LONDON MATHEMATICAL SOCIETY LECTURE NOTE SERIES

Managing Editor: Professor J.W.S. Cassels, Department of Pure Mathematics and Mathematical Statistics, University of Cambridge, 16 Mill Lane, Cambridge CB2 1SB, England

The books in the series listed below are available from booksellers, or, in case of difficulty, from Cambridge University Press.

4 Algebraic topology, J.F. ADAMS
17 Differential germs and catastrophes, Th. BROCKER & L. LANDER
20 Sheaf theory, B.R. TENNISON
27 Skew field constructions, P.M. COHN
34 Representation theory of Lie groups, M.F. ATIYAH *et al*
36 Homological group theory, C.T.C. WALL (ed)
39 Affine sets and affine groups, D.G. NORTHCOTT
40 Introduction to H_p spaces, P.J. KOOSIS
43 Graphs, codes and designs, P.J. CAMERON & J.H. VAN LINT
45 Recursion theory: its generalisations and applications, F.R. DRAKE & S.S. WAINER (eds)
46 p-adic analysis: a short course on recent work, N. KOBLITZ
49 Finite geometries and designs, P. CAMERON, J.W.P. HIRSCHFELD & D.R. HUGHES (eds)
50 Commutator calculus and groups of homotopy classes, H.J. BAUES
51 Synthetic differential geometry, A. KOCK
57 Techniques of geometric topology, R.A. FENN
58 Singularities of smooth functions and maps, J.A. MARTINET
59 Applicable differential geometry, M. CRAMPIN & F.A.E. PIRANI
60 Integrable systems, S.P. NOVIKOV *et al*
62 Economics for mathematicians, J.W.S. CASSELS
65 Several complex variables and complex manifolds I, M.J. FIELD
66 Several complex variables and complex manifolds II, M.J. FIELD
68 Complex algebraic surfaces, A. BEAUVILLE
69 Representation theory, I.M. GELFAND *et al*
72 Commutative algebra: Durham 1981, R.Y. SHARP (ed)
76 Spectral theory of linear differential operators and comparison algebras, H.O. CORDES
77 Isolated singular points on complete intersections, E.J.N. LOOIJENGA
78 A primer on Riemann surfaces, A.F. BEARDON
79 Probability, statistics and analysis, J.F.C. KINGMAN & G.E.H. REUTER (eds)
80 Introduction to the representation theory of compact & locally compact groups, A. ROBERT
81 Skew fields, P.K. DRAXL
83 Homogeneous structures on Riemannian manifolds, F. TRICERRI & L. VANHECKE
84 Finite group algebras and their modules, P. LANDROCK
85 Solitons, P.G. DRAZIN
86 Topological topics, I.M. JAMES (ed)
87 Surveys in set theory, A.R.D. MATHIAS (ed)
88 FPF ring theory, C. FAITH & S. PAGE
89 An F-space sampler, N.J. KALTON, N.T. PECK & J.W. ROBERTS
90 Polytopes and symmetry, S.A. ROBERTSON
91 Classgroups of group rings, M.J. TAYLOR
92 Representation of rings over skew fields, A.H. SCHOFIELD
93 Aspects of topology, I.M. JAMES & E.H. KRONHEIMER (eds)
94 Representations of general linear groups, G.D. JAMES
95 Low-dimensional topology 1982, R.A. FENN (ed)
96 Diophantine equations over function fields, R.C. MASON
97 Varieties of constructive mathematics, D.S. BRIDGES & F. RICHMAN
98 Localization in Noetherian rings, A.V. JATEGAONKAR
99 Methods of differential geometry in algebraic topology, M. KAROUBI & C. LERUSTE
100 Stopping time techniques for analysts and probabilists, L. EGGHE
101 Groups and geometry, ROGER C. LYNDON
103 Surveys in combinatorics 1985, I. ANDERSON (ed)

104 Elliptic structures on 3-manifolds, C.B. THOMAS
105 A local spectral theory for closed operators, I. ERDELYI & WANG SHENGWANG
106 Syzygies, E.G. EVANS & P. GRIFFITH
107 Compactification of Siegel moduli schemes, C-L. CHAI
108 Some topics in graph theory, H.P. YAP
109 Diophantine Analysis, J. LOXTON & A. VAN DER POORTEN (eds)
110 An introduction to surreal numbers, H. GONSHOR
111 Analytical and geometric aspects of hyperbolic space, D.B.A.EPSTEIN (ed)
112 Low-dimensional topology and Kleinian groups, D.B.A. EPSTEIN (ed)
114 Lectures on Bochner-Riesz means, K.M. DAVIS & Y-C. CHANG
115 An introduction to independence for analysts, H.G. DALES & W.H. WOODIN
116 Representations of algebras, P.J. WEBB (ed)
117 Homotopy theory, E. REES & J.D.S. JONES (eds)
118 Skew linear groups, M. SHIRVANI & B. WEHRFRITZ
119 Triangulated categories in the representation theory of finite-dimensional algebras, D. HAPPEL
121 Proceedings of *Groups - St Andrews 1985*, E. ROBERTSON & C. CAMPBELL (eds)
122 Non-classical continuum mechanics, R.J. KNOPS & A.A. LACEY (eds)
123 Surveys in combinatorics 1987, C. WHITEHEAD (ed)
124 Lie groupoids and Lie algebroids in differential geometry, K. MACKENZIE
125 Commutator theory for congruence modular varieties, R. FREESE & R. MCKENZIE
127 New Directions in Dynamical Systems, T.BEDFORD & J. SWIFT (eds)
128 Descriptive set theory and the structure of sets of uniqueness, A.S. KECHRIS & A. LOUVEAU
129 The subgroup structure of the finite classical groups, P.B. KLEIDMAN & M.W. LIEBECK
130 Model theory and modules, M. PREST
131 Algebraic, extremal & metrical combinatorics, M-M. DEZA, P. FRANKL & I.G. ROSENBERG (eds)
132 Whitehead groups of finite groups, ROBERT OLIVER

London Mathematical Society Lecture Note Series. 127

New Directions in Dynamical Systems

Edited by T. Bedford and J. Swift
King's College, Cambridge

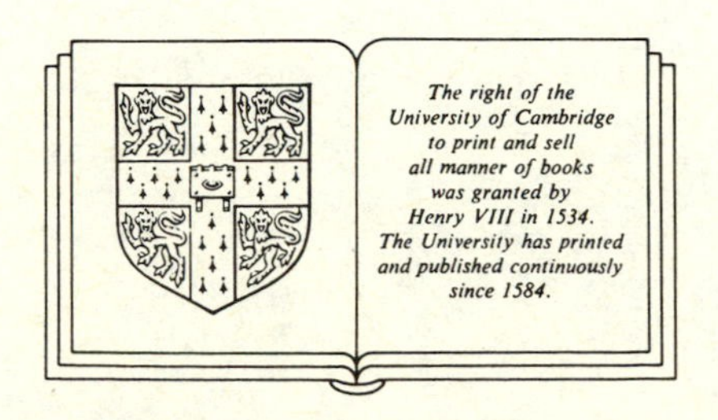

CAMBRIDGE UNIVERSITY PRESS
Cambridge
New York New Rochelle Melbourne Sydney

Published by the Press Syndicate of the University of Cambridge
The Pitt Building, Trumpington Street, Cambridge CB2 1RP
32 East 57th Street, New York, NY 10022, USA
10, Stamford Road, Oakleigh, Melbourne 3166, Australia

First published 1988

Printed in Great Britain at the University Press, Cambridge

Library of Congress cataloging in publication data

New Directions in Dynamical Systems
(London Mathematical Society lecture note series; 127)
Bibliography: p.
1. Differentiable dynamical systems
I. Bedford, T. II. Swift, J. III. Series

QA614.8.N49 1987 003 87-27758

British Library cataloguing in publication data

New Directions in Dynamical Systems
(London Mathematical Society lecture note series; 127)
1. System analysis
I. Bedford, T. II. Swift, J.W. III. Series
003 QA402

ISBN 0 521 34880 3

CONTENTS

PREFACE

This collection of review articles is aimed at those with some background knowledge of Dynamical Systems Theory. It will be useful to graduate students and researchers wishing to familiarize themselves with current research, as well as to those currently working in the field.

Each of the authors has given a survey of an active research topic. The aim is to provide a useful review of the directions in which particular lines of research are going, together with a wide list of references for further reading, and to provide the reader with a number of open problems.

This book is loosely associated with the conference on "Theoretical and Numerical Problems in the Study of Chaotic Ordinary Differential Equations" held at King's College, Cambridge in June and July 1986. The conference was funded by the S.E.R.C. and the Dynamical Systems Project of King's College Research Centre.

The Editors would like to thank David Tranah and Martin Gilchrist at C.U.P., and Ben Mestel and Colin Sparrow of the King's College Research Centre for their help and advice. We are also grateful to Klaus Schmidt and Ian Stewart for letting us use the mathematical fonts they developed for the Apple Macintosh Computer.

T.J.B.
J.W.S.

INTRODUCTION

In recent years Dynamical Systems has attracted attention from workers in diverse fields. The use of powerful computers and computer graphics in numerical simulations has led to growing interest in "chaos". A wide range of scientists including theoretical physicists, engineers, biologists and ecologists have raised interesting problems which provide new sources of "applied" motivation beyond the traditional questions from classical mechanics. Their interaction with mathematicians has stimulated new lines of research and has been particularly important in determining the new directions taken by Dynamical Systems in the last decade.

The approaches to these new problems have several themes in common. Complicated structures are modelled by deterministic systems with a few variables. The bifurcation patterns of parametrized families of systems are studied. Flows are reduced to Poincaré maps and all systems are modelled by one-dimensional maps whenever possible. In experimental systems, attractors are reconstructed from time series.

Typical questions of interest are to prove the existence of numerically observed "strange attractors" such as that in the Hénon map and to describe the structure of such strange attractors. We would like to understand how these complicated sets can be created from dynamically simple ones through a series of bifurcations. Different kinds of scaling behaviour in strange sets can be found and must be explained. In low dimensional systems the possible range of dynamical behaviour is restricted and so, in principle, should be capable of classification.

The articles in this book are primarily research texts and do not provide a systematic introduction to Dynamical Systems theory. Some books giving a more elementary background are Palis and de Melo [1982], Guckenheimer and Holmes [1983], Collet and Eckmann [1980] and Devaney [1986].

Much of the pure mathematical work in Dynamical Systems over the last twenty years has concentrated on Axiom A systems (first defined by Smale [1967] in his famous survey article) and they are now well understood. (Smale's horseshoe map is the best known example of an Axiom A system.) The use of symbolic dynamics has given a nice description of the nonwandering set in terms of subshifts of finite type and has also led to a good understanding of the ergodic theory of these maps. Smale conjectured that Axiom A diffeomorphisms are open and dense in the space of all diffeomorphisms (so that a typical system would satisfy Axiom A) but this was soon found to be false. For example, neither the Hénon map nor the Lorenz equations are structurally stable (and do not therefore satisfy Axiom A) in the parameter regions where strange attractors have been observed numerically. One of the main problems in Dynamical Systems is to find ways of describing such structurally unstable chaotic systems. The Smale programme has bequeathed a variety of techniques for approaching this problem, for example the usual method of showing that a particular system is chaotic is to prove that it contains a horseshoe. Many systems contain important hyperbolic sets (generalized horseshoes). Glendinning's article in this book describes hyperbolic subsets near to homoclinic orbits. In van Strien's paper on one dimensional maps he describes a result of Mañé which says, roughly, that on compact invariant sets away from the critical point a map of the interval is hyperbolic. Applications of the ergodic theory of Axiom A systems arise in the study of scaling spectra in the renormalisation strange sets discussed by Rand.

In order to obtain detailed descriptions of the behaviour of non-Axiom A systems it is necessary to impose restrictive hypotheses. The most natural kind of assumption to make is to specify the phase space in which the dynamics are taking place because the topology of the phase space is such a severe constraint on possible behaviour. This is why the dynamics of maps of the circle and of the interval are relatively well understood. The development of kneading theory and the use of the Schwarzian derivative, together with the observation of universal behaviour, gave the study of unimodal maps of the interval a great boost in the late 1970's. Van Strien's article in this book gives an account of some of the most sophisticated methods available in the study of one dimensional maps. The constraint of dimension is used to good effect elsewhere. For example, Holmes uses the fact that periodic orbits in three dimensional flows are knots whose knot type is invariant as a parameter is changed. By contrast there are apparently simple questions in low dimensions that remain open. Lloyd's article on the number of limit cycles in a polynomial vector field of the plane is a case in point. Such systems are simple enough that one can prove quite a lot about them, but the original problem (Hilbert's 16th) has remained intractable.

Bifurcation Theory is one of the most powerful techniques in the study of dynamical systems and is used in each of the articles in this collection. Here one attempts to understand a family of systems by concentrating on those parameter values which are at the boundary between different classes of structurally stable systems. The analysis has usually been local in a neighbourhood around a fixed or periodic point, although it has often been observed (experimentally or numerically) that the local analysis holds far from the bifurcation. If two or more parameters are varied then highly degenerate systems can be found which are "organising centres" that enable one to describe a wide range of behaviour with a local analysis. Lloyd's article on limit cycles in polynomial systems uses bifurcation techniques to create new limit cycles by bifurcating from the fixed points at zero and infinity, and from homoclinic loops. The article by Stewart considers the physically important problem of bifurcations in systems with symmetries; symmetry leads to more complicated behaviour, yet at the same time the presence of symmetry can simplify the analysis. Rand's article reviews recent work in which one combines the local approach together with rescaling or renormalisation. The renormalisation group was imported from Physics by Feigenbaum for his analysis of the cascade of period doublings. Renormalisation methods can describe sequences of bifurcations (as opposed to isolated bifurcations) that all occur as a result of some simple mechanism. Sequences of bifurcations also occur (for slightly different reasons) near to homoclinic orbits in 3 or higher dimensional flows. Work on such bifurcations was pioneered by by Shil'nikov and is described in Glendinning's article. A related use of Bifurcation Theory is the attempt to understand complicated dynamics in a strange set by understanding the sequence of bifurcations that occur as it is created in a one-parameter family of systems, starting with something well understood. Holmes aims at this by describing some of the knot types that arise as periodic orbits in strange attractors.

An important part of Hamiltonian Systems is the theory of area preserving twist maps of the annulus, which model the dynamics in a neighbourhood of an elliptic fixed point. The celebrated K.A.M. theorem implies that in such a neighbourhood a set of points with positive Lebesgue measure lie on invariant circles. A major problem here is to understand how these invariant circles break up, and Rand's article discusses an approach to this problem using a renormalisation group analysis. Many long-studied Hamiltonian systems, such as the Kepler problem, have symmetries and these are the subject of the penultimate section of Stewart's article. An introduction to the theory of Hamiltonian Systems can be found in the new book compiled by MacKay and Meiss [1987].

Numerical "experiments" are increasingly important in the study of Dynamical Systems. The pioneering studies of Lorenz [1963], Hénon and Heiles [1964], and Hénon [1978] have inspired a huge amount of numerical work which has aided our intuition and motivated numerous theoretical studies. Numerical investigations to discover new theorems are now standard, and the results have been dramatic. Computer experiments with the logistic map pointed the way to many results which are described in van Strien's and Rand's articles. It is modern folklore that Feigenbaum's discovery of the universal scaling in period doubling was made using a hand calculator. Glendinning's paper here discusses several problems arising from numerical observations of bifurcations in the Lorenz equations. Stewart's article is motivated in part by physical systems such as Taylor-Couette flow, and numerical simulations of partial differential equations which model these symmetric systems. A different use of computers - untiring and accurate algebraic manipulation - is used extensively in Lloyd's study of planar polynomial systems.

We hope that the articles in this book will give a flavour of some of the new directions in Dynamical Systems. A whole range of techniques have been adopted to provide descriptions of chaotic and non-chaotic systems. For chaotic systems these include scaling properties of fractal invariant sets, descriptions of persistent structures such as hyperbolic subsets and knotted periodic orbits, and finding sequences of bifurcations that create chaotic behaviour. None of these techniques provide a single satisfactory description of a chaotic system, for the moment a piecemeal approach is all that is possible. For non-chaotic systems the open problems are just as difficult. The techniques of Bifurcation Theory are of some help, but do not really provide a view of the global behaviour of these systems. The wealth of behaviour in even the most simple systems is enough to keep us all intrigued for many years to come.

References

Collet, P. and Eckmann, J.-P. (1980). *Iterated maps of the Interval as Dynamical Systems*, Progress in Physics, Vol 1, Birkhäuser-Boston.

Devaney, R.L. (1986). *An Introduction to Chaotic Dynamics*, Benjamin/Cummings, Menlo Park, CA.

Guckenheimer, J. and Holmes, P. (1983). *Nonlinear Oscillations, Dynamical Systems and Bifurcations of Vector Fields*. Appl. Math. Sci. Series, Vol. 42, Springer-Verlag, New York.

Hénon, M. (1976). A Two-Dimensional Mapping with a Strange Attractor. *Comm. Math. Phys.*, **50**, 69-77.

Hénon, M. and Heiles, C. (1964). The Applicability of the Third Integral of Motion: Some Numerical Experiments. *Astron. J.*, **69**, 73-79.

Lorenz, E.N. (1963). Deterministic Non-periodic Flows. *J. Atmos. Sci.*, **20**, 130-141.

MacKay, R.S. and Meiss, J.D. (1987). *Hamiltonian Dynamical Systems*. Adam Hilger, Bristol.

Palis, J. and de Melo, W. (1982). *Geometric theory of Dynamical Systems: An Introduction*. Springer-Verlag, New York.

Smale, S. (1967). Differentiable Dynamical Systems. *Bull. Amer. Math. Soc.* **73**, 747-817.

UNIVERSALITY AND RENORMALISATION IN DYNAMICAL SYSTEMS.

David Rand.

Mathematics Institute, Warwick University, Coventry CV4 7AL, UK.

CONTENTS

INTRODUCTION.

The renormalisation group formalism has lead to a number of fruitful developments in our understanding of the "transition to chaos". The best known examples concern the quantitative universality of period-doubling cascades and the breakdown of invariant circles in dissipative and area-preserving maps. This paper is meant to be an introduction to, and biased review of, these ideas.

On period-doubling, I just give a relatively brief review of the basic ideas for unimodal maps of the interval. I do not touch upon period-doubling in area-preserving maps because the theory is so similar. The interested reader is referred to Bountis (1981), Benettin, Cercignani, Galgani & Giorgilli (1980), Benettin, Galgani & Giorgilli (1980), Collet, Eckmann & Koch (1981a), Eckmann, Koch & Wittwer (1982), and Greene, MacKay, Vivaldi & Feigenbaum (1981).

After dealing with period-doubling, I discuss the theory of critical circle maps, especially those with golden-mean rotation number. This leads in turn to a theory for the breakup of invariant circles of dissipative maps with golden-mean rotation number. A natural extension is then to the case of area-preserving twist maps and in Section 4 the theory in Sections 2 and 3 is applied to study the breakdown of the last homotopically non-trivial invariant circle. There is an interesting related theory due to Manton and Nauenberg (1983) which deals with the universal small-scale structure found in the boundaries of Siegel domains of rational maps of the Riemann sphere. I shall not discuss it here because of lack of space and because the underlying ideas are similar. The interested reader is referred to Manton & Nauenberg (1983) and Widom (1983).

The results of the next section are not so well-known. In it I describe a general formalism of renormalisation strange sets which, for example, handles the case of general irrational rotation numbers. Predictions about the universal fractal structure of fractal bifurcation sets in parameter space come from this theory. I believe it will be of much more general use in physical problems. Some open problems are discussed in Section 6.

1. PERIOD-DOUBLING CASCADES.

This section contains some introductory remarks on maps of the interval, and a discussion of the Feigenbaum conjectures about period-doubling cascades in unimodal maps of the interval and the associated renormalisation group formalism. The explanation of how to use the renormalisation structure constructed to get similar results for diffeomorphisms and flows is similar to that given in Section 3.2 (see Collet & Eckmann (1980)).

1.1 Unimodal maps of the interval.

1.1.1 Unimodal maps of $I = [-1,1]$.

Definition. A map $f : I \to I$ is unimodal if (a) f is C^1, (b) $f(0) = 1$, (c) f is strictly increasing on $[-1,0]$ and strictly decreasing on $[0,1]$, and (d) f is even.

Remark 1.1. Condition (d) is not really necessary for what follows, but its assumption will make a number of computations considerably easier.

Important Example 1.1 Consider the one-parameter family of unimodal maps given by $f_\mu(x) = 1 - \mu x^2$, $0 < \mu \le 2$. Clearly, when $\mu > 0$ is small the fixed point is a global attractor. On the other hand, if $y = \phi(x) = (4/\pi)(\sin^{-1}\sqrt{(x + 1)/2}) - 1$ then $F = \phi \circ f_2 \circ \phi^{-1} = 1-2|y|$. The Lebesgue measure dx is an invariant ergodic measure for F whence $dx/\pi\sqrt{(1 - x^2)}$ is one for f_2.

$\|f\|_1 = \sup_{x \in I}(|f(x)| + |f'(x)|)$. This topology is called the C^1*–topology*.

Definition. (a) f is p-superstable if 0 is in a p–cycle of f. (b) f is p-filling if there exists p disjoint closed sub-intervals $I_1, \ldots, I_p$ such that f is a homeomorphism from I_j to I_{j+1} for $1 \le j < p$, $f(I_p) \subseteq I_1$ and $g = f^p|I_1$ is a unimodal map of I_1 such that $g(I_1) = I_1$ (i.e. with respect to I_1, g is 1-filling).

Remark 1.4. If f is 1-superstable, points near the orbit of 0 converge to it at a quadratic rate, hence the term *superstable*. The map $f_2(x) = 1 - 2x^2$ is 1-filling.

Definition. A C^1 *family* is a 1-parameter family f_μ in $C^1(I, I)$, $\mu \in (\alpha, \beta)$, which is continuous in the C^1-topology. It is *full* on (α, β) if $f_\mu(1) \to 1$ (resp. $\to -1$) as $\mu \to \alpha^+$ (resp. $\to \beta^-$). For example, $1 - \mu x^2$ is full on $(0, 2)$.

Theorem 1.1. If f_μ is a full C^1-family on (α, β) then there exists $\alpha < \alpha_1 < \alpha_2 < \cdots < \beta_2 < \beta_1 < \beta$ such that 1. f_{α_i} is 2^i-superstable, 2. f_{β_i} is 2^i-filling, and 3. $\lim_{i \to \infty} \alpha_i = \lim_{i \to \infty} \beta_i$.

Remark 1.5. Clearly, there must be a bifurcation point between α_i and α_{i+1} Typically, this will be a generic period-doubling bifurcation as described in the next subsection.

Proof of Theorem 1.1. Consider a full C^1 family on (α, β). Let $\alpha_1 = \sup\{\mu : f_\mu(1) = 0\} = \sup\{\mu : a(f_\mu) = 0\}$. Then there exists $\varepsilon > 0$ such that if $\mu \in (\alpha, \alpha + \varepsilon)$ then $f_\mu \in D(T)$. Let $\beta_1 = \inf\{\mu > \alpha : f_\mu \in D(T)\} < \beta$. Then $\alpha(f_{\beta_1}) \neq 0$ so $f^2_{\beta_1}(\alpha_{\beta_1}) = \alpha_{\beta_1}$. It easily follows that $\mu \to Tf_\mu$ is a full family on (α_1, β_1). Repeating this construction one deduces the existence of sequences $\alpha < \alpha_1 < \alpha_2 < \cdots < \beta_2 < \beta_1 < \beta$ such that for $\alpha_i < \mu < \beta_i$, $f_\mu \in D(T^i)$ and $\mu \to T^i f_\mu$ is full on (α_i, β_i). These have been constructed so that $T^{j-1} f_{\alpha_j}(1) = 0$. Therefore $T^{j-1} f_{\alpha_j}$ is 2-superstable, i.e. f_{α_j} is 2^j-superstable. Moreover, $T^j f_{\beta_j}(1) = -1$ so $T^j f_{\beta_j}$ is 1-filling whence f_{β_j} is 2^j-filling. A simple calculation shows that α_i and β_i satisfy 3. ■

Since $T^{j-1} f_{\alpha_j}$ is 2-superstable, $T^{j-1} f_{\alpha_j + \varepsilon}$ has a 2-sink for $|\varepsilon|$ sufficiently small. Let $\mu_j = \sup\{\mu \in (\alpha_{j-1}, \alpha_j) \mid T^{j-1} f_\mu \text{ has a 2–sink }\}$. Then μ_j is the bifurcation value associated with the bifurcation 2^j-sink $\to 2^{j+1}$-sink, though this is not necessarily a generic bifurcation.

1.2 The period-doubling bifurcation.

Proposition 1.1. If $f : I \times [-1, 1] \to I$ is a family of maps, not necessarily unimodal and $f_\mu(x) = f(x, \mu)$ is such that: $f_0(p) = p$; $f'_0(p) = -1$; $(f_0^2)'''(p) < 0$; $\partial^2 f / \partial\mu\partial x|_{x = p, \mu = 0} < 0$; then there exist intervals $(\nu_1, 0)$ and $(0, \nu_2)$ and $\varepsilon > 0$ such that: (a) if $\mu \in (\nu_1, 0)$ then f_μ^2 has exactly one fixed point in $(p - \varepsilon, p + \varepsilon)$, and this is a sink; (b) if $\mu \in (0, \nu_2)$ then f_μ^2 has three fixed points in $(p - \varepsilon, p + \varepsilon)$, and the largest and smallest are sinks, the middle a source.

Remark 1.6. To deal with $2^n \to 2^{n+1}$ bifurcation, replace f by f^{2^n}.

To get a better picture of how f_μ bifurcates to get from the simple picture for small μ to the complex one for $\mu = 2$ let us concentrate on those aspects brought out by the doubling transformation. For other results the reader should consult Guckenheimer (1977) and (1979) and Jonker & Rand (1981).

1.1.2 Doubling transformation.

Throughout this section f denotes a unimodal map. Let $a = a(f) = -f(1), b = b(f) = f(a)$. Let $D(T)$ denote the set of f's such that (i) $a > 0$ (ii) $b > a$, and (iii) $f(b) \leq a$. For $f \in D(T)$ define the *doubling transformation* T by

$$Tf(x) = -a.f^2(-ax).$$

Remark 1.2. Figure 1 shows the graphs of f and f^2 when $f = 1 - 1.52x^2 + 0.104x^4$. To get Tf one just scales the coordinate using $x_{new} = -ax_{old}$ to scale the small box to have length one and turn the graph upside down. Note that f^2 will look as shown if $f \in D(T)$ because then

$$[-a\ ,\ a] \rightarrow [b\ ,\ 1] \rightarrow [-a = f(1)\ ,\ f(b)] \subseteq [-a\ ,\ a].$$

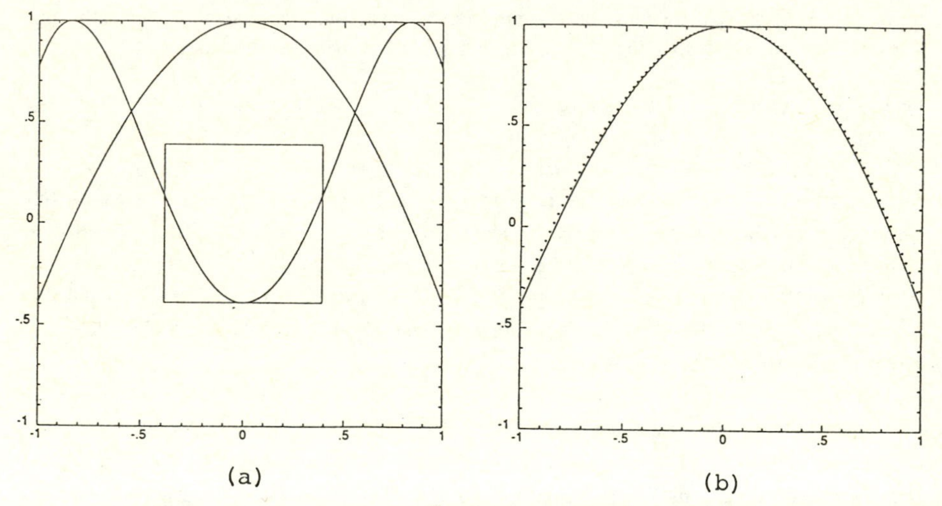

Figure 1. (a) the graphs of f and f^2 where $f = 1 - 1.52x^2 + .104x^4$; the box is $-a \leq x,y \leq a$; (b) points from the graph of Tf superimposed on the graph of f.

Remark 1.3. The boundary of $D(T)$ consists of the surfaces $a = 0$ and $f^2(a) = a$ because, in moving from $D(T)$ to a point where (ii) fails, $b = a$ is encountered i.e. $f^2(a) = a$. Condition (iii) can fail when (ii) is true.

1.1.3 2^n-superstable and 2^n-filling functions.

To get some feeling for the doubling transformation I discuss here an elementary application which is taken from Collet, Eckmann and Lanford (1980). Let $C^1(I\ ,I)$ denote the space of C^1 maps of I to I endowed with the topology defined by the C^1-norm :

Proof. The proof is a simple calculation using the implicit function theorem. It can be found in Guckenheimer (1977) or Whitley (1983) for example. ■

1.3 Feigenbaum conjectures.

1.3.1 Motivation.

I give a heuristic introduction to the Feigenbaum conjectures. Consider the 1-parameter family $f_\mu = 1 - \mu x^2$, $0 < \mu \le 2$, discussed in the previous section. Recall the meaning of the parameter values α_i, β_i of Theorem 1.1. Using a pocket calculator one finds

$$\lim_{n\to\infty} (\alpha_n - \alpha_{n-1})/(\alpha_{n+1} - \alpha_n) = \lim_{n\to\infty} (\beta_n - \beta_{n-1})/(\beta_{n+1} - \beta_n) = \delta = 4.669...$$

and using something a bit more powerful, it appears that there exists $\lambda(\sim -.3995...)$ such that if $\alpha_\infty = \lim_{n\to\infty}\alpha_n$,

$$\zeta = \lim_{n\to\infty} \lambda^{-n} f_{\alpha_\infty}^{2^n} \lambda^n$$

exists and is an analytic function of x^2. Moreover, if one takes any other 1-parameter family close to this one (and, in practice, many not so close) one gets the same experimental values for δ and λ and, (up to a scale change), the same function ζ. This is an example of *universality*.

1.3.2 Feigenbaum Conjectures.

The explanation (essentially proposed by Feigenbaum (1978, 1979) and independently by Coullet and Tresser (1979)) goes as follows: Consider the doubling transformation defined on some suitable subspace of $D(T)$ consisting of analytic functions. Assume that the following facts are true.

Conjecture 1. T has a fixed point f_* with the property that $f_*'(0) = 0$ and $f_*''(0) \neq 0$.

Conjecture 2. The only element of the spectrum of $dT(f_*)$ outside the disk $|z| < 1$ is a simple eigenvalue $\delta = 4.669...$. The rest of the spectrum is contained inside a disk of radius strictly less than 1.

Conjecture 3. The unstable manifold of f_* intersects and is transverse to the submanifolds Σ_n and Λ_n of bifurcation and superstable maps defined as

$$\Sigma_n = \{ f : \text{for some } p \text{ in a } 2^n\text{–cycle of } f,\ (f^{2^n})'(p) = -1 \text{ and } (f^{2^n})'''(p) \neq 0 \}$$

$$\Lambda_n = \{ f : f^{2^n}(0) = 0 \text{ and } f^m(0) \neq 0 \text{ for } 0 < m < 2^n \}.$$

These imply :

1. Since $Tf_* = f_*$,

$$f_*^2(\lambda x) = \lambda f_*(x) \tag{1.1}$$

where $\lambda = -a(f_*) = f_*(1)$. Equation (1.1) is sometimes known as the Cvitanovic-Feigenbaum functional equation.

2. $T^n f \to f_*$ as $n \to \infty$ implies there exists $\beta > 0$ such that

$$\lambda^{-n} f^{2^n}(\lambda^n x) \to \beta^{-1}(f_*(\beta x))$$

uniformly in x as $n \to \infty$.

3. Conjecture 2 implies that, with respect to T, f_* is a saddle point with a 1-

dimensional unstable manifold W^u and a stable manifold W^s of codimension one (see Figure 2); W^u defines a universal 1-parameter family of maps $f_{*,\mu}$. For a 1-parameter family f_μ near f_* with $f_0 \in W^s$ one has

$$\lambda^{-n} f^{2^n}_{\mu\delta^{-n}} \circ \lambda^n \to \beta^{-1} f_{*,\mu} \circ \beta$$

for some $\beta > 0$.

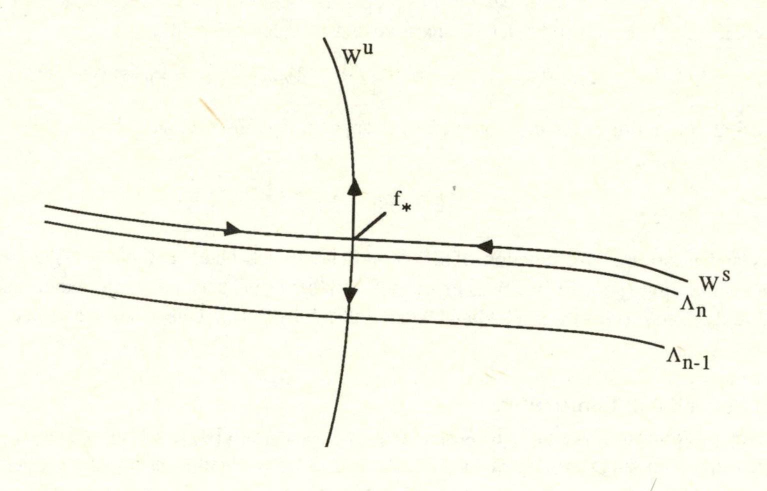

Figure 2. The saddle-point structure of the fixed point and the accumulation of the sets Λ_n.

4. Obviously, $T(\Sigma_n) \subseteq \Sigma_{n-1}$ and $T(\Lambda_n) \subseteq \Lambda_{n-1}$. Thus, Conjecture 3 implies that the Σ_n and Λ_n accumulate on W^s exponentially fast with the distances from W^s decreasing like δ^{-n} (to prove this one one uses the fact that T can be linearised along the unstable direction). If f_μ is a 1-parameter family near $f_{*,\mu}$ and transverse to W^s with say $f_0 \in W^s$ then $f_{\alpha_i} \in \Lambda_i$ and therefore

$$\lim_{i \to \infty} (\alpha_i - \alpha_\infty)/(\alpha_{i+1} - \alpha_\infty) = \delta,$$

if $\alpha_\infty = \lim_{n \to \infty} \alpha_n$. In the same fashion, if μ_n denotes the parameter value at which a $2^n \to 2^{n+1}$ period-doubling occurs then

$$\lim_{i \to \infty} (\mu_i - \mu_\infty)/(\mu_{i+1} - \mu_\infty) = \delta$$

where $\mu_\infty = \lim_{n \to \infty} \mu_n = \alpha_\infty$.

This explains where the simpler universal quantities δ , λ $(= -a(f_*))$ and ζ $(= f_*$ scaled) come from. This sort of a set up is known to physicists as a renormalisation group: a notion which in its most powerful form was invented by Wilson to study problems such as phase transitions in statistical mechanics and quantum field theory (Wilson (1971, 1975)). For this he received the 1983 Physics Nobel Prize.

Lanford has given a proof of Conjectures 1 and 2. His proof makes essential use of rigorous computer-generated estimates.[1] The idea is to cleverly approximate the doubling operator and some related transformations by their action on certain polynomial functions, using the computer to do the more tedious calculations which because of their length are not really accessible to pen and paper, and to keep a rigorous check on the errors. Campanino, Epstein and Ruelle (1981, 1982) previously gave a proof of Conjecture 1 alone which only used very simple numerical estimates and was basically analytical. Very recently, Epstein and Eckmann have proved a result which is essentially, but not quite, Conjecture 1 using the fact that the inverse of (one half of) the fixed point function is a Herglotz function (Epstein (1986), Epstein & Eckmann (1986, 1987)). An attractive aspect of their approach is that it also applies to the related conjecture about golden circle maps (see Section 3) and brings out the similarities between the two problems. However, I should stress at this point that none of these proofs casts any light on the ubiquity of such renormalisation schemes. Why do such low-codimensional renormalisation structures occur so often and in such widely differing situations as 1-dimensional maps and area-preserving twist maps? There must be some general principle underlying this phenomenon which the right proof would reveal and an approach that would unify the seemingly disparate applications. To find these is the main open problem in this area.

An important recent development is a proof by Sullivan that the stable manifold contains all analytic unimodal maps which are topologically conjugate to the fixed point and whose extension to the complex plane is a *quadratic-like map* in the sense of Douady and Hubbard (1985). In particular, this implies that there is a unique quadratic-like fixed point. The proof which is outlined in Sullivan (1987), depends upon the Douady-Hubbard theory for such maps. This theory also gives an important insight into the universality of δ. According to it every quadratic-like map can be quasi-conformally conjugated to a uniquely determined member of the family $f_\mu(x) = 1 - \mu x^2$. In particular, if f_μ is in $D(T)$ then $T(f_\mu)$ can be quasi-conformally conjugated to $f_{h(\mu)}$ for some unique $h(\mu)$. The function h has derivative δ at μ_∞, the parameter value at which the period doublings accumulate.

1.4 The Feigenbaum attractor, the scaling spectrum and Feigenbaum's scaling function.

Firstly, I discuss Feigenbaum's scaling function σ. The definition of σ is from Feigenbaum (1982). Consider a flow $f^t = f^t_\mu$ depending upon a parameter μ. Assume that as μ is increased the system undergoes a cascade of period-doubling bifurcations of the form discussed above with the nth of these occurring at $\mu = \mu_n$ and the μ_n accumulating on μ_∞. Let $f = f_\mu$ be a local Poincaré map for the system obtained as the return map to some 1-codimensional submanifold which is transverse to all the relevant periodic orbits. Let $x(t) = f^t(x)$, $t \in \mathbf{Z}$, denote a solution starting on the appropriate periodic attractor. Then $T_n = 2^{n-1}$ is the period just before the nth period doubling. Thus if $\mu < \mu_n$ is close to μ_n, $x(t + T_n) - x(t) = 0$. Let

$$\psi^{(n)}(t) = x(t) - x(t+T_{n-1}).$$

Then, at least for large n,

$$\psi^{(n+1)}(t) \sim \sigma(t/T_{n+1})\psi^{(n)}(t)$$

where $\sigma(x + 1) = \sigma(x)$ and $\sigma(x + \tfrac{1}{2}) = -\sigma(x)$. Also, defining $r_n(t) = \psi^{(n)}(t)/\psi^{(n-1)}(t)$,

---

[1] **Note added in proof.** A survey of the ideas behind such computer-assisted proofs is contained in Lanford (1987).

$$r_{n+1}(2t) = \frac{\psi^{(n+1)}(2t)}{\psi^{(n)}(2t)} \sim \sigma(2t/T_{n+1}) = \sigma(t/T_n) \sim \frac{\psi^{(n)}(t)}{\psi^{(n-1)}(t)} = r_n(t)$$

which looks as in Figure 3 (at least for large n ; the picture is for $n = 4,5$ in the case of Duffing's equation and is taken from Feigenbaum (1982)). The universal structure of the power spectrum associated with $x(t)$ can be deduced from a knowledge of σ but this is really secondary; as Feigenbaum has pointed out, the sensible thing to look at is σ not the spectrum.

To see in what sense σ is universal I firstly consider the nature of the attractor of the quadratic fixed point $g = f_* \sim 1 - 1.527x^2 + 0.1048x^4 + 0.0267x^6 + \cdots$ of the doubling transformation and then relate it to the *cookie-cutter* F given by (1.2) below.

Let x_0 denote the critical point of g and $x_n = g^n(x_0)$. For $n \geq 1$, let $J_{0,n}$ be the closed interval between x_{2^n} and $x_{2^{n+1}}$ and $J_{i,n}$, $1 \leq i < 2^n$, be the closed interval between x_i and x_{i+2^n}. Then elementary computations using the fixed-point equation $g = \lambda^{-1}.g^2\circ\lambda$ prove:

(a) $g^{2^n}(J_{0,n}) \subseteq J_{0,n}$;

(b) for each $n \geq 1$ the $J_{i,n}$ are pairwise disjoint;

(c) $J_{i,n-1}$ contains $J_{i,n}$ and $J_{i+2^{n-1},n}$ and no other of the $J_{k,n}$;

(d) $J_{2k+1,n} \subseteq J_{1,1}$ and $J_{2k,n} \subseteq J_{0,1}$; and

(e) $|J_{i,n}| \leq (1 + |\lambda|)|\lambda|^n$.

The map g has a unique fixed point which is contained in the gap G between the intervals $J_{0,1}$ and $J_{1,1}$. Since $|g'| > 1$ on $\overline{G}$, if $x \in G$ and $g(x) \neq x$ then $g^m(x) \in J_{0,1}$ for some $m \geq 1$. Using the self-similarity given by $g = \lambda^{-1}.g^2\circ\lambda$, if $x \in G$ where G is any gap between two adjacent cylinders of the form $J_{k,n}$ then either $g^m(x) = x$ for some $m = 2^k$ with $k \leq n$, x is the image of such a point, or $g^m(x) \in J_{0,n}$ for some $m \geq 1$. Thus each $x \in [-1, 1]$ is either an unstable periodic point of period 2^k for some $k \geq 0$, a preimage of such a point, or $g^n(x)$ converges to

$$\Lambda_\infty(g) = \bigcap_{n\geq 1} \bigcup_{i=0}^{2^n-1} J_{i,n}.$$

The set $\Lambda_\infty(g)$ is clearly invariant. Since $g(J_{i,n}) = J_{i+1,n}$ for $0 \leq i < 2^n$, every orbit in $\Lambda_\infty(g)$ is dense in $\Lambda_\infty(g)$. Thus, $\Lambda_\infty(g)$ is *minimal*.

For each $x \in \Lambda_\infty(g)$ there exists a sequence $k(n)$ such that $x \in J_{k(n),n}$ for all $n \geq 1$. But, by (c), $k(n+1) = k(n) + \varepsilon_n 2^n$ where ε_n equals 0 or 1. Thus there exists $\underline{\varepsilon} = \underline{\varepsilon}(x) = \varepsilon_0\varepsilon_1 \cdots \in \{0,1\}^{\mathbb{N}\cup 0}$ such that for all $n \geq 1$, $k(n) = \varepsilon_0 + \cdots + \varepsilon_{n-1}2^{n-1}$. Clearly, x is completely determined by $\underline{\varepsilon}$ and vice versa, and $\underline{\varepsilon}$ gives a homeomorphism from $\Lambda_\infty(g)$ to $\{0,1\}^{\mathbb{N}\cup 0}$ endowed with the usual product topology. But if $x \in J_{k,n}$, $g(x) \in J_{k',n}$ where $k' = k + 1 \bmod 2^n$. Thus, $g|\Lambda_\infty(g)$ is conjugated to the *adding transformation* $(\varepsilon_0, \varepsilon_1, \ldots) \rightarrow (\varepsilon_0', \varepsilon_1', \ldots)$ of $\{0,1\}^{\mathbb{N}\cup 0}$ where the ε_n' are defined by $\varepsilon_0' + \cdots + \varepsilon_{n-1}'2^{n-1} = 1 + \varepsilon_0 + \cdots + \varepsilon_{n-1}2^{n-1} \bmod 2^n$ for all $n \geq 0$.

If f is in the stable manifold of g, then the f-orbit $y_n = f^n(y_0)$ of the critical point y_0 is ordered on the interval exactly as the x_n. Consequently, the intervals $J_{k,n} = [y_k, y_{k+2^n}]$ have properties analogous to (a) to (d) above. Moreover, since f is in the stable manifold, $|J_{k,n}| \leq c.|\lambda|^n$ for some constant $c > 0$. Thus, as above, the attractor for f given by

$$\Lambda_\infty(f) = \bigcap_{n\geq 1} \bigcup_{i=0}^{2^n-1} J_{i,n}.$$

can be conjugated to the adding transformation on $\{0,1\}^{\mathbb{N}\cup 0}$. Consequently, $f\,|\Lambda_\infty(f)$ and $g\,|\Lambda_\infty(g)$ are conjugate.

Before proceeding further it will be useful to introduce a hyperbolic dynamical system (a cookie-cutter) which generates the $J_{i,n}$. Define

$$F(x) \;=\; \begin{cases} \lambda^{-1}x \text{ on } I_0 = J_{1,0} = [x_2, x_4] \\ \lambda^{-1}g(x) \text{ on } I_1 = J_{1,1} = [x_3, x_1] \end{cases} \tag{1.2}$$

Then $F(I_0) = F(I_1) = [x_2, x_1]$ and $|F'| > 1$. Moreover, the n-cylinders of the cookie-cutter F which are defined by

$$I_{a_0,\ldots,a_{n-1}} = \{x : F^j(x) \in I_{a_j}\}\ ,\ a_0,\ldots,a_{n-1} \in \{0,1\}$$

are the 2^n intervals $J_{i,n}$ whose end-points are x_i and x_{i+2^n}, $i = 1,\ldots,2^n$. Thus if

$$\Lambda = \bigcap_{n\geq 0}\ \bigcup_{a_0,\ldots,a_{n-1}} I_{a_0,\ldots,a_{n-1}}$$

then $\Lambda = \Lambda_\infty(g)$ the attractor for g. Of course, the dynamics of $F|\Lambda$ and $g\,|\Lambda_\infty(g)$ are completely different.

Now suppose that f is in the stable manifold of g and $\|T^n(f) - g\| < c.\tau^n$ for some constant $c > 0$ and some $0 < \tau < 1$. Again let y_n denote the f-orbit of the critical point of f. As was already noted the intervals whose end-points are y_i and y_{i+2^n} have the same ordering as and are in one-to-one correspondence with those given by the x_i. Denote the interval corresponding to $I_{a_0,\ldots,a_{n-1}}$ by $I^f_{a_0,\ldots,a_{n-1}}$. Let $f_i = T^i(f)$ and $\alpha_i = a(f_i)^{-1}$. Then, since the map $\alpha_{n-1}f_{n-1}^{a_{n-1}}\circ\cdots\circ\alpha_0 f_0^{a_0}$ sends $I^f_{a_0,\ldots,a_{n-1}}$ injectively onto $I^{f_n}_{a_0,\ldots,a_{n-1}}$ it is not too difficult to deduce that there exists a $\kappa > 0$ depending only upon f such that

$$\kappa^{-1} < |I^f_{a_0,\ldots,a_{n-1}}|/|I_{a_0,\ldots,a_{n-1}}| < \kappa \tag{1.3}$$

independently of n and $a_0,\ldots,a_{n-1}$. This implies that the conjugacy from $\Lambda_\infty(f)$ to the universal attractor $\Lambda_\infty(g)$ is Lipschitz and has a Lipschitz inverse. Geometric universality follows from this and we see in what sense σ is universal.

In particular, this result implies that the Hausdorff dimension d of $\Lambda_\infty(f)$ and $\Lambda_\infty(g)$ are the same. In fact, if $P(\beta)$ denotes the growth rate

$$\lim_{n\to\infty} \log \sum_{a_0,\ldots,a_{n-1}} |I_{a_0,\ldots,a_{n-1}}|^\beta \tag{1.4}$$

then the Hausdorff dimension d is determined by the equation $P(d) = 0$.

Now, $|I_{a_0,\ldots,a_{n-1}}| = (x_1 - x_2)|(F^n)'(x)|^{-1}$ for some $x \in I_{a_0,\ldots,a_{n-1}}$ because $F^n(I_{a_0,\ldots,a_{n-1}}) = [x_2, x_1]$ and $F^n|I_{a_0,\ldots,a_{n-1}}$ is a homeomorphism. Therefore if $r = \lambda^{-1}$ and s (resp. t) denotes the infimum (resp. supremum) of $|a^{-1}g'(x)|$ on $I_1 = [x_2, x_1]$, then $r^{-(n-k)}t^{-k} < |I_{a_0,\ldots,a_{n-1}}| < r^{-(n-k)}s^{-k}$ where $k = a_0 + \cdots + a_{n-1}$. Thus the sum in (1.4) is bounded below and above by $(r^{-\beta} + t^{-\beta})^n$ and $(r^{-\beta} + s^{-\beta})^n$ respectively. Consequently, $(r^{-\beta} + t^{-\beta}) < P(\beta) < (r^{-\beta} + s^{-\beta})$. Inserting estimates for r, s and t one finds

$$0.5345 < d < 0.5544,$$

the estimate given in Falconer (1985). This estimate can be successively improved by estimating the derivatives of F^n on the intervals $I_{a_0,\ldots,a_{n-1}}$.

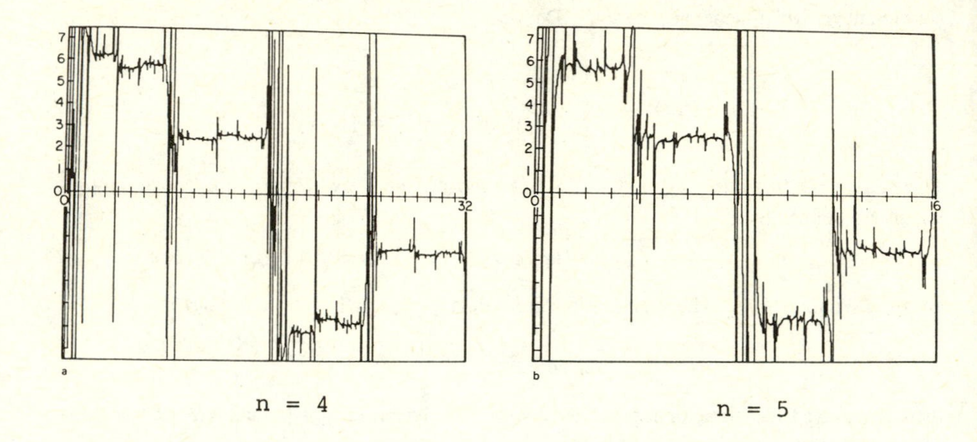

Figure 3. A numerical approximation of r_n obtained from Duffing's equation for $n = 4,5$. (From Feigenbaum (1982).)

The problem with the scaling function σ is that it is not smooth and therefore difficult to specify. Now I introduce a related invariant which is analytic and therefore easier to specify. In this context it was introduced by Halsey, Jensen, Kadanoff, Procaccia & Shraiman (1985), but the form presented here is closer to Bohr & Rand (1987). For $i = 0, \ldots, 2^{n-1} - 1$ consider the lengths

$$l_{i,n} = dist\ (x(T_{n-1} + i), x(i))$$

and their rates of decrease

$$\zeta_i = \lim_{n \to \infty} \log l_{i,n}.$$

There are functions s and p such that if the cascade of bifurcations is in the Feigenbaum universality class then the following are true.

(i) Let $N_n(\zeta)$ be the number of $l_{i,n}$ in $[\zeta, \zeta + d\zeta]$. Then $N_n(\zeta) \sim e^{ns(\zeta)}$.

(ii) Let $p(\beta)$ denote the growth rate of the sums $\sum_{i=1}^{2^{n-1}} l_{i,n}^{\beta}$. Then $p(\beta)$ is real-analytic and strictly convex and s and p are the Legendre transforms of each other.

(iii) If $d(\zeta) = -s(\zeta)/\zeta$, then $d(\zeta)$ is the Hausdorff dimension of the set of points x in the attractor Λ_∞ at $\mu = \mu_\infty$ such that $\lim_{n\to\infty} l_n(x) = \zeta$ where $l_n(x)$ is defined as follows. It follows from an argument similar to that of Proposition 3.2 that the set $\Lambda_\infty' = \overline{\{x(n) : n \in \mathbf{Z}\}}$ is the graph of a Lipschitz function over some coordinate axis. Let π be the projection onto this coordinate axis. Then $l_n(x) = dist\ (x(T_{n-1} + i), x(i))$ where i is such that $\pi(x)$ lies between $\pi(x(T_{n-1} + i))$ and $\pi(x(i))$.

Of course, the definition of the number $N_n(\zeta)$ is somewhat heuristic. This is put on a sounder basis as follows. If J is an interval define $N_n(J)$ to be the number of $l_{i,n}$ in J. Let $s(J)$ be the growth rate of these numbers and let $s(\zeta) = \inf\{s(J) : \zeta \in J\}$.

To see why these results are true for unimodal maps consider the cookie-cutter F defined above. Since, the n-cylinders of the cookie-cutter f are the 2^n intervals whose end-points are x_i and x_{i+2^n}, $i = 1, \ldots, 2^n$, it is easy to see from this construction that the distribution of these scales is given by the exponent entropy function $S(\alpha) = S_F(\alpha)$ for the cookie-cutter F (Bohr and Rand (1986)) and $p(\beta) = P(\beta)$ which is the so-called *pressure* of the function $-\log|f'|$ for the dynamical system $F|\Lambda$ (Bowen (1975)). In fact, $s(\zeta) = S_F(-\zeta)$.

Now, if f is in the stable manifold of g since (1.3) holds the functions $p(\beta)$ and $s(\zeta)$ for f and g are identical. Thus s and p are universal. There is still some work to be done to extend this result to diffeomorphisms, but presumably one can use an argument based on the ideas of Proposition 3.2 below.

A similar construction can be done for the golden fixed point of the renormalisation transformation on circle maps introduced in Section 2 (see Bohr and Rand (1986)).

2. UNIVERSALITY FOR CRITICAL GOLDEN CIRCLE MAPPINGS.

In this section I describe some of the universal properties of critical circle maps with golden-mean rotation number and explain how you prove them using renormalisation group ideas. These ideas easily extend to other rotation numbers with a periodic continued fraction expansion and later, in Section 5, I explain how to extend them to general rotation numbers.

2.1 Circle maps

A continuous map of the circle $\mathbf{T} = \mathbf{R}/\mathbf{Z}$ lifts to a map f of the universal cover $\mathbf{R}$ of $\mathbf{T}$ into itself such that $f(x+1) = f(x) + 1$. This map f is only unique up to addition of an integer; to enforce uniqueness I demand that $0 \le f(0) < 1$. If the original circle map is C^r, $0 \le r \le \omega$, the lift f is C^r. The set of such lifts is denoted $\mathbf{D}^r$. If $x \in \mathbf{R}$ and $f \in \mathbf{D}^0$ the *rotation number* of (f,x) is defined to be

$$\rho(f,x) = \liminf_{n \to \infty} n^{-1}(f^n(x) - x).$$

In general the limit does not exist and $\rho(f,x)$ depends upon x. However, if f is a homeomorphism then the limit exists and $\rho(f,x)$ is independent of x (Arnold (1983)). The number $\rho(f)$ obtained is called the rotation number of f. It depends continuously upon f in the C^0-topology.

Now to bring out some important aspects of circle maps, consider the prototypical 2-parameter family

$$f_{\mu,\nu} = x + \nu - (\mu/2\pi)\sin 2\pi x.$$

If $|\mu| \le 1$ then $f_{\mu,\nu}$ is a homeomorphism; it is a diffeomorphism if $|\mu| < 1$. If $\mu = 0$ then $f_{\mu,\nu}$ is the rotation R_ν so $\rho(f_{0,\nu}) = \nu$ and the dependence of ρ upon ν is trivial. This is not the case if $\mu \ne 0$. To see this fix $0 < |\mu| < 1$ and let f_ν denote $f_{\mu,\nu}$. Let p/q be a rational number expressed in lowest order terms and

$$I_{p/q} = \{\nu : f_\nu^q(x) = x + p \text{ for some } x\}.$$

If $\nu \in I_{p/q}$, f_ν has a periodic orbit of period q (a q-cycle) and $\rho(f_\nu) = p/q$. If $\mu = 0$, $I_{p/q}$ is a point. Suppose this to be the case for $\mu \ne 0$. Then if $\nu < I_{p/q}$, $f_\nu^q(x) < x + p$ for all x and if $\nu > I_{p/q}$ then $f_\nu^q(x) > x + p$ for all x. Thus when $\nu = I_{p/q}$, $f_\nu^q(x) = x + p$ for all x, i.e.

$f_\nu^q = R_p$. Thus $R_{-p} \circ f_\nu^{q-1}$ is the inverse of f_ν and since f_ν is entire it has an entire inverse whence f_ν must be affine which is not the case. Therefore, $I_{p/q}$ is a non-trivial closed interval.

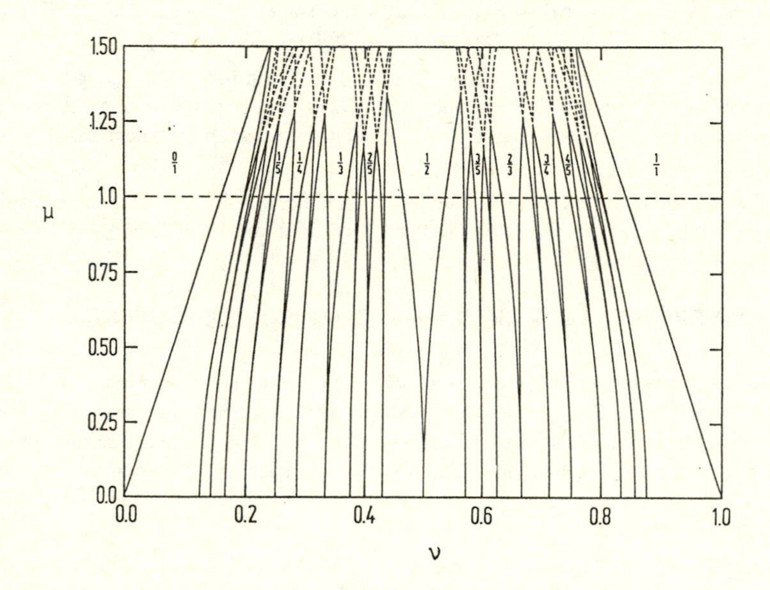

Figure 4. The Arnold tongues of the family $f_{\mu,\nu}$ above. (Courtesy of T. Bohr.)

Consequently, $\rho(f_\nu)$ is constant upon the countably infinite set of intervals $I_{p/q}$ and irrational elsewhere. To see how the intervals $I_{p/q}$ vary as μ changes consider the so-called *Arnold tongues :*

$$A_{p/q} = \{(\mu,\nu) : f_{\mu,\nu}^q(x) = x + p \text{ for some } x\}.$$

A picture of some of these is shown in Figure 4. Notice how fast they taper off as $\mu \to 0$. In fact, as $\mu \to 0$ the Lebesgue measure of the union of the $I_{p/q}$ converges to 0. Moreover as $\mu \uparrow 1$, they fill more and more of the line and it is conjectured that the union of the $I_{p/q}$ has full Lebesgue measure on $\mu = 1$. This is supported by numerical experiments of Jensen, Bak and Bohr (1983) and Lanford (1985a). Indeed Jensen, Bak and Bohr (1983) claim on the basis of their numerical experiments that it, like other families, has the property that the complement of the $I_{p/q}$s in $\mu = 1$ has Hausdorff dimension approximately equal to .87 and that the exact value of this is a universal constant. Later, I will indicate an explanation for this discovery using ideas about renormalisation strange sets from Ostlund, Rand, Sethna and Siggia (1983) and Lanford (1985b). However, for the moment I want to discuss some singular scaling properties of maps with rotation number $\upsilon = (\sqrt{5} - 1)/2$ which modulo 1 is the *golden mean.* Such a map is called *golden.* Later, in Section 5, I will discuss other irrational rotation numbers.

If ω is an irrational in (0,1), its rational approximations p_n/q_n are defined inductively by setting $p_0 = 0$ and $q_0 = 1$, and requiring that q_n is the smallest positive integer such that

$|q_n\omega - p_n| < |q_{n-1}\omega - p_{n-1}|$. These are the rational numbers obtained by truncating the continued fraction expansion $\omega = 1/(a_1 + 1/(a_2 + \cdots)) = [a_1,a_2,\ldots]$ of ω as follows:

$$p_k/q_k = [a_1,\ldots,a_k,a_{k+1} = \infty].$$

Then p_k and q_k satisfy the recursion relations

$$p_{k+1} = a_k p_k + p_{k-1}\ ,\ q_{k+1} = a_k q_k + q_{k-1}.$$

If $\omega = \upsilon$ then $a_k = 1$ for all $k \geq 1$. Thus the rational approximates of υ are ratios of Fibonacci numbers.

Now fix $|\mu| < 1$ and let $f_\nu = f_{\mu,\nu}$, so that f_ν is a diffeomorphism. Let ν_* be the value of ν such that $\rho(f_\nu) = \upsilon$ and let ν_n be the value of ν closest to ν_* such that $\rho(f_\nu) = p_n/q_n$. If $\mu = 0$, $\nu_n = p_n/q_n$ and $\nu_* = \upsilon$. More generally, if $|\mu| < 1$, the following results are consequences of Herman's theory of diffeomorphisms of the circle (Herman (1979), Yoccoz (1984)).

(a) If $f = f_\nu$ then $f^{q_n}(0) - p_n$ decreases as a^n where $a = -\upsilon$.

(b) $a^{-n}(f^{q_n}(a^n x) - p_n)$ converges, up to a scale change, to $x \to x + \upsilon$.

(c) $\lim_{n\to\infty}(\nu_n - \nu_{n-1})/(\nu_{n+1} - \nu_n) = \delta$ where $\delta = -\upsilon^{-2}$.

I now discuss the critical case $\mu = 1$.

2.2 Critical circle maps

Definition A *critical map* of the circle is the lift of an analytic homeomorphism with a single critical point and such that this critical point is cubic.

Let $f_\nu = f_{1,\nu}$. Then f_ν is critical and numerical experiments indicate the following facts:

(a)′ if $f = f_\nu$ is golden then $f^{q_n}(0) - p_n$ decreases as a^n where $a = -.776\ldots = -\upsilon^{.527\ldots}$,

(b)′ $a^{-n}(f^{q_n}(a^n x) - p_n)$ converges to an analytic function ζ of x^3 as $n \to \infty$, and

(c)′ if ν_n is as above with $\mu = 1$ then $\lim_{n\to\infty}(\nu_n - \nu_{n-1})/(\nu_{n+1} - \nu_n) = \delta$ where $\delta = -2.834\ldots = -\upsilon^{-2.164\ldots}$.

Moreover,

(d)′ if $\nu_\infty = \lim_{n\to\infty}\nu_n$ there is a neighbourhood U of $(1,\nu_\infty)$ such that if $(\mu,\nu) \in U$ then $\rho(f_{\mu,\nu}) = \upsilon$ if and only if $\nu = \nu_*(\mu)$ and $\mu \leq 1$ where the function ν_* is C^∞ on $\mu \neq 1$ and C^2 at $\mu = 1$. If $\mu < 1$ then $f_{\nu_*(\mu)}$ is analytically conjugate to the rotation R_υ, while if $\mu = 1$ then it is C^0-conjugate to R_υ.

I will say that any 2-parameter family which satisfies these conditions (a)′-(d)′, perhaps after a change of coordinates in the phase and/or parameter space, is in the *golden mean universality class*. Numerical studies show that there are many families in this universality class. Much of the remainder of this section is devoted to explaining why.

2.3 Further motivation : dissipative diffeomorphisms of the annulus.

Before proceeding to the renormalisation analysis I give some further motivation for the study of critical circle maps related to the breakup of invariant circles in dissipative diffeomorphisms of the multi-dimensional annulus. The theory for these is explained in detail in Section 3. It is this aspect that has the real physical significance because, by taking Poincaré sections, such results immediately imply analogous results for the breakdown of invariant tori in dissipative differential equations. When these ideas were first discussed in 1982 the link between critical circle maps and the breakdown of invariant circles was very surprising.

Definition. An invariant circle with rotation number υ is called *golden*.

To describe the structure for invariant circles of diffeomorphisms, consider as prototype the following family of embeddings of the annulus $\mathbf{T} \times [R^{-1}, R]$ where $R > 1$:

$$F_{\mu,\nu}(\phi,r) = (\phi + r + \nu \,,\, 1 + \lambda(r - 1) - (\mu/2\pi)\sin 2\pi(\phi + r))$$

where $\lambda > 0$ is small. This map contracts areas by a factor of λ. Let $\tilde{F}_{\mu,\nu}$ denote the lift to the universal cover $\mathbf{R}^2$ of the annulus. For $\mu = 0$, $F_{\mu,\nu}$ leaves invariant the circle $r = 1$. In fact, this circle is normally hyperbolic and therefore persists near to $r = 1$ for $|\mu|$ small. Let $f_{\mu,\nu}$ denote the restriction of $F_{\mu,\nu}$ to this invariant circle when it exists.

Let $A_{p/q}$, $p/q \in \mathbf{Q}$, denote the set of parameter values (μ,ν) such that $\tilde{F}^q_{\mu,\nu}(\phi, r) = (\phi + p, r)$ for some (ϕ, r). Let $\nu = \nu_*(\mu)$ denote the curve on which $\rho(f_{\mu,\nu}) = \upsilon$ and $\nu = \nu_{p/q}(\mu)$ denote the boundary curve of $A_{p/q}$ nearest $\nu = \nu_*(\mu)$ in the sense of Figure 4.

There is strong numerical evidence for the following facts (Feigenbaum, Kadanoff and Shenker (1982)):

(a) There exists μ_* (= .978837778... when $\lambda = 0.5$) such that if $\mu < \mu_*$ and $\nu = \nu_*(\mu)$ then $F_{\mu,\nu}$ has an analytic invariant circle and $f_{\mu,\nu}$ is analytically conjugate to a rotation. Moreover, on $[0,\mu_*)$, ν_* is analytic.

(b) F_{μ_*,ν_*} has a golden invariant circle, but it is not analytic. The map f_{μ_*,ν_*} can be conjugated to a rotation.

(c) For $\mu > \mu_*$ there is no golden invariant circle.

(d) F has the following scaling structure. (i) If $\nu_n = \nu_{p_n/q_n}(\mu_*)$ then F_{μ_*,ν_n} has a unique (p_n/q_n)-periodic orbit (a saddle node). Let a_n denote the shortest distance between two points on this orbit. Then $\lim_{n\to\infty}(a_n/a_{n+1}) = |a|$ and $\lim_{n\to\infty}(\nu_n - \nu_{n-1})/(\nu_{n+1} - \nu_n) = \delta$ where the numbers a and δ are exactly as for critical circle maps.

2.4 Renormalisation analysis

I now discuss how the results concerning critical circle maps can be understood in terms of a renormalisation transformation T. Let f be a diffeomorphism or critical map of the circle whose rotation number is approximately υ, and represent f as a map of the interval $[b, c] = [f(0) - 1, f(0)]$ as in Figure 5(a). Note that $[b, 0]$ gets mapped into $[0, c]$. Now construct a new map $\tilde{f}$ of the circle in the following way (see Figure 5(b)). Let $\hat{f} : [b, f(b)] \to [b, f(b)]$ be defined by $\hat{f} = f^2$ on $[b, 0]$ and $\hat{f} = f$ on $[0, f(b)]$. Take

$$\tilde{f}(x) = a^{-1}\hat{f}(ax)$$

where $a = a(f) = -(f(b) - b)$. Note that a is negative.

It is shown below in Section 2.5 that $\rho(\tilde{f}) = \rho(f)^{-1} - 1$; thus ρ is invariant under this process if and only if $\rho = \upsilon$, and a map with rotation number p_n/q_n is sent to one with rotation number p_{n-1}/q_{n-1}.

The problem is how to set this up as a renormalisation scheme so that the scaling properties etc. can be deduced from the structure of the dynamics near a fixed point in a similar fashion to the Feigenbaum theory for period-doubling. Firstly, note that even though f was analytic the map $\tilde{f}$ is not; it has discontinuities in its derivatives at 0 and at the end-points of the interval. A convenient and natural way to deal with this problem is to enlarge the space of

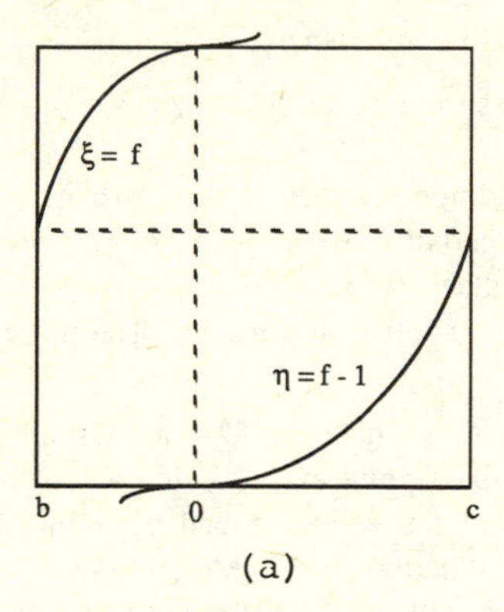

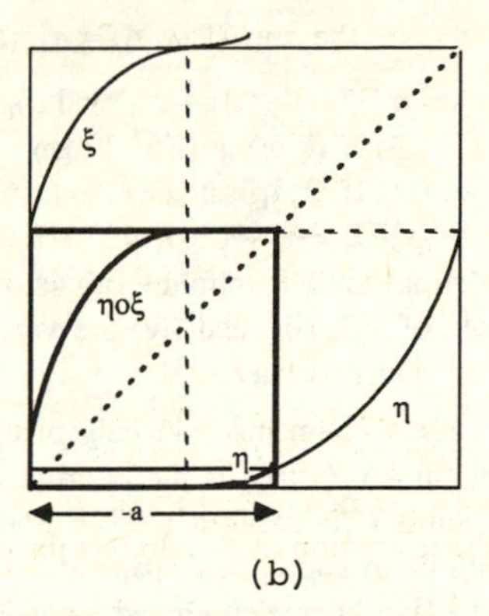

Figure 5.(a) Representation of a circle map as pair of commuting maps.
(b) Construction of the renormalised map for T_1.

maps and work with pairs (ξ, η) of analytic maps of the line which when glued together define a map of the circle.

To be specific consider the set $\mathbf{E}$ of pairs (ξ, η) which satisfy the following conditions:

(a) $0 < \xi(0) = \eta(0) + 1 < 1$;

(b) $\xi''(0) = 0$, $\xi'(\eta(0)) \neq 0 \neq \eta'(\xi(0))$ and $\xi''(\eta(0)) \neq 0 \neq \eta''(\xi(0))$; and

(c) $(\xi \circ \eta - \eta \circ \xi)^{(i)}(0) = 0$ for $0 \leq i \leq 3$.

Actually the condition (c) is only chosen for technical reasons. The natural and important condition is commutation i.e. (c)′ $\xi \circ \eta = \eta \circ \xi$ near 0, but this presents technical difficulties and (c) is adequate. The pair (ξ, η) is said to be *critical* if $\xi'(0) = 0 = \eta'(0)$.

Let R_α denote the rotation $x \to x + \alpha$. Note that if $f \in \mathbf{D}^\omega$ then the pair $u_f = (f, R_{-1} \circ f)$ satisfies the above conditions (a), (b) and (c)′ so that $\mathbf{D}^\omega$ is naturally embedded in $\mathbf{E}$. Moreover, if $(\xi, \eta) \in \mathbf{E}$ then the map $f = f_{\xi,\eta}$ defined by $f = \xi$ on $x \leq 0$ and $f = \eta$ on $x > 0$ defines a continuous map of the circle. Let $\mathbf{E}_{crit}$ denote the subset consisting of those (ξ, η) which are critical. Let $\mathbf{E}_0$ consist of those pairs in $\mathbf{E}$ such that (d) $\xi(\eta(0)) > 0$.

These condition (d) is introduced to ensure that T is well-defined and (c) ensures that the fixed points of T are hyperbolic.

The map $T : \mathbf{E}_0 \to \mathbf{E}$ corresponding to $f \to \tilde{f}$ is given by

$$T(\xi, \eta)(x) = (a^{-1}\eta(ax), a^{-1}\eta(\xi(ax)))$$

where $a = a(\xi, \eta) = -(\xi(\eta(0)) - \eta(0))$.

The transformation T has the following properties:

(i) Define $\rho(\xi, \eta) = \rho(f_{\xi,\eta})$ if $f_{\xi,\eta}$ is a homeomorphism. Then $\rho(T(\xi, \eta)) = \rho(\xi, \eta)^{-1} - 1$. Thus ρ is fixed under T if and only if $\rho = \upsilon = (\sqrt{5} - 1)/2$ and T sends $\rho = p_n/q_n$ to $\rho = p_{n-1}/q_{n-1}$.

(ii) If $f \in \mathbf{D}^\omega$ and $(\xi, \eta) = (f, R_{-1} \circ f)$ then

$$T^n(\xi,\eta) = a_{(n)}^{-1}(f^{q_n} - p_n, f^{q_{n+1}} - p_{n+1}) \circ a_{(n)}$$

where $a_{(n)} = a(T^{n-1}(\xi,\eta)).a_{(n-1)}$ and $a_0 = 1$. This uses the relation $q_{n+1} = q_n + q_{n-1}$.

(iii) If (ξ,η) is critical then so is $T(\xi,\eta)$.

(iv) If (ξ,η) satisfies the commutation condition (c)′ then so does $T(\xi,\eta)$.

(v) $\xi = x + \upsilon$, $\eta = x + \upsilon - 1$ is a hyperbolic fixed point of T with a 1-dimensional unstable manifold (with associated eigenvalue $-\upsilon^{-2}$).

The proofs of (ii), (iii) and (iv) are very simple and are left to the reader. The proofs of (i) and (v) are sketched later.

The main claim made in Feigenbaum, Kadanoff & Shenker, (1982), and Ostlund, Rand, Sethna & Siggia, (1983) is that a critical hyperbolic fixed point exists. Using a computer and methods similar to Lanford (1982), Mestel has proved the existence and hyperbolicity of a fixed point in an appropriate space of pairs of analytic functions of x^3 (Mestel (1985)). Lanford and de la Llave have also proved the existence in work which is as yet unpublished. It follows from Mestel's results that:

(i) There exist suitable and natural choices of **E** and **O** for which the map $T : \mathbf{O} \to \mathbf{E}$ is well-defined and C^∞.

(ii) T has a unique fixed point $u_* = (\xi_*, \eta_*)$ in **O** with $a = a(\xi_*, \eta_*)$ in $[-.78, -.77]$.

(iii) The spectrum of $dT(u_*)$ has two simple eigenvalues; $\delta \in [-2.9, -2.8]$ and $\gamma = a^{-2}$, with the rest of the spectrum inside the circle $|z| = .875$. The corresponding eigenvectors X_δ and X_γ have the properties described below.

(iv) The local unstable manifold of u_* associated with X_δ and X_γ is contained within the space of pairs satisfying the commutation condition (c)′ and the local strong unstable manifold associated with X_δ consists of pairs which are analytic functions of x^3.

(v) $\mathbf{O}_{crit}$, the set of critical pairs in **O**, is a 1-codimensional submanifold. Both the local stable manifold of u_* and the strong unstable manifold associated with X_δ are contained in $\mathbf{O}_{crit}$.

(vi) X_δ is positive on $[\eta_*(0), \xi_*(0)]$.

Some elementary consequences of these results are as follows:

1. Suppose that $f \in \mathbf{D}^\omega$ is such that u_f lies in the stable manifold of $u_* = (\xi_*, \eta_*)$. Then $T^n(u_f) \to u_*$ as $n \to \infty$. Since $|a(T^n(\xi,\eta)) - a_*| \le c.\nu^{-n}$ for some $c > 0$ and $0 < \nu < 1$, $a_{(n)}/a_*^n$ converges to some constant $\tau > 0$ as $n \to \infty$ so if $a = a_*$, $a^{-n}.(f^{q_n} - p_n) \circ a^n \to \tau^{-1}.\xi_* \circ \tau$. In particular, $a^{-n}.(f^{q_n}(0) - p_n) \to \tau^{-1}.(\xi_*(0))$ so $f^{q_n}(0) - p_n$ decreases as a^n.

2. Consider the submanifold Σ_n consisting of those critical u such that if f is the continuous lift of f_u to **R** then $f^{q_n}(0) = p_n$. Then $T(\Sigma_n) \subseteq \Sigma_{n-1}$ and, using the fact that near to u_* in $\mathbf{E}_{crit}$ the map T can be linearised in the direction of the unstable manifold in $\mathbf{E}_{crit}$, it follows that the Σ_n accumulate on the stable manifold $\Sigma^s = \Sigma^s(u_*)$ of u_* at the rate δ^n. A similar result holds for the sets Λ_n consisting of those u such that $f^{q_n}(x) - x \ge p_n$ for all x and $f^{q_n}(x) - x = p_n$ for some x.

Let f_ν be an analytic 1-parameter family of critical circle maps and $u_\nu = (f_\nu, R_{-1} \circ f_\nu)$. Suppose that $u_0 \in \Sigma^s$ and that $\nu \to u_\nu$ is transverse to Σ^s at $\nu = 0$. If ν_n denotes the value of ν closest to 0 such that if $f = f_\nu$ then $f^{q_n}(0) = p_n$, then ν_n corresponds to the last point where u_ν intersects Σ_n. Consequently, the ν_n accumulate on 0 at rate δ^n i.e.

$$\lim_{n \to \infty} (\nu_n - \nu_{n-1})/(\nu_{n+1} - \nu_n) = \delta.$$

Similarly this result is true if ν_n is the last value of ν such that $u_\nu \in \Lambda_n$ (i.e. ν is on the edge of I_{p_n/q_n}).

Some other consequences such as the existence of a C^0 conjugacy from $f = f_{\xi,\eta}$ to the rotation R_υ for all $(\xi,\eta) \in \Sigma^s$ are discussed below. However, before proceeding to these, I want to discuss the main theorem of this section. Part of this theorem claims the existence of an open set of families f_ν in the golden-mean universality class. Such a result is non-trivial for this case because it follows from the fixed-point equations $\xi_* = a^{-1}.\eta_* \circ a$, $\eta_* = a^{-1}.\eta_* \circ \xi_* \circ a$ where $a = a(\xi_*,\eta_*)$, that $\eta_*'(\xi_*(0)) = a^{-4}$ and $\xi_*'(\eta_*(0)) = a^{-2}$. Thus f_{u_*} is not analytic and there is a neighbourhood of u_* which contains no pairs of the form u_f with f in $\mathbf{D}^\omega$. Although this is not a problem in principle it makes it difficult to deduce that there is any such family f_ν in this universality class. This contrasts with the situation for period-doubling where the fixed point is an analytic map of the class of interest. I get over the difficulty by showing that there is a sense in which f_{u_*} is analytic. The construction involved which is described in Section 3.3 will be even more important for the study of the breakdown of invariant tori of diffeomorphisms and provides a basic connection between commuting pairs of maps and smooth maps of the circle and annulus. Before discussing this construction, I prove that T acts on rotation numbers in the way claimed above.

2.5 The action of T on ρ.

Suppose that f is the lift of a homeomorphism of the circle with $0 \le f(0) < 1$. As above represent f as a map $\hat{f}$ of the interval $[b,c] = [f(0) - 1, f(0)]$. If $x \in [b,c]$ then $f^n(x) = \hat{f}^n(x) + k_n$ where $k_n = k_n(\hat{f})$ is the number of times $x, \dots, \hat{f}^n(x)$ lies in $[0,c)$. Thus the rotation number $\rho = \rho(f)$ of f is given by

$$\rho \;=\; \lim_{n\to\infty} n^{-1}(f^n(x) - x) \;=\; \lim_{n\to\infty} k_n/n.$$

Assume that $d = \hat{f}^2(0) > 0$ so that $T(f, f-1)$ is well-defined. The interval $[b,0]$ is mapped onto $[d,c]$.

Proposition 2.1. If $d = \hat{f}^2(0) > 0$ then $\rho(T(f,f-1)) = \rho^{-1} - 1$.

Proof. Consider the orbit $x_i = \hat{f}^i(0)$. Let $x_0, \dots, x_{n-1}$ be a long orbit segment and r_n and s_n be respectively the number of points in $[b,0)$ and $[d,c)$. Then $|r_n - s_n| \le 1$ since $\hat{f}$ maps $[b,0]$ onto $[d,c]$. Let $y_i = \tilde{f}^i(0)$, $0 < i < n - s_n - 1$ where $\tilde{f} = f_{T(f,f-1)}$. Then the y_i correspond one-to-one to those x_i lying outside $(d,c]$. Thus, if $\hat{k}_n = k_n(\tilde{f})$, $\hat{k}_{n-s_n} = r_n$ and $k_n = n - r_n$. Therefore,

$$\rho(\tilde{f}) = \lim_{n\to\infty} \hat{k}_{n-s_n}/(n-s_n) = \lim_{n\to\infty} r_n/(n - r_n) = \lim_{n\to\infty} (n-k_n)/k_n = \rho^{-1} - 1$$

as required. ∎

Note. In Section 5 I shall introduce renormalisation transformations for general rotation numbers by defining transformations T_n analogous to $T = T_1$. A similar proof shows that these have the property that $\rho(T_n(u)) = \rho(u)^{-1} - n$.

2.6 Commuting pairs and circle maps

Clearly if $f \in \mathbf{D}^\omega$ then the pair $(f, R_{-1} \circ f)$ commutes i.e. satisfies condition (c)′. Hence, if p is a diffeomorphism of some appropriate interval into $\mathbf{R}$ then the pair $(p^{-1} \circ f \circ p, p^{-1} \circ R_{-1} \circ f \circ p)$ commutes. I show in this section that all the commuting pairs (ξ, η) of interest are of this sort.

The construction which proves this depends upon the existence of an analytic map g which sends some neighbourhood U of $\eta(0)$ diffeomorphically onto a neighbourhood of $\xi(0)$ and is such that

$$g \circ \eta = \xi \text{ near } 0 \text{ and } \eta \circ g = \xi \text{ near } \eta(0). \tag{2.1}$$

Then (ξ, η) will be of the form claimed. There are three special cases to which I shall want to apply this result, namely: (a) when ξ and η commute, η has an analytic inverse near 0 and $g = \xi \circ \eta^{-1}$; (b) when ξ and η commute, $\xi = e(x^3)$, $\eta = f(x^3)$, f has an analytic inverse near 0 and $g = e \circ f^{-1}$, and (c) to the multidimensional fixed point as explained in Section 3. In these cases the fact that g satisfies the relations (2.1) above follows directly from the commutativity.

The idea behind the proof that such a pair (ξ, η) is of the form $(p^{-1} \circ f \circ p, p^{-1} \circ R_{-1} \circ f \circ p)$ is as follows. Let Δ be a neighbourhood of $[\eta(0), \xi(0)]$ which contains U and gU and such that $f_{\xi,\eta}$ is defined on Δ and sends Δ into itself. Then if points in Δ are identified by g, the quotient manifold $\Sigma = \Delta / g$ is clearly a topological circle. Moreover, Σ acquires the structure of an analytic manifold via the canonical projection $\pi : \Delta \to \Sigma$. It is easy to see that $f = f_{\xi,\eta} : \Delta \to \Delta$ induces a continuous map $\hat{f} : \Sigma \to \Sigma$ such that $\hat{f} \circ \pi = \pi \circ f$. Clearly, f is analytic away from $\pi(0)$ and $\pi(\eta(0))$ since ξ and η are. But $\hat{f}(\pi(x)) = \pi(\xi(x))$ for x near 0 and for $\pi(x)$ near $\pi(\eta(0))$, so $\hat{f}$ is analytic. Since Σ is a circle its universal cover is $\tilde{\pi} : \mathbf{R} \to \Sigma$ and $\hat{f}$ lifts to an analytic f in $\mathbf{D}^\omega$. Since Δ is simply connected there is a diffeomorphism $p : \Delta \to \mathbf{R}$ such that $\pi = \tilde{\pi} \circ p$. Using this one sees that $\xi = p^{-1} \circ f \circ p$ and $\eta = p^{-1} \circ R_{-1} \circ f \circ p$.

I want to emphasise that this argument did not really depend upon the fact that ξ and η were defined on intervals, and works equally well in higher dimensions to give analytic maps of the multi-dimensional annulus $A^n = \mathbf{T}^1 \times \mathbf{R}^{n-1}$.

2.7 Main theorem for circle maps

Theorem 2.1. There is a 2-codimensional submanifold Σ in $\mathbf{D}^\omega$ such that if $f_{\mu,\nu}$ is a 2-parameter family in $\mathbf{D}^\omega$ which intersects Σ transversally at $(\mu,\nu) = (0,0)$ then $f_{\mu,\nu}$ is in our universality class.

Outline of proof. The first point is that by the result described in Section 2.6 there exists p_* and f_* in $\mathbf{D}^\omega$ such that $\xi_* = p_*^{-1} \circ f_* \circ p_*$ and $\eta_* = p_*^{-1} \circ R_{-1} \circ f_* \circ p_*$. For f in $\mathbf{D}^\omega$ let $z(f)$ be defined by $f''(z(f)) = 0$. Then $z(f)$ is uniquely defined and C^∞ on some neighbourhood of f_* in $\mathbf{D}^\omega$. If f lies in this neighbourhood let $\bar{f} = f \circ R_{-z(f)}$. Consider the map $\Phi(f) = (p_*^{-1} \circ \bar{f} \circ p_*, p_*^{-1} \circ R_{-1} \circ \bar{f} \circ p_*)$ defined on some neighbourhood U of f_*. This sends f_* to u_*. Moreover, if $X_1(x) = \sin 2\pi x$ then $d\Phi(f_*).X_1$ is clearly transverse to the 1-codimensional subspace of critical pairs $\mathbf{O}_{crit}$. A calculation also shows that the tangent vector X_2 to the image under Φ of $t \to (p_*^{-1} \circ R_t \circ f \circ p_*, p_*^{-1} \circ R_{t-1} \circ f \circ p_*)$ at $t = 0$ is transverse to the stable manifold Σ^s of u_* in $\mathbf{O}$ (see Rand (1984)). But this last vector is tangent to $\mathbf{O}_{crit}$. Thus $T_{u_*}\mathbf{O}$ is spanned by $T_{u_*}\Sigma^s$ and the images of X_1 and X_2 i.e. Φ is transverse to Σ^s at u_*. Thus

$\Sigma = \Phi^{-1}(\Sigma^s)$ is a submanifold. Any 2-parameter family $f_{\mu,\nu}$ transverse to Σ is sent by Φ to a family $u_{\mu,\nu}$ transverse to Σ^s. The universality for the family $u_{\mu,\nu}$ follows from the usual arguments for such renormalisation saddle points as above and in Rand (1984), and follows for $f_{\mu,\nu}$ because this only differs from $u_{\mu,\nu}$ by the conjugacy p_*. ■

2.8 Existence of a conjugacy to a rotation

Theorem 2.2. Suppose that $T^n(\xi,\eta) \to (\xi_*,\eta_*)$ as $n \to \infty$. Then $f = f_{\xi,\eta}$ is C^0-conjugate to the notation R_υ.

Proof. A calculation shows that for $f = f_{u_*}$, $|f'| > a_*^{-2}$ on $[f^2(0), f(0)]$. Now consider the case where $u = (\xi,\eta)$ is so close to $u_* = (\xi_*,\eta_*)$ that if $u_n = (\xi_n,\eta_n) = T^n(\xi,\eta)$ and $f_n = f_{u_n}$ then $f'_n > .9a_*^{-2}$ on $[f_n^2(0), f_n(0)]$ and $|a_{u_n}^{-1}| > .9|a_*^{-1}|$ for all $n \geq 0$. If $f = f_0$ is not C^0-conjugate to a rotation then there exists a non-trivial closed interval I such that $f^n I \cap I = \varnothing$ for all $n > 0$. Let I_f denote a largest such interval.

Note that 0, $f(0)$ and $f^2(0)$ cannot lie in I_f because $f^{q_n}(0) \to 0$ from both sides as $n \to \infty$. This is because $T^n(\xi,\eta) \to (\xi_*,\eta_*)$ as $n \to \infty$ and $|a_*| < 1$. Also $I_f \cap (f^2(0), f(0)) = \varnothing$ because otherwise $I_f \subset (f^2(0), f(0))$ and then $|fI_f| > |I_f|$ contradicting the maximal length of I_f. Thus if $J = a_f^{-1} I_f$, $(Tf)^n J \cap J = \varnothing$ for all $n > 0$ whence $|I_{Tf}| > .9a_*^{-1}|I_f|$. This implies that on iterating T, $|I_{T^m f}| > 1$ for some $m > 0$ which gives a contradiction.

Finally, note that the assumption that u was close to u_* was not necessary since $f = f_{\xi,\eta}$ is C^0-conjugate to a rotation if and only if $f_{T(\xi,\eta)}$ is. This is because one has a non-trivial interval I_f if and only if the other does. If $u = (\xi,\eta)$ is any element in a stable manifold of u_* then for large n, $T^n(u)$ is so close to u_* that the above argument can be applied. Thus $f_{T^n u}$ is C^0-conjugate to R_υ. Therefore, f is C^0-conjugate to R_υ. ■

Note. MacKay (1987) has given a new and very simple proof of Denjoy's Theorem along these lines by exploiting the way in which the renormalisation transformation acts upon the variation of $\log |f'|$.

2.9 C^1-uniqueness of the conjugacy.

I now describe some numerical results about these conjugacies (see Ostlund, Rand, Sethna & Siggia (1983) for a detailed discussion). Let $f = f_{\xi,\eta}$ with (ξ,η) in the stable manifold of (ξ_*,η_*). Let $h = h(f)$ denote the associated conjugacy to a rotation and let $\chi = h - id$. Then $\chi(\theta + 1) = \chi(\theta)$. Let the kth Fourier coefficient of χ be denoted by $\hat{\chi}(k)$. Then $q_n \hat{\chi}(q_n)$ converges to a positive constant as $n \to \infty$. In fact, if Q_n is any sequence of integers satisfying $Q_{n+1} = Q_n + Q_{n-1}$ then $Q_n\, \chi(Q_n)$ also converges to a positive constant as $n \to \infty$ with the constant depending upon the sequence (but not on the map f). A good and also experimentally useful way of expressing this is as follows. Let

$$\tilde{f}(\omega) = \lim_{L \to \infty} L^{-1} \sum_{l=0}^{L-1} e^{2\pi i l \omega}(f^l(0) - l\upsilon).$$

Then if $\omega = n\upsilon \bmod 1$,

$$\tilde{f}(\omega) = \lim_{L \to \infty} L^{-1} \sum_{l=0}^{L-1} e^{2\pi i n l \upsilon} \chi(l\upsilon) = \int_0^1 e^{2\pi i n \theta} \chi(\theta) d\theta = \hat{\chi}(n).$$

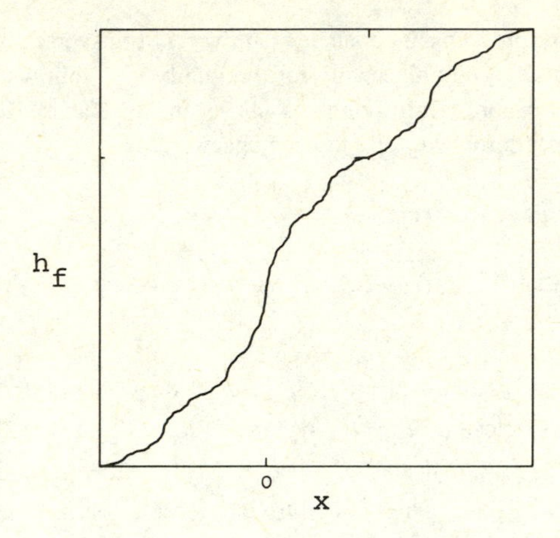

Figure 6. A numerical approximation to the conjugacy between the map $f : x \to x + \nu - (1/2\pi)\sin 2\pi x$ and the rotation R_υ.

One finds numerically that $\tilde{f}$ is universal as $\omega \to 0$. In fact, if g is another critical map $\tilde{f}(\omega) - \tilde{g}(\omega) = O(\omega \tilde{g}(\omega))$. Also, $\tilde{f}$ satisfies the scaling law

$$\tilde{f}(\omega) = -\upsilon \tilde{f}(\omega/\upsilon) + O(\omega^2).$$

The universal form of $\tilde{f}$ is shown in Figure 7.

A proof of these facts, based on the following conjectures,[2] which were also found numerically, is given in Ostlund, Rand, Sethna & Siggia (1983).

Conjecture 1. If f and g are critical maps with $\rho(f) = \rho(g) = \upsilon$, then f and g are C^1-conjugate.

Conjecture 2. If $f = f_{\xi_1,\eta_1}$ and $g = f_{\xi_2,\eta_2}$ where each (ξ_i, η_i) is critical and $\xi_i \circ \eta_i = \eta_i \circ \xi_i$, then f and g are conjugated by a homeomorphism h such that h and h^{-1} are C^1 except at the endpoints $\eta_i(0)$ and $\xi_i(0)$, $i = 1,2$.

Another way of saying this is as follows. If f and g are two golden critical circle maps in the stable manifold and h_f and h_g are the corresponding conjugacies to the rotation R_υ, then $h_f^{-1} \circ h_g$ is C^1. The numerical evidence was that this conjugacy was not in general C^2.

Universality of the low-frequency power spectrum and the scaling spectrum (which is analogous to that described in Section 1.4 for period-doubling) follows from this. It is therefore more fundamental as a test of phase-space universality, and rather than constructing subsidiary objects such as the power spectrum one should directly construct the conjugacy from a

[2] **Note added in proof.** I now have a proof of a stronger version of these conjectures.

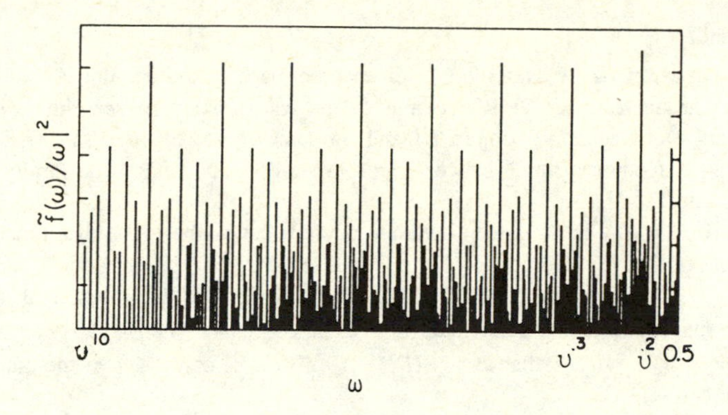

Figure 7. $\tilde{f}$ for the map f of Figure 6.

time-series to a standard example of a golden critical map and attempt to determine whether or not this is C^1.

3. UNIVERSALITY FOR THE BREAKDOWN OF DISSIPATIVE GOLDEN INVARIANT TORI.

Consider a flow with an attracting invariant torus T on which the flow has no stationary points. Let S be a smooth circle on T which is transverse to the flow on T. Then there is a small annular transversal Σ to T which meets T in S such that the flow is transverse to Σ. If $x \in \Sigma$ let $f(x)$ denote the point where the orbit through x first returns to Σ. This defines a return map $f : \Sigma \to \Sigma$ which has S as an invariant circle. The map f is an embedding (i.e. a diffeomorphism onto its image) because of the way it comes from a differential equation.

My aim in this section is to study the way in which such a torus or the associated invariant circle breaks down as some parameter on which the equations depend is varied. This will give results about the structure of the complex bifurcation sets around and organised by points in parameter space where invariant tori break down. In fact, I obtain some global self-similarity from the renormalisation structure developed in Section 5 for general rotation numbers. The aim in this section will be to prove a version of Theorem 2.1 for golden invariant circles of dissipative diffeomorphisms of the annulus $A^n = \mathbf{T}^1 \times \mathbf{R}^{n-1}$ which proves the observations described in Section 2.3 above. Firstly, I construct a renormalisation transformation on pairs of commuting maps in $\mathbf{R}^n$. Then I use an argument similar to that of Theorem 2.1 and results about renormalisation and the existence and non-existence of invariant circles to apply this to analytic annulus maps. The construction of the renormalisation transformation uses ideas developed by Collet, Eckmann and Koch (1981b) for period-doubling. The proof uses the structure of the 1-dimensional problem in a fundamental way.

3.1 Universality.

I shall say that a 2-parameter family $F_{\mu,\nu}$ of analytic maps of the annulus $A^n = \mathbf{T}^1 \times \mathbf{R}^{n-1}$ or their lifts to the universal cover $\mathbf{R}^n$ belongs to the *golden-mean universality class* if for the same numbers a and δ as in Theorem 2.1 and for some choice of the parametrisation μ,ν and some analytic diffeomorphism P between appropriate domains of $\mathbf{R}^2$, the family has the following properties:

(i) $F = F_{0,0}$ has a golden invariant circle which is the graph of a function which is Lipschitz but not C^1.

(ii) There is a sequence B_n of coordinate changes of the polynomial form given in (3.1) below and converging exponentially fast to the coordinate change B_* given in (3.3) such that if $B_{(n)} = B_1 \circ \cdots \circ B_n \circ P$ then $B_{(n)}^{-1} \circ (F^{q_n} - (p_n, 0)) \circ B_{(n)}$ converges to the map E_* defined in (3.2) as $n \to \infty$.

(iii) If $\nu = \nu_n$ is such that $F_{0,\nu}$ has a p_n/q_n-periodic point on $x = 0$ and ν_n is minimal with this property, then $\lim_{n \to \infty} \delta^n (\nu_{n+1} - \nu_n)$ exists and is non-zero.

(iv) There is a neighbourhood U of (0,0) such that if $(\mu,\nu) \in U$ then $F_{\mu,\nu}$ has a golden invariant circle if and only if $\nu = \nu_*(\mu)$ and $\mu \leq 0$ where the function ν_* is C^∞ on $\mu < 0$ and C^2 at $\mu = 0$. If $\mu < 0$ then this circle is C^∞.

The rest of this section is an outline of the ideas behind the proof of the following theorem. However, some of the ideas of these later sections are of more general interest. For example, one can use analogues of Proposition 3.2 to prove existence of invariant circles in other contexts and Hoidn (1985) has used an extension of these ideas to prove a version of the Moser Twist Theorem.

Theorem 3.1. There is a 2-codimensional submanifold Σ in the space of analytic diffeomorphisms of the annulus such that if $F_{\mu,\nu}$ is a 2-parameter family which transversely intersects Σ at $(\mu,\nu) = (0,0)$ then $F_{\mu,\nu}$ is in the golden-mean universality class.

3.2 Renormalisation structure in higher dimensions.

Firstly, note an important difference between the 1-dimensional and multi-dimensional cases. In the 1-dimensional case I was able to easily pick out which point to scale about. For critical maps it was the critical point while for diffeomorphisms it did not matter. There is no such obvious choice in the higher dimensional case since I want to treat diffeomorphisms and one must only work about special points on the invariant circle. In the 1-dimensional case I could have ignored the choice of origin and allowed transformations of the form $b(x) = \tau_0 + \tau_1 x$ in place of the simple scale change $x \to ax$. In one dimension it was clear that these are the only coordinate changes to worry about. However, in higher dimensions it is necessary to allow a slightly larger class of coordinate changes. In fact I use a renormalisation transformation of the form

$$T(E, F) = B^{-1} \circ (F, F \circ E) \circ B$$

where $B = B_{E,F}$ belongs to the set $\mathbf{P}$ of polynomial maps of the form

$$B(x,y) = (\tau_0 + \tau_1 x,\ \underline{\tau}_0 + \underline{\tau}_1 x + \underline{\tau}_2 x^2 + \underline{\tau}_3 x^3 + \underline{\mu}.y), \tag{3.1}$$

where $(x,y) \in \mathbf{R} \times \mathbf{R}^{n-1}$, $\tau_0, \tau_1 \in \mathbf{R}$, $\underline{\tau}_0, \underline{\tau}_1, \underline{\tau}_2, \underline{\tau}_3 \in \mathbf{R}^{n-1}$ and $\underline{\mu} \in \mathrm{Lin}(\mathbf{R}^{n-1}, \mathbf{R}^{n-1})$.

Fix $\alpha \in \mathbf{R}^{n-1}$ such that $|\alpha| = 1$ and define

$$E_*(x,y) = (\xi_*((x^3 - \alpha.y)^{1/3}),\ 0) \text{ and } F_*(x,y) = (\eta_*((x^3 - \alpha.y)^{1/3}),\ 0). \tag{3.2}$$

(Recall that ξ_* and η_* are analytic functions of x^3.) Then clearly $U_* = (E_*, F_*)$ is a fixed point of the map T if

$$B_{E_*,F_*}(x,y) = B_*(x,y) = (ax, a^3 y) \tag{3.3}$$

where $a = a(\xi_*, \eta_*)$. The commutation conditions that are imposed are $(E \circ F - F \circ E)(0) = 0$ and $(\partial^3/\partial^3 x)(E \circ F - F \circ E)(0) = 0$.

Theorem 3.2. There is a affine map $(E, F) \to B_{E,F}$ into **P** such that the transformation T has the following properties. (i) $B(U_*) = B_*$. (ii) It is well-defined and C^∞ on an appropriate space and $dT(U_*)$ is compact. (iii) The eigenvalues of $dT(U_*)$ of modulus > 1 are δ and $\gamma = a^{-2}$. The other eigenvalues are contained strictly inside $|z| = 1$. (iv) The eigenvectors associated with δ and γ are of the form $P_X = (X((x^3 - \alpha.y)^{1/3}), 0)$ where X is the corresponding eigenvector for the 1-dimensional problem.

The proof of this Theorem is in Rand (1984), and I restrict my comments here to trying to give some more insight into the choice of $B_{E,F}$ and further explaining the point made at the beginning of this section about the need to factor out extra coordinate changes in order to obtain a renormalisation transformation with stable and unstable manifolds of the appropriate dimension and codimension. Instead of T one could consider $S(E,F) = B_*^{-1} \circ (F, F \circ E) \circ B_*$. Then U_* is a fixed point of S. It turns out that the only eigenvalues of $dS(U_*)$ of modulus > 1 are as follows: (i) δ which is simple and whose eigenvector is of the form P_X; (ii) a^{-2} whose spectral subspace is spanned by an eigenvector of the form P_X and an $(n-1)$-dimensional subspace corresponding to coordinate changes of the form $(0, \underline{\tau}_1 x)$; (iii) a^{-3}, a^{-1} and 1 whose spectral subspaces are of dimensions $n-1$, n and $n^2 - n + 1$ respectively and which correspond to coordinate changes of the form (3.1) with $\underline{\tau}_1 = 0$.

These extra eigenvalues of modulus > 1 correspond to the extra coordinate changes that have to be factored out. One constructs $(E, F) \to B_{E,F}$ using the spectral projection of $dS(U_*)$ associated with the eigenvalues a^{-3}, a^{-2}, a^{-1} and 1 to pick out the appropriate coordinate change. Equivalently, B can be fixed by imposing a number of normalisation conditions on (E, F) such as $E(0,0) = (1,0)$.

3.3 Commuting pairs and invariant circles.

Suppose that (E, F) is close to $U_* = (E_*, F_*)$. As in the 1-dimensional case, if there exists a diffeomorphism G from a neighbourhood V of $F(0,0)$ to a neighbourhood of $E(0,0)$ such that $G \circ F = E$ near (0,0) and $F \circ G = E$ near $F(0,0)$ then there exists a diffeomorphism P between appropriate domains and an analytic map Φ of the annulus into itself such that $E = P^{-1} \circ \Phi \circ P$ and $F = P^{-1} \circ R_{-1} \circ \Phi \circ P$ where $R_{-1}(x,y) = (x - 1, y)$. The particular cases that I shall be interested in are (a) when (E, F) commutes and E and F are invertible on the appropriate domains so that one can take $E \circ F^{-1}$ for G and (b) when $(E, F) = (E_*, F_*)$ and $G(x,y) = ((g((x^3 - \alpha.y)^{1/3})^3 + \alpha.y)^{1/3}, y)$ where $g = e_* \circ f_*^{-1} = \xi_* \circ \eta_*^{-1}$. In the latter case I denote the associated annulus map by Φ_*.

Given (E, F), let $K = K_{E,F}$ denote the map defined as follows: $K = E$ on $x \geq 0$ and $K = F$ on $x < 0$. I say that (E, F) has an invariant circle if $K_{E,F}$ does. This is called a C^r, circle, $0 < r \leq \omega$, if it is C^r in the annulus obtained by factoring out G.

Proposition 3.1. $T(E, F)$ has a C^r golden invariant circle if and only if (E, F) does.

Proof. If σ is a golden invariant circle for (E, F) and $B = B_{E,F}$ then $B^{-1}\sigma$ is one for $T(E, F)$. On the other hand if σ is a golden invariant circle for $T(E, F)$ and $B = B_{E,F}$ then $\sigma_1 \cup \sigma_2$ is one for (E, F) where $\sigma_1 = B\sigma$ and $\sigma_2 = E\sigma_1$. ■

3.4 Existence of golden invariant circles.

I now want to indicate how to prove the existence of golden invariant circles for pairs (E, F) that are asymptotic to a fixed point U_* of T. Here the main application of this result is to the case $U_* = (E_*, F_*)$ but one can also apply it to pairs converging to the trivial fixed point. The same ideas work for arbitrary rotation numbers ω satisfying a Diophantine condition of the following form: there exist constants $C > 0$ and $\tau > 0$ such that for all relatively prime $p, q \in \mathbf{Z}$ with $q \neq 0$,

$$|\omega - p/q| > C/|q|^{2+\tau}.$$

In the area-preserving case, an extension of these ideas has been used by Hoidn (1985) to deduce a version of the Moser Twist Theorem.

Definition. A *cyclic orbit segment* γ *of length* N of $K = K_{E,F}$ is a sequence of points $z_0, \ldots, z_{N-1}$ such that $z_i = K(z_{i-1})$ and such that if $z_i = (x_i, y_i)$ then $x_0 = 0$ and $x_0, \ldots, x_{N-1}$ have the same ordering on $\mathbf{R}$ as they would for the rotation R_υ. The Lipschitz constant $L(\gamma)$ associated with γ is the least $k > 0$ such that for each i, γ is contained in the cone $C_i(k)$ given by $|y - y_i| \leq k|x - x_i|$.

Lemma 3.1. Suppose that (E, F) is invertible. If $T(E, F)$ has a cyclic orbit segment γ_1 of length q_n then (E, F) has one γ of length q_{n+1}. Moreover, if (E, F) is sufficiently close to (E_*, F_*), $L(\gamma) \leq L(\gamma_1)$.

Proof. Let $\gamma' = B\gamma_1$ where $B = B_{E,F}$. Then q_{n-1} of the points of γ' lie in $x \leq 0$. Let γ'' be the image of these points under E. Then for γ take $\gamma' \cup \gamma''$. To see the last fact note that since γ_1 is contained in the cones $C_i(k)$, $k = L(\gamma_1)$ then γ' is contained in the image $C_i'(k)$ of each of these under B and γ'' is in the images $C_i''(k)$ under E of the $C_i'(k)$ corresponding to points with $x_i \leq 0$. Thus a necessary and sufficient condition for a set of cones $C_j(k')$ to suffice is that the one through $E(F(0))$ (the point at which γ' and γ'' join) contains all the $C_i'(k)$ and the $C_i''(k)$ corresponding to points with $x_i \leq 0$. ■

Proposition 3.2. Suppose that (E, F) is invertible and lies in the stable manifold of (E_*, F_*). Then (E, F) has a golden invariant circle of the form $y = \psi(x)$ where ψ is a Lipschitz function.

Proof. Since $(E_n, F_n) = T^n(E, F) \to (E_*, F_*)$ as $n \to \infty$ one can construct a cyclic orbit segment $\gamma^{(n)}$ of length q_n for (E, F). Let z_0 be a limit point of the set of initial points of the $\gamma^{(n)}$. Then the orbit $z_0, z_1, \ldots$ of z_0 under $K = K_{E,F}$ is ordered as for R_υ. Moreover, $L(\gamma^{(n)})$ is bounded since by Lemma 3.1, $L(\gamma^{(n+1)}) \leq L(\gamma^{(n)})$ for large n. Thus if $z_i = (x_i, y_i)$ then there is a Lipschitz function ψ such that $y_i = \psi(x_i)$. Using T, this then gives similar orbits $y_i^{(n)} = \psi^{(n)}(x_i^{(n)})$ for (E_n, F_n), $n > 0$.

It remains to check that the x_i are dense in the interval. If this not the case there exists a *complementary interval* J for (E, F) i.e. a subinterval which does not contain an x_i, and hence complementary intervals for each of the (E_n, F_n). Let J_n be the longest such. J_n cannot contain $x_0^{(n)} = 0$ and hence not $x_1^{(n)} < 0$ and $x_2^{(n)} > 0$. This is the case because the convergence to the fixed point implies that x_0 is a limit point of the x_n, $n > 0$. Moreover, since $|\partial F_n/\partial x| > a_*^{-1}$ for n large (since the same is true of (E_*, F_*)) the longest such interval J_n must be contained in $[x_1^{(n)}, x_2^{(n)}]$. Thus (E_{n+1}, F_{n+1}) has a complementary interval J which is essentially the image under the B for (E_n, F_n) of J_n. But if (E_n, F_n) is very close to (E_*, F_*), B is close to B_* and $|J|$ is approximately $a_*^{-1}|J_n|$. Letting J_{n+1} denote the largest such interval J, it follows that $|J_{n+1}| > 0.9a_*^{-1}|J_n|$ for all large n which is clearly a contradiction since $0.9|a_*^{-1}| > 1$ then implies that $|J_n| > 1$ for large n whereas, in fact, $|J_n| < 1$ for all $n \geq 0$. ■

Note. With very small changes the above argument can be used to prove that if (E_n, F_n) converges to the *simple fixed point* $U_{*,0}$ given by $E_{*,0}(x,y) = (x + \alpha.y + \upsilon, 0)$ and $F_{*,0}(x,y) = (x + \alpha.y + \upsilon - 1, 0)$ then (E, F) has an golden invariant circle. A more sophisticated argument shows that in this case the golden invariant circle is C^∞.

3.5 Outline of proof and nonexistence of golden invariant circles.

1. The unstable manifold of U_* defines a 2-parameter family $U_{\mu,\nu}$ of retractions onto the circle $\mathbf{T}^1$ whose parametrisation by $(\mu,\nu) \in \mathbf{R}^2$ can be chosen so that if $\mu < 0$ then $u_{\mu,\nu} = U_{\mu,\nu}|\mathbf{T}^1$ is a diffeomorphism of $\mathbf{T}^1$ (after gluing by g) and $\rho(f_{\mu,\nu}) = \upsilon$ if and only if $\nu = 0$. Using Herman's Theorem, (Herman (1979), Yoccoz (1984)) $u_{\mu,0}$ can be conjugated to a rotation whence $T^n(U_{\mu,0}) \to U_{*,0}$ as $n \to \infty$ if $\mu < 0$.

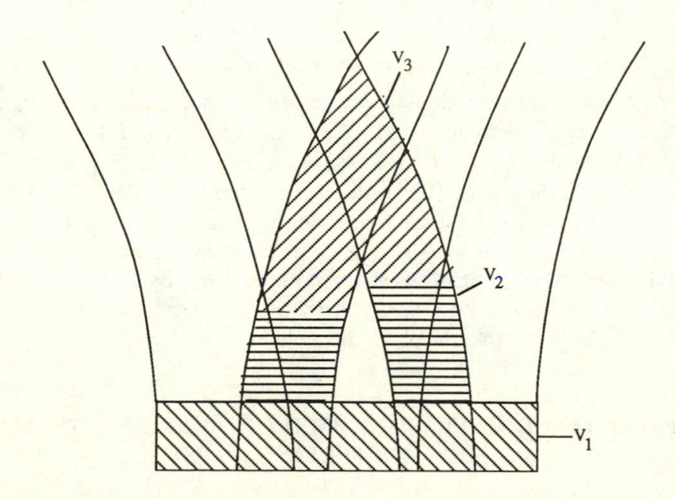

Figure 8. The annular region $A = V_1 \cup V_2 \cup V_3$ in the unstable manifold $U_{\mu,\nu}$

2. Now choose an annular region $A = V_1 \cup V_2 \cup V_3$ in the unstable manifold $U_{\mu,\nu}$ of the form shown in Figure 8 and with the following properties:

(i) If (μ,ν) is close to $(0,0)$ then $T^n(U_{\mu,\nu}) \in A$ for two consecutive n.

(ii) If $(\mu,\nu) \in V_1$ then $u_{\mu,\nu}$ is a diffeomorphism and $\rho(u_{\mu,\nu}) = \upsilon$ if and only if $\nu = 0$.

(iii) If $(\mu,\nu) \in V_2$ then $U_{\mu,\nu}$ has no orbit with rotation number υ.

(iv) If $(\mu,\nu) \in V_3$ then $u = u_{\mu,\nu}$ is such that there is a small subinterval I of $\mathbf{T}^1$ such that $u^m(I) = \mathbf{T}^1$ for some $m \geq 0$.

The existence of such regions follows from the theory of circle maps (see Rand (1984) for references).

But then one can extend A to an open subset $\mathbf{A} = \mathbf{V}_1 \cup \mathbf{V}_2 \cup \mathbf{V}_3$ of the full function space such that if U is near the stable manifold Σ^s of U_* then $T^n(U) \in \mathbf{A}$ for some $n \geq 0$ and such that:

(i) There is a 1-codimensional submanifold W of $\mathbf{V}_1$ such that if $U \in \mathbf{V}_1$ then U has a golden invariant circle if and only if $U \in W$. To see that one can extend V_1 to get $\mathbf{V}_1$ note that it follows from the results of Rand (1984) and arguments analogous to those of the previous section that there is a neighbourhood $\mathbf{W}$ of $U_{*,0}$ and a 1-codimensional submanifold W' of $\mathbf{W}$ such that U in $\mathbf{W}$ has a golden invariant circle if and only if $U \in W'$. Thus, by 1, there is a neighbourhood $\mathbf{V}$ of the curve in the unstable manifold given by $\mu < 0$, $\nu = 0$ in which the set of U with a golden invariant circle forms a 1-codimensional C^∞ submanifold.

(ii) If $U \in \mathbf{V}_2$ then U has no orbit with rotation number υ.

(iii) If $U \in \mathbf{V}_3$ then U has an overflowing strip i.e. there exists a strip I in the annulus of the form $\theta_0 \leq \theta \leq \theta_0 + c$, $c > 0$ small, such that for some $m > 0$ the projection of $U^m(I)$ into $\mathbf{T}^1$ covers $\mathbf{T}^1$.

3. Thus if $U \in \mathbf{A}$ then U has a golden invariant circle if and only if $U \in W$. It follows from Proposition 3.1 that if U is near Σ^s then U has a golden invariant circle if and only if $T^n(U) \in W$ for some $n \geq 0$. With a little work it can now be shown that if $U \in \Sigma^u$ then there is a coordinate system (x,y,z) at U with $x,y \in \mathbf{R}$ and z in the unit ball of some Banach space, such that $U_{(x,y,z)}$ has a golden invariant circle if and only if $x = 0$ and $y \leq 0$.

4. The rest of the proof is much the same as that of Theorem 2.1 (with Φ_* playing the role of f_*). As there, one uses conjugation by P_* to construct a map Ξ from a neighbourhood of Φ_* to a neighbourhood of U_* such that $\Xi(\Phi_*) = U_*$ and such that Ξ is transverse to Σ^u at Φ_*. Let Σ' be the intersection of $\Xi^{-1}(\Sigma^u)$ with this neighbourhood. Then Σ' is a submanifold, at least if one reduces the neighbourhood sufficiently. One then shows that Σ' contains a diffeomorphism Φ. Next one applies 3 to deduce the existence of the coordinate system (x,y,z) at $U = \Xi(\Phi)$. Using the transversality and pulling back by Ξ, gives a similar coordinate system (x',y',z') at Φ with the property that $\Phi(x',y',z')$ has a golden invariant circle if and only if $x' = 0$ and $y' \leq 0$. This completes the outline of the proof. ■

4. PERSISTENCE AND DESTRUCTION OF INVARIANT CIRCLES IN AREA-PRESERVING MAPS.

This section is about the persistence and destruction of invariant tori in Hamiltonian systems with two degrees of freedom. The main general references for what is covered are Arnold (1978), Moser (1973) and Siegel & Moser (1971). Specific references are Greene (1968, 1979, 1980), Shenker & Kadanoff (1982) and MacKay (1982, 1983). Other recent important references on this topic are Mather (1982, 1984, 1986), Aubry (1983), Aubry & Le Daeron (1983) and Herman (1983) but I do not discuss their results here. You should also read

Kolmogorov's address (Kolmogorov (1954)).

4.1 Introduction and motivation

Recall that a n-degree of freedom Hamiltonian system with Hamiltonian H is completely integrable if there exist n integrals of the motion $H = f_1, f_2, \ldots, f_n$ which are in involution. Then if the energy surface $H = cst.$ is compact, say a $(2n-1)$-sphere, the level surfaces of the map $f = (f_1, \ldots, f_n)$ corresponding to regular values are n-dimensional tori and the flow induced by H on these tori is linear. Thus locally the phase space can be represented as $\mathbf{T}^n \times \mathbf{R}^n$. The associated system of canonical coordinates $(\theta, r) \in \mathbf{T}^n \times \mathbf{R}^n$ are the so-called angle-action coordinates. In this coordinate system the equations of motion reduce to

$$\dot{I} = 0\,,\quad \dot{\theta} = \omega(I)$$

Call such a system nonresonant if the derivative of ω is invertible at each value of I. The Kolmogorov-Arnold-Moser (KAM) Theorem states that most of these invariant tori persist when a non-resonant integrable system is perturbed by a small amount. There are also versions of this result for certain special resonant systems. This sort of result has a number of practical implications, for example, in the following fields.

Celestial mechanics. The Earth's motion about the Sun is a completely integrable dynamical system if one considers it to be merely a two-body system. However, the other planets introduce perturbations which could possibly lead to instability of the Earth's orbit. Similar remarks hold for many other celestial systems. The Sun, Jupiter and the Trojan group of asteroids form an approximate equilateral triangle which is reminiscent of Lagrange's equilateral triangle solution of the 3-body problem (Siegel & Moser (1971), p. 91). In fact, in a problem with 3 planar bodies P_1, P_2 and P_3 whose masses satisfy $m_1 >> m_2 >> m_3$, the effect of P_2 is to perturb the simple Kepler orbit of P_3 about P_1. Thus orbits on those invariant tori which persist under the perturbation will remain bounded but some of those in regions of phase-space where the tori have been destroyed will be unstable. It has been proposed that this is related to the observation by Kirkwood in the 1860s of gaps in the distribution of asteroids ($P_1 = \text{Sun}, P_2 = \text{Jupiter}, P_3 = \text{asteroid}$) corresponding to resonances.

Statistical Mechanics. The statistical ensemble of states of Boltzmann's hard-sphere gas tend to equilibrium for any initial state and it was expected that any sufficiently "complex" system would do the same. The KAM theorem shows that this is certainly not the case for, for example, a system of weakly coupled oscillators as in this case invariant tori persist and there will be no equipartition of energy. More precisely, for this system the Liouville measure will not be ergodic. However, in this situation the strength of the coupling allowed in the hypotheses of the KAM theorem will tend to zero as the number n of oscillators grows. Thus if n is of the order of Avogadro's number the KAM Theorem is not likely to be practicable.

Plasma confinement. The simplest model for a plasma is to regard it as a conducting fluid. Then the magnetic field is tangent to the surface of constant pressure, at least in equilibrium. These surfaces are called the magnetic surfaces. To confine the plasma in, say, a solid torus it is therefore necessary that some of the magnetic surfaces are tori. Now, although this system is not strictly speaking a Hamiltonian system, it gives rise (by following the field lines once round the solid torus) to an area-preserving return map in certain coordinates. The confining toroidal surfaces will correspond to the invariant circles of the map.

In each of these areas, and more generally, one would like to go beyond the KAM theorem and ask how do stability and confinement break down. In particular, how do invariant tori, which persist for small perturbations λH_1, break down as λ increases? I will only consider area-preserving maps here. These arise as the Poincaré maps associated with systems with two degrees of freedom. The invariant tori for the flow correspond to invariant circles for the map. In this particular case the invariant circles separate the phase-space so breakdown really does correspond to the loss of stability and confinement in the strongest sense. On the other hand, for systems with more degrees of freedom the invariant tori do not separate the phase-space.

4.2 Twist maps

Let θ, r denote polar coordinates in the plane and consider the map $(\theta, r) \rightarrow (\theta_1, r_1)$ defined by

$$\theta_1 = \theta + \alpha(r), \quad r_1 = r \tag{4.1}$$

of the annulus $A : a_0 \leq r \leq b_0$ into itself. This corresponds to the angle-action equations for an integrable system. The map (4.1) leaves each circle about the origin invariant, rotating it through an angle $\alpha(r)$ which is assumed to increase with r i.e. $\alpha'(r) > 0$. My aim is to study small and large perturbations of such maps. These will have the form

$$\theta_1 = \theta + \alpha(r) + \tilde{f}(\theta, r), \quad r_1 = r + \tilde{g}(\theta, r) \tag{4.2}$$

where $\tilde{f}$ and $\tilde{g}$ are periodic in θ. If $\partial\theta_1/\partial r = \alpha' + \partial\tilde{f}/\partial r$ is bounded away from 0, this is called a *twist map*. Putting $x = \theta$ and $y = \alpha(r)/\kappa$ where $\kappa = |\alpha(b_0) - \alpha(a_0)| > 0$ then (4.2) becomes

$$x_1 = x + \kappa y + f(x,y), \quad y_1 = y + g(x,y) \tag{4.3}$$

where f and g are periodic in x. It is this particular class of maps which I shall look at.

Example 4.1. Consider an analytic area preserving map $P : \mathbf{R}^2 \rightarrow \mathbf{R}^2$. Suppose that $P(0) = 0$ and that $dP(0)$ is an elliptic matrix i.e. its eigenvalues λ, λ^{-1} have modulus 1 but are not equal to ± 1. Assume that $\lambda^k \neq 1$ for $k = 3, \ldots, 2l + 1$. Then (Siegel & Moser (1971), Section 23), in suitable coordinates near 0

$$P(u,v) = (u\cos w - v\sin w + O_{2l+2},\ u\sin w + v\cos w + O_{2l+2})$$

where

$$w = a + b(u^2 + v^2)^l, \quad b > 0$$

and O_{2l+2} stands for a power series in u and v of order $\geq 2l + 2$.

Now consider P in the punctured disk $0 < u^2 + v^2 < \varepsilon^2$ and introduce coordinates x and y by

$$u = \varepsilon y^{1/2l} \cos 2\pi x, \quad v = \varepsilon y^{1/2l} \sin 2\pi x.$$

In the new coordinates the map takes the form

$$(x,y) \rightarrow (x + a + b\varepsilon^{2l} y + O(\varepsilon^{2l+1}),\ y + O(\varepsilon^{2l+1}))$$

where the terms are analytic functions in (x,y) for $0 < y < 1$ and have period 1 in x. Making the coordinate change $(x,y) \rightarrow (x, y - ab^{-1}\varepsilon^{-2l})$ puts P in the form

$$x_1 = x + b\varepsilon^{2l} y + O(\varepsilon^{2l+1}), \quad y_1 = y + O(\varepsilon^{2l+1}) \tag{4.4}$$

Example 4.2. Consider the class of perturbed twist maps defined as follows. Let $\phi : \mathbf{T} \rightarrow \mathbf{R}$

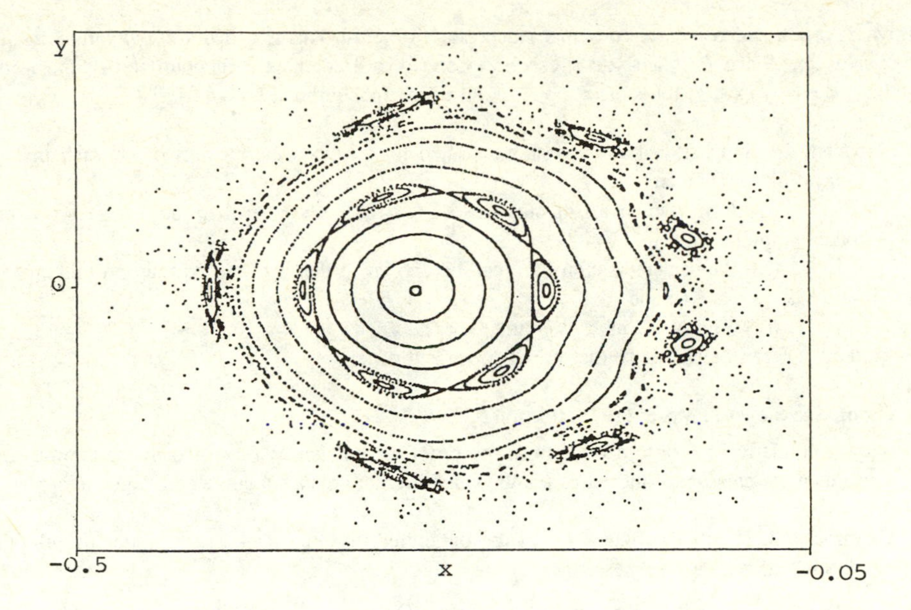

Figure 9. Some orbits surrounding an elliptic fixed point

be analytic and P_ϕ be defined by

$$P_\phi(\theta, r) = (\theta + r,\ r + \phi(\theta + r)). \tag{4.5}$$

With $\phi = k\sin$, (4.5) is the much studied standard map. Assume that $\int_0^1 \phi = 0$. Then $P = P_\phi$ has the following *intersection property:* if C be a closed curve of the form $r = r(\theta)$, then $P(C) \cap C \neq \emptyset$.

Suppose that the curve $C : r = \psi(\theta)$ is invariant under P. Since $P_\phi(C) = C$ and $P_\phi^{-1}(C) = C$,

$$(id + \psi)^{-1} + (id + \psi) = 2id + \phi \tag{4.6}$$

and conversely. Thus given a diffeomorphism h of the circle, by setting $\psi = h - id$, one can construct a twist map with an invariant circle on which the induced map is h.

4.3 Moser Twist Theorem

Consider the twist map P given by

$$x_1 = x + \kappa y + f(x,y),\ y_1 = y + g(x,y)$$

with $\kappa > 0$, f and g periodic in x, and $y \in [a,b]$. I assume that P possesses the intersection property. For $0 < s_0 < 1/4$ choose ω so that $\kappa^{-1}\omega$ lies in $[a + s_0, b - s_0]$ and for all $p/q \in \mathbf{Q}$ in lowest order terms,

$$|\omega - p/q| \geq c/|q|^{2+\sigma}$$

where $c > 0$ and $\sigma > 0$ are constants independent of p/q. This is called a *Diophantine condition* on ω. Since P is analytic it can be extended to a complex neighbourhood of the form $|\text{im} x| < r, |y - \omega| < s$, for some $r, s > 0$. Call this neighbourhood U.

Theorem 4.1 Suppose that P is as above. Then for all $\varepsilon > 0$ there exists $\delta > 0$ such that if $|f| + |g| < \delta\kappa$ on U, then

(i) P has an invariant curve $x = \xi + u(\xi), y = v(\xi)$ with u and v analytic and periodic on $|\text{im}\xi| < r/2$;

(ii) the parametrisation by ξ can be chosen so that the induced map on this invariant curve is $\xi \to \xi + \omega$; and

(iii) $|u| + |v - \kappa^{-1}\omega| < \varepsilon$ on $|\text{im}\xi| < r/2$.

Moreover, δ can be chosen so that it is independent of κ.

Proof. See Siegel & Moser (1971) Section 32.

It is relatively easy to deduce from the proof of this that as $f, g \to 0$ in the appropriate sense then the measure of the set of points which are not on such a curve converges to zero.

Example 4.3. Recall Example 4.1. In the punctured disk $0 < u^2 + v^2 < \varepsilon$ about the elliptic fixed point the map was expressed as

$$P : x_1 = x + b\varepsilon^{2l} y + O(\varepsilon^{2l+1}),\ y_1 = y + O(\varepsilon^{2l+1}).$$

The error terms f and g are estimated by $(|f| + |g|)/\kappa = O(\varepsilon^{2l+1})/b\varepsilon^{2l} = O(\varepsilon)$. Moreover, P possesses the intersection property because if a closed curve $y = \psi(x)$ for x and y real does not intersect its image then the two regions enclosed by these curves (in the (u,v)-plane) would have different areas. The existence of invariant circles in the punctured disk $0 < u^2 + v^2 < \varepsilon^2$ follows immediately. Moreover, the relative measure of these in the disk converges to 1 as $\varepsilon \to 0$.

4.4 Empirical results on the breakdown of golden invariant tori

As prototypes consider the families

$$P_k : x_1 = x + y\ ,\ y_1 = y + (k/2\pi)\sin 2\pi(x + y) \tag{4.7}$$

and

$$Q_p : x_1 = p - y - x^2\ ,\ y_1 = x. \tag{4.8}$$

From the Twist Theorem it follows that, for $k > 0$ small, P_k has an analytic invariant circle with golden rotation number $\upsilon = (\sqrt{5} - 1)/2$. Careful numerical studies show that this breaks down when $k = k_c = 0.9716354...$(Greene (1968, 1979), Shenker & Kadanoff (1982)). Similarly the golden invariant torus for Q_p breaks down at $p = p_c = 2.382194...$. At the breakdown there is a golden invariant torus, but it does not appear to be analytic and the conjugacy h between the induced map and a rotation is certainly not differentiable. A picture of the critical invariant torus for Q_{p_c} is given in Figure 10 and Figure 11 shows the Fourier coefficients $\hat{\chi}(k)$ of $\chi = h - id$ from which one sees that $\sum_{k \geq 0} k\hat{\chi}(k)$ certainly does not converge.

I am going to claim here that the structure of this breakdown has certain quantitatively universal features similar to those of the dissipative case discussed in Section 3. This was discovered numerically by Shenker and Kadanoff (1982) following work of Greene (1979). Here I closely follow the work of MacKay (1982, 1983) who was the first person to

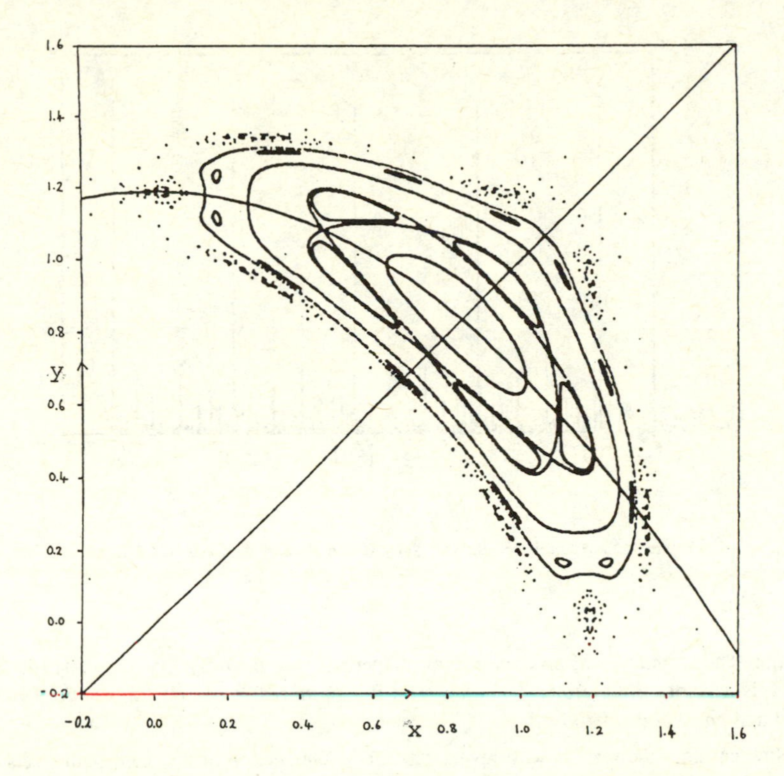

Figure 10. Birkhoff orbits in the quadratic map Q_{P_c}, converging to a critical circle (the outermost one) whose rotation number is in the same $SL_2(\mathbf{Z})$-orbit as υ. Also shown are two symmetry lines. From MacKay (1982).

successfully implement an exact renormalisation scheme for this problem. This approach closely follows that for circle maps described in Section 2. Previously, Escande and Doveil (1981a,b) had invented an approximate scheme where the renormalisation transformation acted on the Hamiltonian of a model system. I now describe the empirical scaling results in more detail as these give some insight into what is happening.

4.5 Reversibility.

A map P is said to have be *reversible* if it possesses a *reversor* S by which I mean an involution S ($S^2 = id$) such that $(PS)^2 = id$. Then $SPS^{-1} = P^{-1}$ and P^nS is also a reversor for P. A *symmetry line* of P is a component of the fixed point set of some reversor S of P. Near a fixed point of S one can choose coordinates so that $S(x,y) = (-x,y)$. These are called *symmetry coordinates.*

Note. $S(x,y) = (-x+y,y)$ is a reversor for the map given by $x_1 = x + y_1$, $y_1 = y + g(x)$. If $g(x) = k\sin 2\pi x$ then this gives a reversor for the family P_k above.

A periodic orbit is said to be *symmetric* with respect to S if it is its own reflection under S. It is then easy to see that a periodic orbit is symmetric if and only if it has a point on some

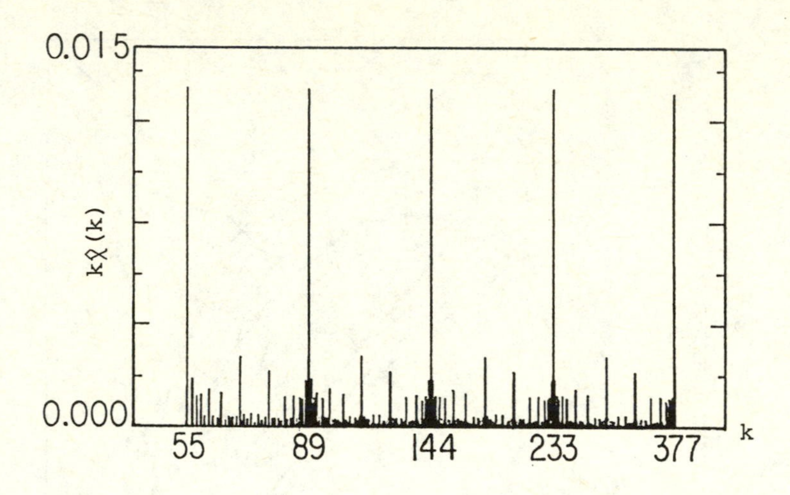

Figure 11. The Fourier coefficients $\hat{\chi}(k)$. (From Shenker & Kadanoff (1982).)

symmetry line.

Definition The *residue* R of a periodic orbit of period q is given by $R = (2 - \text{trace}\, dP^q(x))/4$ where x is a point on the orbit. Then the orbit is *elliptic* if $0 < R < 1$, *hyperbolic* if R is not in $[0,1]$ and *parabolic* otherwise.

Now consider the map P with reversor S. The intersection of the fixed points sets of the reversors S and PS give a fixed point of P. If this is removed four half-lines are left. Shenker and Kadanoff (1982) and Greene (in unpublished work) made the observation that for (4.7) and most other examples studied, on one of these half-lines there is a periodic point with non-negative residue and rotation number p/q for all $p/q \in \mathbf{Q}$. Following MacKay this half-line and the periodic points on it are called *dominant.*

Now I can discuss the scaling behaviour for (4.7) and (4.8) in detail. Let μ_n denote the (k or p) parameter value for which there is a p_n/q_n-periodic point on the dominant half-line with residue equal to 1. Then in both cases as $n \to \infty$

$$(\mu_n - \mu_{n-1})/(\mu_{n+1} - \mu_n) \to \delta = 1.632...$$

and $\mu_\infty = \lim_{n\to\infty} \mu_n$ is the value at which the golden torus breaks down. Let (X, Y) be symmetry coordinates with respect to the dominant half-line and let (O, Y_n) denote the coordinates of the above dominant p_n/q_n-periodic point. Then as $n \to \infty$,

$$(Y_n - Y_{n-1})/(Y_{n+1} - Y_n) \to \beta = -3.06688....$$

Let X_n be the X-coordinate of the point in the orbit of the dominant periodic point nearest to the dominant half-line. Then as $n \to \infty$,

$$(X_n - X_{n-1})/(X_{n+1} - X_n) \to \alpha = -1.41483...$$

Choose the origin of the (X, Y)-coordinate system to be at the limit of the dominant p_n/q_n-periodic points. Let $\tilde{P}$ be the lift of P to the universal cover $\mathbf{R}^2$. Then from the above

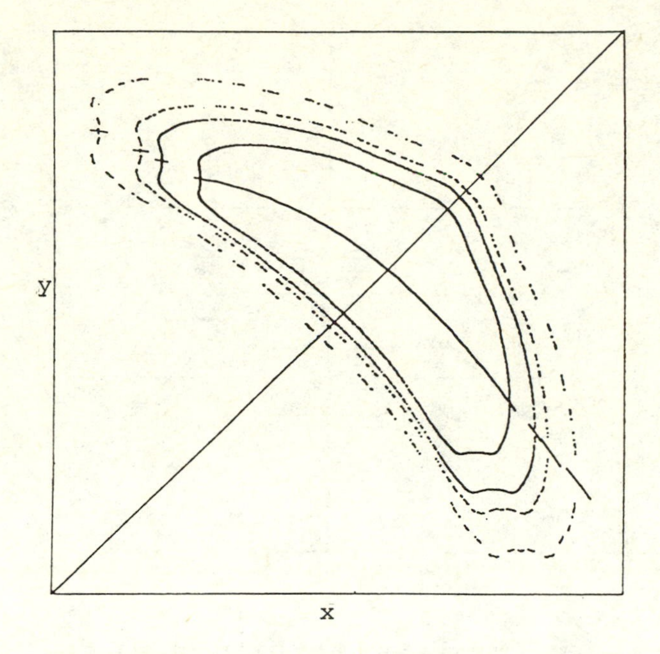

Figure 12. An orbit of rotation number υ for four parameter values in the quadratic map, (4.8) one subcritical, one critical, and two supercritical. From MacKay (1982).

one expects that in the (X,Y)-coordinates $P_n = \Lambda^n \circ R_{-1}{}^{p_n} \circ P^{q_n} \circ \Lambda^{-n}$ converges as $n \to \infty$ to an analytic area-preserving twist map if $\Lambda = diag(\alpha,\beta)$. To convince you of this MacKay's beautiful picture of P_n for $n = 11$ is included as Figure 13. Note that after a reflection everything repeats itself on a smaller scale in the smaller box obtained by scaling down by α^{-1} and β^{-1} in the X and Y directions respectively. It is a non-trivial task to get the right symmetry coordinates here (see MacKay (1982) for an explanation of how to do it).

I will say that any area-preserving twist map which satisfies the above lies in the *golden area-preserving universality class.*

4.6 Renormalisation.

I outline the details and results for the renormalisation scheme for this problem.

4.6.1 Action representation

Given a function $\tau(x,x_1)$, the relations

$$y_1 = \tau_2(x,x_1)\ ,\ y = -\tau_1(x,x_1)$$

generate an area-preserving map $P : (x,y) \to (x_1,y_1)$. (τ_i denotes the derivative of τ with respect to the ith argument.) One says that τ is a generating function for P, and I write this symbolically as $\tau \to P$. (See Arnold (1978) Section 47 for a discussion of generating functions.) I shall also assume that $\tau_{12} < 0$ because then

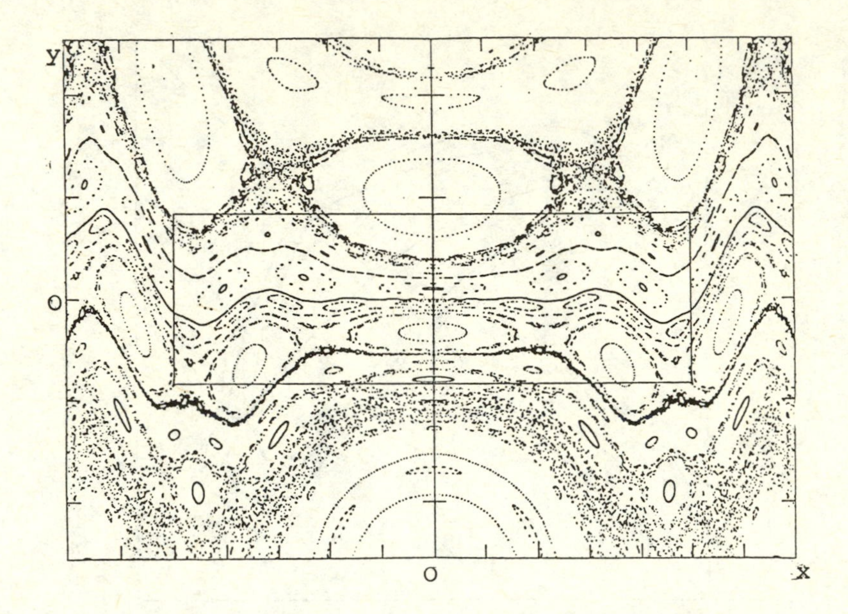

Figure 13. Some orbits of the universal map for the neighbourhood of critical golden circles. From MacKay (1982).

$$\partial x_1/\partial y = -1/\tau_{12}(x,x_1) > 0$$

so that P satisfies the twist condition.

Given two generating functions ν and τ, define $\tau\circ\nu$ as follows

$$\tau\circ\nu(x,x_2) = \nu(x,x_1) + \tau(x_1,x_2) \tag{4.9}$$

where $x_1 = x_1(x,x_2)$, is chosen so as to make the sum (4.9) stationary with respect to perturbations in x_1 i.e. $\nu_2(x,x_1) + \tau_1(x_1,x_2) = 0$. I ignore any problems about uniqueness in this definition as they do not occur in the present context. It is easy to see that $\nu \to E$ and $\tau \to F$ implies $\tau\circ\nu \to F\circ E$. Moreover, if $\tau \to P$ and $S(x,y) = (-x,y)$ is a reversor for P, then

$$\tau(x,x_1) = \tau(-x_1,-x) \tag{4.10}$$

4.6.2 Renormalisation transformation

The idea is to define a renormalisation transformation as in the dissipative case which is of the form

$$T(E,F) = \Lambda^{-1}\circ(F,F\circ E)\circ\Lambda \tag{4.11}$$

where in some suitably chosen symmetric coordinate system Λ is a diagonal matrix of the form $\Lambda(x,y) = (ax,by)$. In fact, one chooses a and b so that the following conditions are satisfied

(a) the x-coordinate of $F(0,0)$ is a, and

(b) the twist $\partial x_1/\partial y\,(0,0)$ of F at (0,0) is b/a.

This ensures that the renormalised maps $(\tilde{E},\tilde{F}) = T(E,F)$ satisfy the following: (i) the x-coordinate of $\tilde{E}(0,0)$ is 1 and (ii) the twist $\partial x_1/\partial y$ associated with $\tilde{E}$ at (0,0) is normalised to 1. At the fixed point (E_*, F_*) (if it exists) one will have $a = \alpha^{-1}$ and $b = \beta^{-1}$. A schematic representation of this is shown in Figure 14.

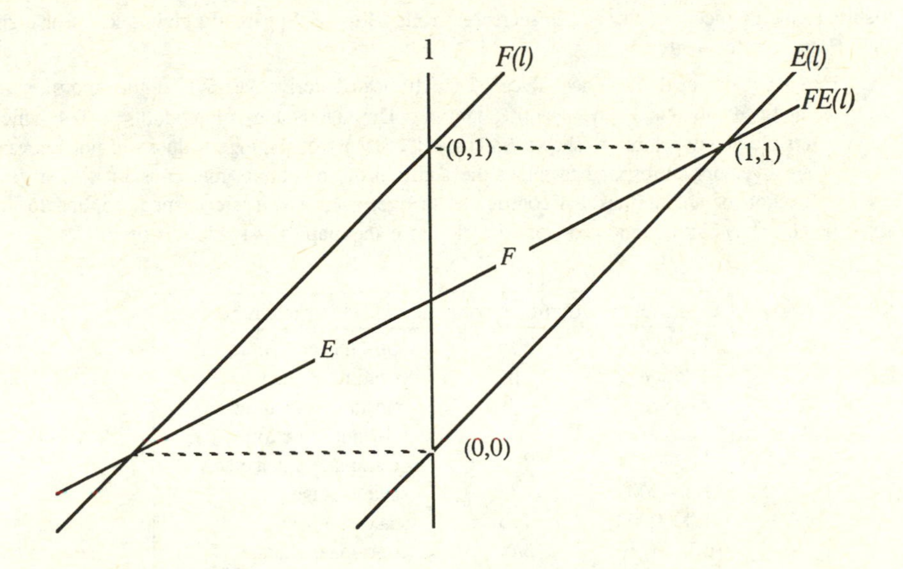

Figure 14. A schematic representation of the construction of T.

It is convenient to use the action representation because this is a convenient way to force the maps considered to be area-preserving which, for example, is otherwise difficult to enforce in a computerised scheme involving truncations. In this representation the renormalisation transformation (4.11) becomes

$$N(\nu,\tau)(x,x_1) = a^{-1}b^{-1}(\tau(ax,ax_1)\,,\ \tau\circ\nu(ax,ax_1)) \tag{4.12}$$

where a and b are chosen so that $\tau(0,a) = 0$ and $b/a = \tau_{12}(0,a)$. If $(\tilde{\nu},\tilde{\tau}) = N(\nu,\tau)$ this normalisation implies $\tilde{\nu}(0,1) = 0$ and $\tilde{\nu}_{12}(0,1) = 1$.

It turns out to be very difficult to find good domains in $\mathbb{C}^2$ on which to define E and F so that the composition, scaling etc. have the right properties i.e. so that ν,τ analytic on these domains implies $\tilde{\nu},\tilde{\tau}$ analytic on slightly bigger domains. However, MacKay discovered that the situation is simpler if one concentrates on reversible maps and looks for a reversible fixed point of N. The reversor $S(x,y) = (-x,y)$ is assumed. Then for the renormalisation transformation take

$$N_S(\nu,\tau)(x,x_1) = a^{-1}b^{-1}(\tau(-ax_1,-ax),\tau\circ\nu(ax,ax_1)). \tag{4.13}$$

By (4.10), N_S and N agree on the space of reversible functions. Moreover, a simple calculation shows that if E and F are reversible then $\tilde{E}$ and $\tilde{F}$ are reversible if and only if E and F

commute. Thus N and N_S preserve and agree on the space of reversible commuting maps.

With this new renormalisation N_S one can find good domains in $\mathbf{C}^2$ for ν and τ, for example, for τ use $|x - c| < r$ and $|x_1 - c_1| < r_1$ where $c = .050707985$, $r = .502060282$, $c_1 = -.655406307$ and $r_1 = .329680205$ and for ν use the rescaled and reflected domains. MacKay then used Newton's method to find an approximate fixed point of N_S truncating at various degrees. The scaling factors α and β can be read off from this. The results converge nicely as the degree increases from 14 to 18, apparently giving seven or eight figure accuracy for degree 18.

He then calculated the eigenvalues of the truncated derivative dN_S at the approximate fixed point by diagonalising the resulting matrix. The eigenvalues of modulus > 0.4 which have reversible eigenvectors are shown in Table 4.1. Most of the eigenvalues are not relevant as their eigenvectors correspond to either the addition of an infinitesimal constant to the generating function or an infinitesimal coordinate change, such as a scale change, applied to the fixed point. They can be removed from the picture if the map (4.14) below is used.

Eigenvalue	Compare with:	Interpretation
7.0208826	$\alpha\beta/\upsilon$	constant term in action
-3.0668882	β	coordinate change
-2.6817385	$-\alpha\beta\upsilon$	constant term in action
1.6279500	δ	relevant direction
-1.5320951	β/α^2	coordinate change
1.0000001	α/α	scale change
1.0000000	β/β	scale change
-0.7653736	β/α^4	coordinate change
-0.6108303	δ'	essential convergence rate
0.4995593	α/α^3	coordinate change

Table 4.1.

The non-reversible eigenvalues which have been omitted may, of course be non-commuting. From the table it can be seen that there is only one relevant eigenvalue of modulus > 1, the others corresponding to infinitesimal coordinate changes and the addition of constants to the generating functions ν and τ. The values of these non-relevant eigenvalues are calculated as in Section 3.2 (for details see Rand (1984)).

The eigenvalue δ' has an interesting meaning. Let R_n^+ and R_n^- be the residues of the (p_n/q_n)-periodic orbit with non-negative and non-positive residues respectively and assume that the system is critical (i.e. $k = k_c$ or $p = p_c$). Then as $n \to \infty$,

$$R_n^+ \to 0.250088\ldots \text{ and } R_n^- \to -0.255426\ldots$$

and the convergence is at rate $\delta' = -0.6108\ldots$. This convergence rate was first discovered by Greene (1979).

There is no reason in principle why these calculations cannot be carried out to arbitrary precision and a proof given along the lines of Lanford's proof of the Feigenbaum Conjectures. However, it is clear that the numerical estimates are much more delicate in this case and the eigenvectors etc. relatively ill-conditioned, so much more computing time and power would be involved.

Finally, note that the eigenvalues corresponding to addition of a constant to the action will not be present for the transformation T given by (4.11) on pairs of area-preserving maps. Moreover, if instead of the transformation T given in (4.11) one uses

$$T(E,F) = (B^{-1}\circ F\circ B\,,\, B^{-1}\circ F\circ E\circ B) \tag{4.14}$$

with $B(x,y) = (ax, by - c - dx^2)$ chosen so as to preserve the (four) normalisation conditions implied by $E(0,0) = (0,0)$, $F(0,1) = (0,1)$, $E(0,1) = (1,1)$ and the positive twist, then the only eigenvalue of $dT(E_*, F_*)$ of modulus > 1 is $\delta = 1.62795...$. The use of B to provide a coordinate change to impose the normalisations removes the four eigenvalues β, β/α^2, α/α and β/β corresponding to the four basic relevant coordinate changes.

4.7 The boundary of existence and non-existence of invariant circles.

The existence of the fixed-point $U_* = (E_*, F_*)$ has very interesting consequences for the structure of the boundary of the set of area-preserving maps with an invariant torus. As in the dissipative case there exists a diffeomorphism P_* such that

$$E_* = P_*^{-1}\circ f_*\circ P_* \text{ and } F_* = P_*^{-1}\circ R_{-1}\circ f_*\circ P_*$$

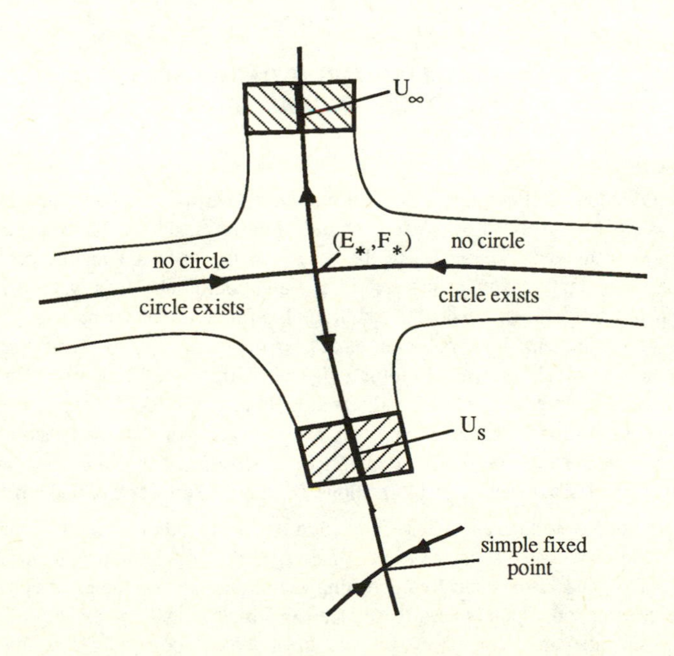

Figure 15. Schematic representation of the action of T near the fixed point and the fundamental neighbourhoods defined below.

where f_* is the lift of an area-preserving diffeomorphism of the annulus $\mathbf{T}^1 \times \mathbf{R}$ to its universal cover $\mathbf{R}^2$. Then, as in the dissipative case, the map defined on a neighbourhood of f_* by

$$\Xi(f) = (P_*\circ f\circ P_*^{-1}\,,\, P_*\circ R_{-1}\circ f\circ P_*^{-1})$$

sends f_* to U_* and is transverse to the stable manifold Σ^s of U_* at U_*. Thus $\Sigma = \Xi^{-1}(\Sigma^s)$ is a 1-codimensional submanifold near to f_* and if f_k is a 1-parameter family transversally intersecting Σ at $k=0$ then f_0 has the universal self-similar structure in phase-space described above and the family f_k has the universal scaling structure in parameter space. To deduce that for $k \le 0$, f_k has a golden invariant circle I assume that one has a situation as shown in Figure 15. In particular, assume (i) that the unstable manifold of U_* contains a fundamental neighbourhood U^s which converges to the *simple line* of integrable maps given by the pairs $((x+y/\omega-\omega\ ,y)\ ,(x+1+y\ ,y))$ under iteration of the transformation T of (4.14) and (ii) that on the other side of Σ^s it contains a fundamental neighbourhood U^∞ such that no f in U^∞ has a golden invariant circle. Then if (E,F) lies just "below" Σ^s it converges to the simple line and one deduces, as in Rand (1984) and Section 3.4, that it has a golden invariant circle. If it is just above, under iteration of T it gets sufficiently close to U^∞ that non-existence of a golden invariant circle can be deduced. From this it would follow that in a neighbourhood of f_* the boundary of existence of golden invariant circles is exactly given by the submanifold Σ. Thus, for the family f_k above, the golden invariant circle breaks down at $k=0$. The interesting problem thrown up by this argument is the conjecture (ii) above. It is very important to develop techniques for proving such a result. This is the one major part of the argument which we do not know how to do.

5. RENORMALISATION STRANGE SETS AND FRACTAL BIFURCATION SETS.

5.1 Introduction.

The theorems and conjectures that I have discussed above use the traditional renormalisation formalism which was invented by Wilson to study phase transitions. Its use in dynamical systems to study the universal properties of the transition to chaos has mainly been restricted to the study of single bifurcation points and their neighbourhoods. However, many bifurcation processes in dynamics involve complicated fractal bifurcation sets whose global structure is more physically important. Two important examples of this are (a) the Cantor set of points in parameter space at which an invariant torus of general irrational rotation number breaks down and (b) the Cantor structure of the 2-codimensional set of period-doublings encountered, for example, inside an Arnold tongue. These sets act as organising centres for complicated bifurcation sets which are often encountered in physical systems, for example in many of those experiments involving a sequence of bifurcations from a phase-locked periodic orbit to chaos.

The traditional formalism relies upon finding a hyperbolic saddle point. Then the geometrical and dynamical structure of the saddle point and its stable manifold is used to deduce physically and mathematically interesting consequences. In the extension I want to discuss in this section and which is used to deduce the structure and universality of such complex fractal structures in parameter space, the role of the fixed point is played by a hyperbolic strange set Λ which can be a strange saddle (e.g. a horseshoe) or a strange attractor. I call such sets *renormalisation strange sets*. One of the basic observations will be that the fractal bifurcation sets within a given universality class are all *lipeomorphic* to each other provided either the unstable manifolds of the renormalisation strange set are 1-dimensional or the expansion and contraction rates in the renormalisation strange set sayisfy certain conditions. A *lipeomorphism* is a Lipschitz homeomorphism which has a Lipschitz inverse so lipeomorphic sets have the same scaling properties. I believe that the ideas involved here are of wider interest than dynamical systems because such fractal structures are very likely to abound in

Physics.

In discussing various examples one of my main aims will be to describe the sort of interesting universal objects that can be deduced from a renormalisation strange set. For renormalisation strange sets, the objects corresponding to the eigenvalues of the linearisation at the fixed point and the functional structure of the fixed point will be dimensions, scaling functions and scaling spectra derived from the detailed dynamical structure of the strange set Λ. They have analogous interpretations in terms of the fractal bifurcation set. The scaling spectra mentioned above are invariants of the fractal structure and give a representation of the spectrum of scales in the fractal bifurcation sets and their relation with dynamical quantities. Universality of the bifurcation sets up to lipeomorphism implies universality of these subsidiary objects.

Because of limitations of space, I shall only discuss the following two examples here:

1. Critical circle maps with general rotation number which also gives the universal structures associated to the the breakdown of invariant tori in strongly dissipative systems.

2. The structure of the cascade of breakdowns of the invariant circles of area-preserving maps.

Other examples of renormalisation strange sets in dynamical systems includes:

3. A theory for unimodal maps of the interval with general kneading invariant which generalises the usual period-doubling theory (Rand (1987)).

4. An application to deduce self-similarity of the boundary of chaos in some 2-parameter unfoldings of homoclinic orbits (Rand (1987)) and the more general, but symbolic, formalism of Procaccia, Thomae and Tresser (1987) for renormalising the bifurcation sets of Lorenz-like maps.

5. The renormalisation structure developed by van Zeijts in his Ph.D. thesis to describe the 2-codimensional Cantor set of 2-codimensional period-doublings in maps with two generic critical points. This has also been analysed on the symbolic level by Gambaudo, Los and Tresser (1987).

6. The work of MacKay and Percival (1986) on the structure of the boundaries of Siegel domains of general rotation number.

Another early example, which, strictly speaking is not within dynamical systems and which I will also not discuss is the application to the structure of quasi-periodic Schrödinger operators due to Kohmoto, Kadanoff and Tang (1983) and Ostlund, Pandit, Rand, Schellnhuber and Siggia (1982). This was perhaps the first example where a renormalisation strange set was properly worked out and an excellent rigorous treatment is given by Casdagli (1986).

The ideas presented here are very speculative and much remains to check even the various conjectures let alone provide rigorous proofs. Proofs in this area have tended to rely upon computers to rigorously check certain estimates and although all the hypotheses I make can certainly be checked in this way in principle, at the present time it is beyond our computing facilities. However, I believe that the most important aspect of the sort of analysis presented here is that it provides a powerful mathematical picture which can be justified subject to a small number of precise hypotheses and which leads to a number of new and useful insights and conjectures about the transition to chaos which would not be seen without this picture.

5.2 Critical circle maps.

Consider once again the following prototypical 2-parameter family of circle maps

$$f_{\mu,\nu}(x) = x + \nu - (\mu/2\pi)\sin 2\pi x.$$

The family $f_\nu = f_{1,\nu}$ is critical in the sense defined in Section 2. Now, as above, if p/q is rational, let $I_{p/q} = \{\nu : f_\nu^q(x) = x + p\}$ denote the phase-locked ν-interval of rotation number p/q. Consider the complement of the phase-locked intervals,

$$M_f = \{\nu : \rho(f_\nu) \text{ is irrational }\}.$$

Computer experiments by Jensen, Bohr & Bak (1983) indicate that this has zero Lebesgue measure and Hausdorff dimension $HD(M_f)$ approximately equal to 0.87. They also noted that $HD(M_f)$ appears to be independent of f. Following Lanford (1985b), I will explain why this is the case: if g_ν is a family close to f_ν then $HD(M_g) = HD(M_f)$. In fact, M_f and M_g are lipeomorphic. This is the basic statement of universality and is relatively easily checked, but I will also introduce and explain universal functions associated with f which are universal and are more sensitive invariants than Hausdorff dimension.

The renormalisation transformations for this case, invented by Ostlund, Rand, Sethna & Siggia (1983) essentially act on the space **E** of pairs of functions (ξ,η) introduced in Section 2. For each n in **N** define a renormalisation transformation as follows:

$$T_n(\xi,\eta) = (a^{-1}.\xi^{n-1}\circ\eta\circ a \ , \ a^{-1}.\xi^{n-1}\circ\eta\circ\xi\circ a)$$

where $a = \xi^{n-1}(\eta(0)) - \xi^{n-1}(\eta(\xi(0)))$. The transformation appropriate to the golden-mean case is $T = T_1$.

For T_n to be well-defined each of the compositions must be compatible. This will be the case if $1/(n+1) < \rho(\xi,\eta) < 1/n$ where the rotation number $\rho(\xi,\eta)$ is as defined in Section 2.4.1. Therefore one takes the set D_n of pairs satisfying this condition for the domain of T_n. Then, since the D_n are disjoint, the T_n can be put together into a single transformation T whose domain is $\bigcup_{n\geq 1} D_n$ and which is defined by $T|D_n = T_n$.

Now, by adapting the proof of Proposition 2.1 one can easily show that T_n sends a pair with rotation number ρ to one with rotation number $\rho^{-1} - n$. Thus $T_n(D_n)$ contains pairs of every rotation number in (0,1). It is known that at the golden-mean fixed point one direction (roughly corresponding to ρ) is expanded and the rest are contracted. Thus, following Lanford (1985b), I assume the picture shown in Figure 16. In particular, I assume that the geometry of the various intersections $T_n(D_n)\cap D_m$ is as shown and that there is uniform expansion in the 1-dimensional "vertical" direction and uniform contraction in the 1-codimensional "horizontal" direction.

It follows from this that **E** contains a T-invariant set Λ homeomorphic to the space $\mathbf{N}^\mathbf{Z}$ of biinfinite sequences of positive integers, and that the homeomorphism carries over the action of T on Λ to the shift σ on $\mathbf{N}^\mathbf{Z}$: $\sigma(\cdots a_{-1}a_0a_1\cdots) = \cdots b_{-1}b_0b_1\cdots$ where $b_i = a_{i+1}$. Given $\underline{a}^+ = a_1a_2\cdots$ in $\mathbf{N}^\mathbf{N}$, define

$$H_{a_1,\ldots,a_n} = \{u : T^{j-1}u \in D_{a_j} \text{ for } 1 \leq j \leq n\}.$$

Then $H_{a_1,\ldots,a_{n+1}}$ is contained in $H_{a_1,\ldots,a_n}$ and

$$H_{\underline{a}^+} = \bigcap_{n\geq 1} H_{a_1,\ldots,a_n}$$

defines a "horizontal" 1-codimensional submanifold. Moreover, given $\underline{a}^- = a_0a_{-1}a_{-2}\cdots$ in $\mathbf{N}^\mathbf{N}$ define "vertical" strips $V_{a_0,\ldots,a_{-n}}$ inductively by

$$V_{a_0} = T(H_{a_0}), \quad V_{a_0,\ldots,a_{-n}} = T(V_{a_0,\ldots,a_{-(n-1)}} \cap H_{a_{-n}}).$$

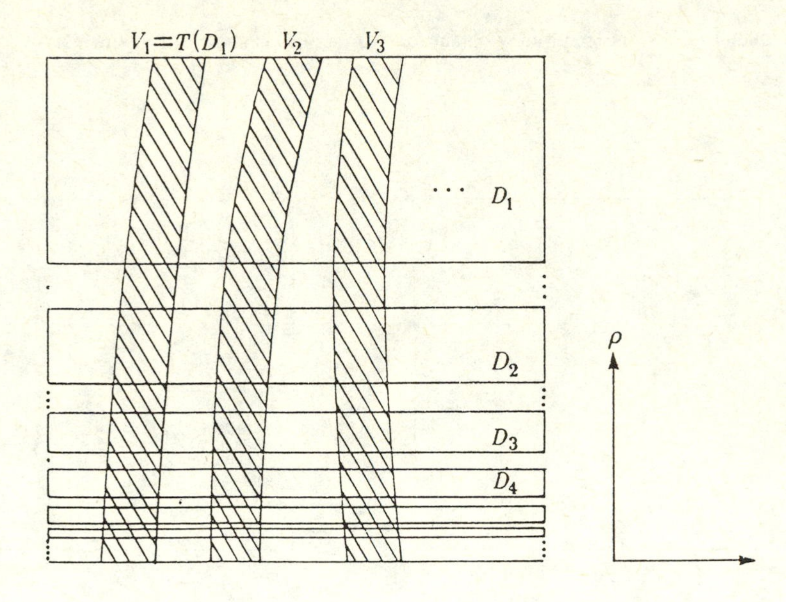

Figure 16. Schematic representation of the action of T on $\bigcup_n D_n$.

Then $V_{\underline{a}^-} = \bigcap_{n \geq 0} V_{a_0, \ldots, a_{-n}}$ is a "vertical" 1-dimensional curve. The homeomorphism between $\mathbf{N}^{\mathbf{Z}}$ and Λ sends $\cdots a_{-1}a_0a_1 \cdots$ to $V_{\underline{a}^-} \cap H_{\underline{a}^+}$.

If $u \in H_{\underline{a}^+}$ then $1/(a_j + 1) < \rho(T^{j-1}(u)) < 1/a_j$ so $\rho = \rho(u)$ has continued fraction $[a_1, a_2, \ldots] = 1/(a_1 + 1/(a_2 + \cdots))$. Thus if u_ν is a 1-parameter family close to Λ then

$$\rho(u_\nu) = [a_1, a_2, \ldots] \text{ if and only if } u_\nu \in H_{\underline{a}^+}$$

and the set $M_u = \{\nu : \rho(u_\nu) \text{ is irrational}\}$ corresponds to the intersection of the various $H_{\underline{a}^+}$ with the curve in $\mathbf{E}$ given by u_ν. The fact that the pairs in Λ and those close to Λ come from analytic circle maps follows from the glueing construction given in Rand (1984) and described in Section 2.6 above. From the theory of hyperbolic sets with 1-dimensional unstable manifolds this implies that if u_ν is transverse to the $H_{\underline{a}^+}$, then the fractal properties of the set M_u are the same as those for the families given by the $V_{\underline{a}^-}$ and the latter are independent of $\underline{a}^-$. Thus these fractal properties are independent of u near Λ. In fact, a more basic observation is that it also follows that if f and g are two families transverse to the $H_{\underline{a}^+}$, then there is a lipeomorphism from M_f to M_g which is obtained by sliding along the stable manifolds of Λ from M_f to M_g. Lippeomorphism of M_f and M_g implies that they have the same scaling structures and fractal properties. For the above case, the existence of this lipeomorphism can be tested numerically in the following way. For each $p/q \in \mathbf{Q}$ let $I_{p/q}(f) = [l_{p/q}(f), r_{p/q}(f)]$. For two critical families f and g plot the $l_{p/q}(f)$ and $r_{p/q}(f)$ against the $l_{p/q}(g)$ and $r_{p/q}(g)$. This should give the graph $\gamma_{f,g}$ of a Lipschitz function. After I had conjectured this, Kim and Ostlund informed me that they have observed it in some numerical experiments (see Kim &

Ostlund (1987)). The graph they obtain for two families is shown in Figure 17.

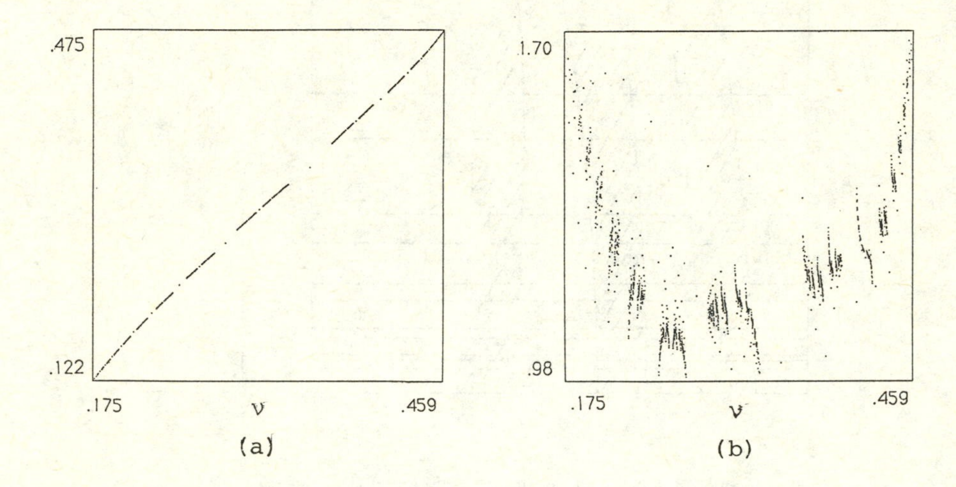

Figure 17. (Courtesy of S. Kim.) (a) The graph $\gamma_{f,g}$ obtained numerically by Kim and Ostlund (1987) for f as above and $g_\nu(x) = x + \nu - (5/16\pi)(\sin\pi x + .2\sin 6\pi x)$.
(b) The numerically computed "derivative" of the graph $\gamma_{f,g}$ of (a).

I now want to use Lanford's construction (Lanford (1985b)) to relate the above picture to the numerical work of Farmer and Satija (1984) and Umberger, Farmer and Satija (1984). Consider 1-parameter families u_ν such that ν ranges from 0 to 1, $\nu \to \rho(u_\nu)$ is non-decreasing, $0 < \rho(u_\nu) < 1$, and $\rho(u_0) = 0$ and $\rho(u_1) = 1$. I call such families *full*. Each unstable manifold $V_{\underline{a}^-}$ can be parameterised by ν so that it is such a family. Now given a full family u_ν define $u_\nu^{[n]}$ to be $T_n(u_{b+\nu(a-b)})$ where (a,b) is the interval of values of μ for which $u_\mu \in D_n$. Lanford (1985b) has shown (i) that all the $V_{\underline{a}^-}$ can be parameterised so that

$$(V_{a_{-1},a_{-2},\ldots})^{[a_0]} = V_{a_0,a_{-1},a_{-2},\ldots}$$

in the obvious sense, and (ii) if $u_\nu^{[a_0,\ldots,a_{-n}]}$ is defined recursively as $(u_\nu^{[a_{-1},\ldots,a_{-n}]})^{[a_0]}$, then for an open set of full families u_ν,

$$u_\nu^{[a_0,\ldots,a_{-n}]} \to V_{a_0,a_{-1},\ldots}$$

as $n \to \infty$, the convergence being uniform in $a_0, a_{-1},\ldots$ and exponentially fast in n. Farmer and Satija had previously implemented the construction of the families $u_\nu^{[a_0,\ldots,a_{-n}]}$ numerically and found reasonably good convergence and a representation of the $V_{\underline{a}^-}$ using a projection into two representative dimensions.

In the above analysis I restricted to critical maps. This was not really necessary and I could have defined the transformation T on a neighbourhood of $\bigcup_n D_n$ in the space of all analytic pairs satisfying the conditions (a) - (c) of Section 2.4 above. This will introduce an extra unstable direction so that the stable manifolds have codimension 2. In the extension to higher dimensional systems which include dissipative diffeomorphisms of the annulus $\mathbf{T} \times \mathbf{R}^{n-1}$ it is necessary to work in this way as the renormalisation transformation $\tilde{T}$ there will act on pairs of maps from $\mathbf{R}^n$ to $\mathbf{R}^n$ in a fashion similar to T (see Section 3 above and Rand (1984)) and for these maps there is no simple criterion corresponding to criticality. These maps contain Λ as they contain the singular maps whose image is 1-dimensional as a subspace and $\tilde{T}$ is an extension of T. If the above conjectures about Λ are correct then using the methods in Rand (1984) outlined in Section 3 above it can be shown that Λ is also the appropriate renormalisation strange set for $\tilde{T}$ because under renormalisation the strongly disspative maps and diffeomorphisms converge to singular 1-dimensional maps. Suppose that $F_{\mu,\nu}$ is a 2-parameter family of dissipative diffeomorphisms of the annulus such that the family of pairs $U_{\mu,\nu} = (F_{\mu,\nu}, R_{-1} \circ F_{\mu,\nu})$ is transverse to the stable manifolds. Using methods similar to those in Rand (1984) for the golden-mean case (see Section 3 above) it is possible to show that

$$K_F = \{(\mu,\nu) : U_{\mu,\nu} \text{ is in a stable manifold}\}$$

is precisely the set of points (μ,ν) at which the invariant circles break up. If $U_{\mu,\nu} \in H_{a_1,a_2,\ldots}$ then the invariant circle of rotation number $[a_1,a_2,\ldots]$ breaks up at (μ,ν). It follows from the hyperbolic structure of Λ that the set K_F will have the same fractal structure as the set M_f for transverse families of critical circle maps. In particular, $HD(K_F) = HD(M_f)$. Bohr has numerical evidence for this (see Figure 18).

5.3 Breakup of the invariant circles of area-preserving maps.

I want consider here the way in which all the homotopically non-trivial invariant circles of an area-preserving twist map of the annulus such as that given by $f_k(x,y) = (x+y \pmod 1), y + (k/2\pi)\sin 2\pi(x+y))$ break down as k is increased. I shall abbreviate the term *homotopically non-trivial invariant circle* to *invariant circle* in what follows. (Of course, with the correct interpretation these results apply equally well to the other invariant circles.) Recall from Section 4 that it is believed that (a) if $0 < k < k_c = .9716354..$ then f_k has a Cantor set of invariant circles, while, (b) if $k > k_c$ there are none, and (c) at $k = k_c$ there is a single circle whose rotation number is $\upsilon = (\sqrt{5}-1)/2$.

To describe the cascade of breakdowns, I introduce the diagram K_f which consists of those (ω,k) such that if $0 < k' \leq k$ then $f_{k'}$ has an invariant circle of irrational rotation number ω. Below I will explain why it follows from the renormalisation picture that, up to lipeomorphism, the diagram K_f should be a universal object.

The basic renormalisation transformations, which are essentially the same as those defined above for circle maps were first used by MacKay (1983). They are analogous to the transformation T ($= T_1$ below) introduced in Section 4.6.2 above. Define

$$T_n(E,F) = (B^{-1} \circ E^{n-1} \circ F \circ B \,,\, B^{-1} \circ E^{n-1} \circ F \circ E \circ B)$$

with $B(x,y) = (ax, by - c - dx^2)$ chosen so as to preserve the normalisations $E(0,0) = (0,0)$, $F(0,1) = (0,1)$ and $E(0,1) = (1,1)$. This acts on ρ as $\rho \to \rho^{-1} - n$. For the space $\mathbf{B}$ I take triples (E,F,ω) where ω is a real number in $(0,1)$. Extend T_n to these triples as $(E,F,\omega) \to (T_n(E,F), \omega^{-1} - n)$ which is defined on the subset D_n given by

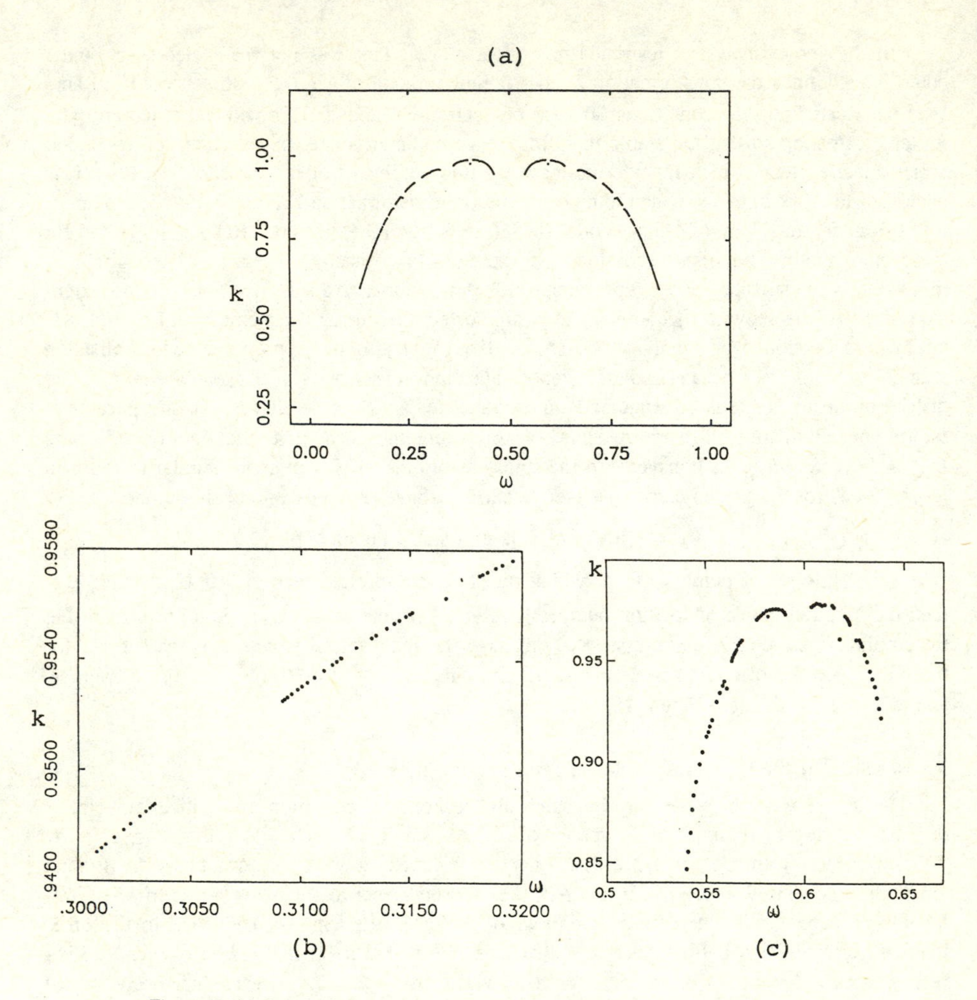

Figure 18. (a) A numerical approximation of the set of points (ω,k) at which the irrational invariant circles break down for the map $(x,y) \to (x + y + \omega, \lambda y + (k/2\pi)\sin 2\pi(x + y))$ for $\lambda = 0.25$. (b) A blowup of part of (a). (c) Same as (a) but $\lambda = 0.5$. (Courtesy of T. Bohr.)

$1/(n + 1) < \omega < 1/n$. In Figure 19(a) I have represented the space of triples picking out the ω direction and one other expanding direction. Assume that these are uniformly expanding and that all the others are uniformly contracting, and also assume that the action of T_n on D_n is geometrically as that shown for T_2 on D_2. MacKay (1986) and Escande (1987) have worked out in detail a similar picture for an approximate renormalisation scheme due to Escande and Doveil (1981a,b), and I believe that they should be able to verify the above conjecture for this model.

The picture is complicated a little by the fact that one cannot isolate the renormalisation strange set from the *simple line* of integrable maps given by $((x + y/\omega - \omega, y), (x + 1 + y, y))$ because circles with irrational rotation number break down arbitrarily close to the simple line. Given $\underline{a} = \cdots a_{-1}a_0a_1 \cdots \in \mathbf{N}^{\mathbf{Z}}$, define

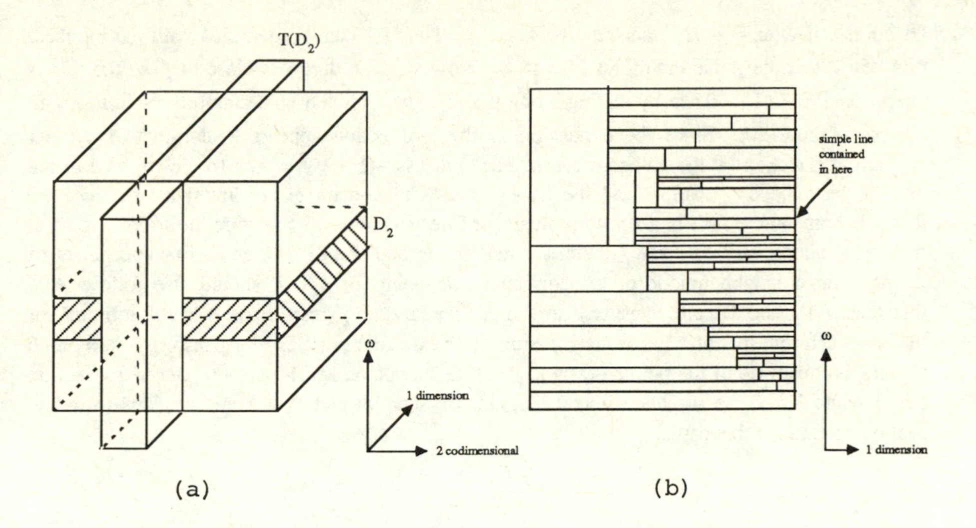

Figure 19. (a) A schematic representation of the action of T on D_2 in the space of triples for area-preserving maps. (b) A schematic representation of the projection of the invariant set obtained by factoring out the stable directions.

$H_{\underline{a}^+} = \{u : T^{j-1}u \in D_{a_j}\}$ as for circle maps. For this case of area-preserving maps it can be deduced from my picture that $H_{\underline{a}^+}$ is locally a 1-codimensional submanifold with boundary (see Figure 19). It is contained in $\omega = [a_1,a_2,\ldots]$. Let $\tilde{H}_{\underline{a}^+}$ be its boundary. Locally this will be a 2-codimensional submanifold. The sets $H_{\underline{a}^+}$ and $\tilde{H}_{\underline{a}^+}$ are distinguished dynamically by the fact that if $u \in H_{\underline{a}^+} - \tilde{H}_{\underline{a}^+}$ then $T^n u$ converges to the simple line as $n \to \infty$ while if $u \in \tilde{H}_{\underline{a}^+}$ then it does not, instead it converges to the renormalisation strange set Λ defined below. For rotation numbers satisfying a Diophantine condition these two cases correspond respectively to the existence of a smooth invariant circle and the existence of a critical invariant circle.

One can also define "vertical strips" $V_{\underline{a}^-}$ as for circle maps. The intersection of the various $\tilde{H}_{\underline{a}^+}$ and $V_{\underline{b}^-}$ gives the strange invariant set Λ. To get a better picture of Λ factor out the stable direction and only consider the two unstable directions. Then a picture as shown in Figure 19(b) is obtained.

Now consider the 2-dimensional surface Γ_f in **B** given by $\Phi(\omega,k) = (f_k, R_{-1} \circ f_k, \omega)$. If $\Phi(\omega,k) \in H_{a_1,a_2,\ldots}$ then $T^n(\Phi(\omega,k))$ converges to Λ or the simple line. Thus using the arguments in Rand (1984) outlined in Section 3 above, $\Phi(\omega,k)$ will have an invariant circle. Assuming that there is a fundamental neighbourhood U "above" Λ such that each $(\xi,\eta) \in U$ has no invariant circles, then it follows that

$$K_f = \{(\omega,k) : \Phi(\omega,k) \in \tilde{H}_{\underline{a}^+} \text{ for some } \underline{a}^+\}.$$

Of course, if $\Phi(\omega,k) \in H_{\underline{a}^+}$ then $\omega = [a_1,a_2,...]$. Thus K_f can be identified with the intersection of Γ_f and the stable manifolds $\tilde{H}_{\underline{a}^+}$ of Λ. Now suppose that g is close to f so that Γ_g is nearly parallel to Γ_f. Then by sliding along the $\tilde{H}_{\underline{a}^+}$ you obtain a homeomorphism from K_f to K_g (see Figure 20). Since the expansion in the ω-direction appears to dominate the other expanding direction (at the golden-mean fixed point it is $-(1+\upsilon)^2 = -2.618...$ compared to the other eigenvalue $\delta = 1.618...$) and the hypersurfaces of contant ω are invariant, I conjecture that this homeomorphism is a lipeomorphism. More generally, I conjecture the following local result: for almost all $(\omega,k) \in K_f$ there is a neighbourhood U of (ω,k) and a lipeomorphism of $K_f \cap U$ onto a neighbourhood of the corresponding points in K_g. It should also follow from this that if Γ_f and Γ_g are transverse to the $\tilde{H}_{\underline{a}}^+$s and $k_f(m)$ (resp. $k_g(m)$) is the infimum of those k such that the total Lebesgue measure of the invariant circles of f (resp. g) is m, then the graph consisting of the points $(k_f(m),k_g(m))$ is monotone and Lipschitz. Below in Section 5.4, I relate K_f to the so-called fractal diagram of Schmidt and Bialek and use this to define scaling spectra for this case.

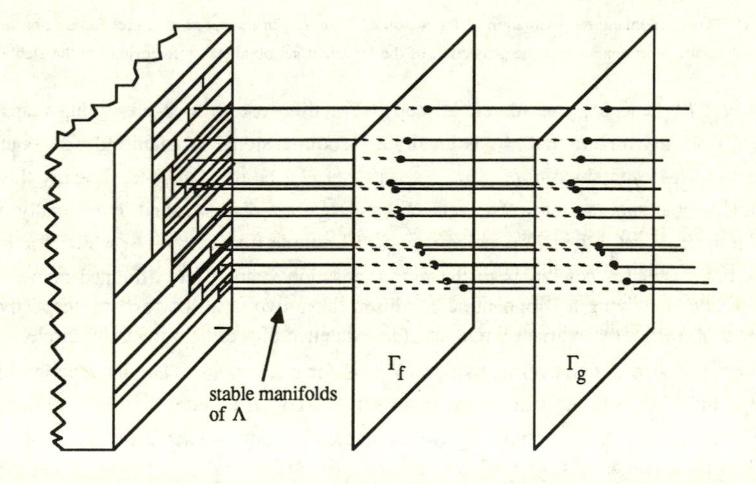

Figure 20. A schematic representation of the way in which one obtains a homeomorphism from K_f to K_g by sliding along the stable manifolds of the renormalisation strange set.

Satija (1985) has carried out a numerical investigation of a set related to Λ. The viewpoint of her work is similar to that of Farmer and Satija (1984) on circle maps and its relation with the picture presented here is similar to the relationship for circle maps described in Section 5.2.

5.4 Scaling and other spectra as universal invariants.

As we have seen above one of the primary consequences of the existence of a renormalisation strange set of the kind discussed is the universality of the associated fractal "bifurcation" set up to lipeomorphism. The scaling spectrum is an invariant of this relation. Similar invariants related to the Farey tree hierarchy have been measured by Cvitanovic (1987).

5.4.1 The scaling spectrum.

For concreteness consider the renormalisation strange set Λ constructed above for circle maps, although it will be clear that the results and approach is of much greater generality. Fix an unstable manifold $\Sigma = V_{\underline{a}^-}$. Let $\Sigma_{a_1,...,a_n}$ be the intersection of $H_{a_1,...,a_n}$ with Σ. This is an open interval. Denote the set of such intervals by $\mathbf{C}_n$. Let λ denote any smooth measure on Σ. Let $N_n(a,b)$ denote the number of C in $\mathbf{C}_n$ such that, $a < n^{-1}\log\lambda(C) < b$, i.e. $\lambda(C) \in e^{n(a,b)}$. Let $S(a,b)$ be the growth rate $\lim_{n\to\infty} n^{-1}\log N_n(a,b)$ and

$$S(\alpha) = \inf\{S(a,b) : a < \alpha < b\}$$

Then $S(\alpha)$ is a continuous concave function which is independent of the choice of Σ and λ. Its Legendre transform is the function $P(\beta)$ which is the growth rate of the sums $\sum_{\mathbf{C}_n} \lambda(C)^{-\beta}$ (Bohr & Rand (1986), Gundlach (1986)). In fact, $S(-\alpha)$ is the entropy function for the characteristic exponents of Λ as defined in Bohr & Rand (1986).

Now let u_ν be a full family as in Section 5.2 which is transverse to all the $H_{\underline{a}^+}$ and so that $u_\nu^{[a_0,...,a_{-n}]} \to V_{a_0,a_{-1},...}$ as $n \to \infty$ as explained. Let $I_{a_1,...,a_n}$ denote the set of ν such that u_ν lies in $H_{a_1,...,a_n}$. Let $S_u(\alpha)$ be defined as $S(\alpha)$ except that the $I_{a_1,...,a_n}$ replace the $\Sigma_{a_1,...,a_n}$. Let $C_{n,\nu}$ denote the interval $I_{a_1,...,a_n}$ containing ν. Then the following result follows from those of Lanford (1985b), Bohr & Rand (1986) and Gundlach (1986).

Theorem 5.1 If u_ν is transverse to the $H_{\underline{a}^+}$s, then $S_u = S$. Hence (i) S_u is independent of u, (ii) S_u is real-analytic on its support, (iii) $D_u(\alpha) = -S_u(\alpha)/\alpha$ is the Hausdorff dimension of the set of points ν such that $\lim_{n\to\infty} n^{-1}\log\lambda(C_{n,\nu}) = \alpha$, (iv) $HD(M_u)$ is the value of α such that $D_u'(\alpha) = 0$ and (v) the total size $\sum_{C \in \mathbf{C}_n} \lambda(C)$ of the complement of the gaps goes to zero like e^{nP} where

$$P = S_u(-\alpha) - \alpha$$

where α satisfies $S'_u(-\alpha) = 1$.

The interpretation of this in terms of parameter space is as follows: Let u_ν be a critical family. Consider the set A_n of rational numbers of the form $[a_1, . . . , a_n]$ where each $a_i \in \mathbf{N}$. The complement of the set of intervals $I_{p/q}$, $p/q \in \bigcup_{i=1}^{n} A_i$ consists of a set K of open intervals. These are precisely the $I_{a_1,...,a_n}$. So the universal function tells us the set of scales in this set as $n \to \infty$. Alternatively, and perhaps more appealingly, the fact that each $I_{a_0,a_1,...,a_n}$ contains the phase-locked tongue $I_{p/q} = I_{[a_0,...,a_n]}$ which is mapped onto $I_{[a_1,...,a_n]}$ by T_{a_0} can be used to deduce that the same result is true and the same spectrum obtained when the $I_{a_1,...,a_n}$ are replaced by the $I_{[a_1,...,a_n]}$.

5.4.2 The q-spectrum.

The most interesting quantity associated with the intervals $I_{p/q}$ is the length q of the associated periodic orbits of the u_ν, $\nu \in I_{p/q}$. Thus it is interesting to construct an invariant which, together with the geometry, takes account of this dynamical information. Let $N_n(a,b)$ be the number of $C = I_{[a_1, \ldots, a_n]}$ such that $a < \log\lambda(C)/\log q < b$, where $q = q(a_1, \ldots, a_n)$ is given by $[a_1, \ldots, a_n] = p/q$ in lowest order terms. The function $\tilde{S}(\alpha)$ is defined in terms of the $N_n(a,b)$ as before so that, roughly speaking, the number of $I_{[a_1, \ldots, a_n]}$ with $\lambda(I_{[a_0, \ldots, a_n]}) = q^\alpha$ grows like $e^{n\tilde{S}(\alpha)}$ as $n \to \infty$.

If the $I_{a_1, \ldots, a_n}$ are used instead of the $I_{[a_0, \ldots, a_n]}$ the same function is obtained, and this in turn is independent of the choice of $V_{\underline{a}^-}$. Thus, if u_ν is transverse to the $H_{\underline{a}^+}$ the function $\tilde{S}$ is independent of u.

This function is analysed using the thermodynamic formalism of Collet and Lebowitz (1986), Rand (1986) and Gundlach (1986). In the latter there is a general treatment for the fluctuation spectra S_{ψ_1,ψ_2} of quantities of the form $\sum_{i=0}^{n-1}\psi_2(T^i(x))/\sum_{i=0}^{n-1}\psi_1(T^i(x))$ where ψ_1 and ψ_2 are Hölder continuous. The function $\tilde{S}$ is the fluctuation spectrum associated with $\psi_1(x) = -\log\|dT(x)|E^u(x)\|$ where $E^u(x)$ is the tangent space to $V_{a_0 a_{-1}} \cdots$ where $x = \cdots a_{-1}a_0a_1 \cdots$ and $\psi_2(x) = [a_0a_{-1}a_{-2}\ldots]$. Then ψ_1 controls the rate at which the length of the $I_{a_1, \ldots, a_n}$ decrease as $n \to \infty$ and $\sum_{i=0}^{n-1}\log\psi_2(T^i(x))/\log q(x)$ is bounded above and below by positive constants independent of x and n, so ψ_2 determines the increase in q as $n \to \infty$.

It follows from these results that $\tilde{S}$ is a real-analytic concave function on its support. Therefore, if $\tilde{S}$ has a maximum it will be unique. Assume that this maximum is achieved at $\alpha = \alpha_{\max}$. Then it follows from the definition of $\tilde{S}$ that for large n, almost all of the phase-locked intervals $I_{[a_0, \ldots, a_n]}$ scale as $q^{\alpha_{\max}}$. I believe that $\alpha_{\max}$ is approximately -3. On the other hand the lengths of the phase-locked intervals $I_{p_n/q_n} = I_{[a_0, \ldots, a_n]}$, $a_j \equiv 1$, corresponding to ratios of Fibonacci numbers, scale as $|\delta|^{-n}$ and the q_n grow as $(1 + \upsilon)^n$ so the α-value corresponding to these is given by $\alpha = \alpha_{golden} = -\log|\delta|/\log(1 + \upsilon) = -2.1647..$. Since these are the largest phase-locked intervals for a given q, α_{golden} is bigger than any other values of α for which $\tilde{S}(\alpha) > 0$.

Note that $S \to \infty$ as $\alpha \to -\infty$ because the associated cookie-cutter (Bohr & Rand (1986)) has infinite topological entropy. To get a finite S one can take the same definition except that the a_i are restricted to be $\le n$. The scaling spectrum S_n obtained by doing this has maximum value $\log n$. Thus for the normalised S one can take $S_{renorm} = \lim_{n\to\infty} S_n/(\log n)$.

5.4.3 Scaling spectra for area-preserving maps.

Scaling spectra can be defined for the set K_f defined in Section 5.3 for area-preserving maps and, as in the circle case, these will be independent of f. In particular, the Hausdorff dimension of K_f should be universal. The set K_f should be approximated by the so-called *fractal diagram* of Schmidt and Bialek (1982). This is the set $\tilde{K}_f = \bigcup_{n\ge1}\tilde{K}_{f,n}$ of points (ω,k) defined as follows: for each $\omega = [a_1,a_2, \ldots, a_n] \in A_n$ let $k_c(\omega)$ denote the value of k at which the minimax orbit with rotation number ω becomes destabilised i.e. bifurcates from elliptic to hyperbolic. Then $\tilde{K}_{f,n} = \{(\omega,k_c(\omega)) : \omega \in A_n\}$. A picture of $\tilde{K}_{f,1}\cup\tilde{K}_{f,2}$ is shown

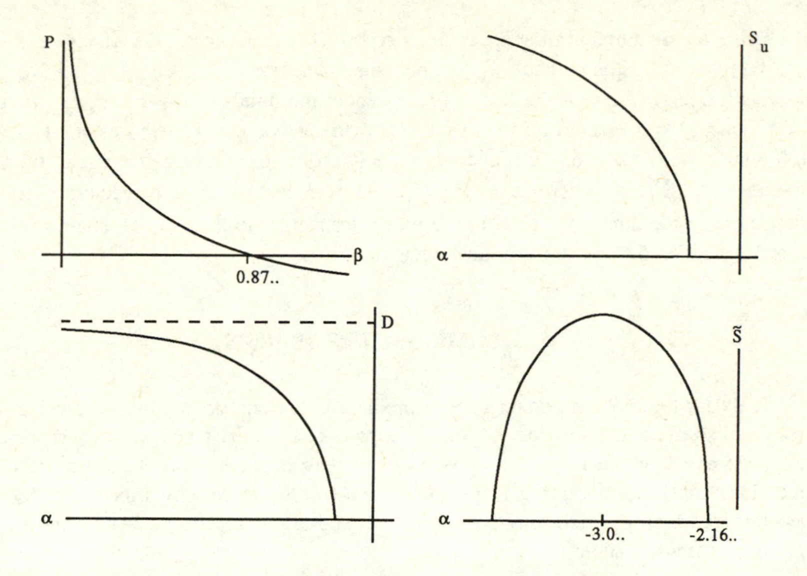

Figure 21. (a) The pressure $P(\beta)$. (b) The scaling spectrum $S_u(\alpha)$.
(c) The dimension function $D(\alpha)$. (e) The q-spectrum $\tilde{S}$.

in Figure 22. This diagram will be self-similar since each T_n maps the subset of $\tilde{K}_{f,m}$ with $a_1 = n$ onto $\tilde{K}_{f,m-1}$. The set K_f should be the boundary of $\tilde{K}_f$ i.e. $K_f = \mathrm{cl}(\tilde{K}_f) - \tilde{K}_f$.

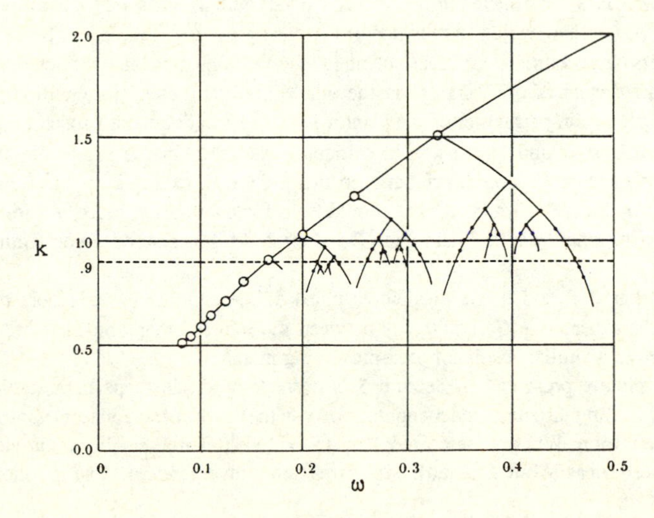

Figure 22. The first two layers of the fractal diagram of Schmidt and Bialek (Courtesy of G. Schmidt).

Using $\tilde{K}_f$ a natural scaling spectrum is constructed as follows. For given $a_1, \ldots, a_n$ let $\omega_{n,m} = [a_1, \ldots, a_{n-1}, m]$ and $\delta_{n,m}$ be the distance between $(\omega_{n,m}, k(\omega_{n,m}))$ and $(\omega_{n,m+1}, k(\omega_{n,m+1}))$ for $m \geq 0$. Let $N_n(a,b)$ denote the number of $a_1, \ldots, a_{n-1}, m$ such that $a < n^{-1}\log\delta_{n,m} < b$ and let $S(a,b)$ be the growth rate of the $N_n(a,b)$ as $n \to \infty$. Then if $S(\alpha) = \inf\{S(a,b) : a < \alpha < b\}$ and $D_f(\alpha) = -S(\alpha)/\alpha$ then under the usual transversality hypothesis that Γ_f is transverse to the $H_{\underline{a}+}$s or even locally so, I conjecture that $D_f(\alpha)$ is independent of the family f. It will have properties similar to those of the analogous functions defined in Section 5.4 above for critical circle maps.

6. SOME OPEN PROBLEMS

1. By far the most important general problem is to understand why these universal structures occur so often and so widely. All of the proofs of the existence and hyperbolicity do not really cast any light on this ubiquity or on why they occur in such differing contexts as 1-dimensional and area-preserving maps. There must be some general principle underlying this phenomenon which the right proof would reveal and an approach that would unify the seemingly disparate applications.

2. A related problem is to understand in a deeper way the nature of the attractors in phase space associated with the various fixed points and renormalisation strange sets. Are these always almost-periodic (in the sense of Bohr) and do they always have a universal geometric structure (i.e. can two in the same universality class always be Lipschitz conjugated)? A good understanding of this question should lead to insight into the problems raised in 1. Some specific problems associated with the existence of this geometric structure[3] are:

(a) Prove that if f and g are in the stable manifold of the fixed point of the doubling operator then there is a $C^{1+\varepsilon}$ diffeomorphism of the interval which sends the minimal Cantor set $\Lambda_\infty(f)$ onto $\Lambda_\infty(g)$ and which conjugates f on $\Lambda_\infty(f)$ to g on $\Lambda_\infty(g)$. (Since the eigenvalues of the periodic orbits are an invariant of Lipschitz conjugacy one cannot hope that this will conjugate the dynamics of f and g on the whole interval.) This proof will have to make use of the universal scaling structure of the Cantor set $\Lambda_\infty(f)$ to deduce a Lipschitz conjugacy.

(b) Prove that if f and g are golden critical circle maps in the stable manifold of the fixed point then there is a C^1 conjugacy between them. Give an example to show that this conjugacy is not C^2 in general. Similar results should hold for commuting pairs and for maps with other quadratic irrational rotation numbers. Do they hold for general Diophantine rotation numbers?

(c) Prove that if f and g are dissipative maps of the annulus in the stable manifold of the fixed point then there is a C^1 conjugacy between the induced maps on their critical golden invariant tori. Prove a similar result for area-preserving maps.

(d) If the picture presented in Section 5 is correct the circle maps in a stable manifold have a universal scaling structure independently of whether or not the rotation number satisfies a Diophantine condition. What is the correct way to express this universality geometrically?

(e) Use these ideas about geometric structures and convergence under the renormalisation

[3] **Note added in proof.** Paluba (1987) has proved that the conjugacy in 2(a) is Lipschitz. Also, in Sullivan (1987) it is announced that in as yet unpublished work, Sullivan and Feigenbaum have proved 2(a). Independently, I have proved the first part of 2(b) and the ideas used in this proof can be used to prove 2(a).

operators to give a new proof of the Herman-Yoccoz theorem on the existence of a smooth conjugacy to a rotation for diffeomorphisms of the circle whose rotation number satisfies a Diophantine condition. Some progress has already been made on this by J. Stark.

(f) Suppose that f_ν and g_ν are full families of critical circle maps and M_f and M_g are the corresponding sets of parameter values ν where the rotation number is irrational. Prove that if the two families intersect the stable manifolds of Section 5.2 transversally then the natural homeomorphism from M_f to M_g (i.e. that which preserves rotation number) extends to a $C^{1+\varepsilon}$ diffeomorphism of the interval. To help prove this show that the map $\chi : M_f \to M_g$ defined by $\rho(\chi(f)) = G(\rho(f))$ where G is the Gauss map, can be extended to a $C^{1+\varepsilon}$ map of the parameter interval.

(g) Prove similar results for the other fractal bifurcation sets associated with renormalisation strange sets (see Section 5).

3. Prove the existence of the golden area-preserving fixed point as described in Section 4.

4. Show how to prove or give plausibility arguments for the existence of a fundamental domain F in the unstable manifold of the golden area-preserving fixed point which is also in the stable manifold of the simple fixed point. (See Section 4.7)

5. Prove the existence of a fundamental domain F in the unstable manifold of the golden area-preserving fixed point such that there is a neighbourhood U of F with the property that every element of U has no golden invariant circle.

6. Investigate the possibly homoclinic behaviour of the stable manifold of the golden area-preserving fixed point near the "boundary" of the analytic maps which perhaps corresponds to the erratic behaviour found by Wilbrink (1987).

7. Develop better and more controlled numerical tests of the conjectures of Section 5, in particular the existence and hyperbolicity of the renormalisation strange sets. Numerically check the conjectures for area-preserving maps.

8. Prove the existence and hyperbolicity of the renormalisation strange sets of Section 5. More simply, prove the existence and hyperbolicity of one for critical circle maps with ε-singularities in the sense of Jonker and Rand (1983).

9. Develop a theory along the lines of that in Section 2 for analytic maps of the torus $\mathbf{T}^2$. In particular, determine what rotation vector plays the role of the golden mean (see Kim and Ostlund (1985)), study the breakdown of smooth conjugacy to a rotation as the nonlinearity is increased and isolate a critical system, study the scaling properties in parameter and phase space $\mathbf{T}^2$ and construct an appropriate renormalisation operator. An entirely analogous set of problems can be posed for the breakup of invariant 2-dimensional tori in 4-dimensional symplectic maps.

10. This problem was suggested by Robert MacKay. Determine the condition for a twist map (not necessarily area-preserving) to lie in the stable manifold of the golden area-preserving fixed point. For example, are the breakups of irrational invariant circles which occur in dissipative systems close to the collision of two invariant circles (see Chenciner (1983)) in this universality class.

Acknowledgements.

I am indebted to Ben Mestel and the referee who suggested a number of corrections and improvements to the original draft, and to Robert MacKay for many useful discussions. I would also like to thank the Applied Mathematics Program of the University of Arizona for its hospitality during the visit when part of this paper was written. This was partially supported by the University of Arizona and the US ONR under grant NOOO14-85-K-0412. Also, I gratefully acknowledge the financial support of the UK Science and Engineering Research Council.

REFERENCES

V. I. Arnold, 1961, *Small denominators I. On the mapping of a circle into itself.* Izvestia Akad. Nauk. Math. Series **25**, (1961), 21-86 [Transl. Amer. Math. Soc., Series 2, **46**, (1965), 213-284].

V.I. Arnold, 1963, *Small denominators and problems of stability of motion in classical and celestial mechanics,* Russian Math. Surveys, **18**, 85-192.

V.I. Arnold, 1978, *Mathematical methods of classical mechanics.* Springer-Verlag.

V.I. Arnold, 1983, *Geometrical methods in the theory of ordinary differential equations.* Springer-Verlag.

S. Aubry, 1983, *The twist map, the extended Frenkel-Kontorova model and the devil's staircase.* Physica **7D**, 240-258.

S. Aubry and P. Y. Le Daeron, 1983, *The discrete Frenkel-Kontorova model and its extensions I: Exact results for the ground states* Physica **8D**, 381-422.

G. Benettin, C. Cercignani, L. Galgani and A. Giorgilli, 1980, *Universal properties in conservative dynamical systems.* Lettere Nuovo Cimento **28**, 1-4.

G. Benettin, L. Galgani and A. Giorgilli, 1980, *Further results on universal properties in conservative dynamical systems.* Lettere Nuovo Cimento **29**, 163-166.

T. Bohr and D. A. Rand, 1986, *The entropy function for characteristic exponents.* Physica **25D**, 387-398.

T. C. Bountis, 1981, *Period doubling bifurcations and universality in conservative systems.* Physica **3D**, 577-589.

R. Bowen, 1975, *Equilibrium states and the ergodic theory of Anosov diffeomorphisms.* Springer Lecture Notes in Math. No. 470.

M. Campanino, H. Epstein & D. Ruelle, 1981 *On the existence of Feigenbaum's fixed point.* Comm. Math. Phys. **79**, 261-302.

M. Campanino, H. Epstein & D. Ruelle, 1982 *On Feigenbaum's functional equation.* Topology **21**, 125-129.

M. Casdagli, 1986, *Symbolic Dynamics for the renormalisation map of a quasi-periodic Schrödinger Equation.* Comm. Math. Phys., **107**, 295-318.

A. Chenciner, 1983, *Bifurcations de diffeomorphismes de* $\mathbf{R}^2$ *au voisinage d'un point fixe elliptique.* in *Chaotic behaviour of deterministic systems.* (Editors: G. Iooss, R. Helleman and R. Stora), North-Holland, 273-348.

P. Collet, J.-P. Eckmann, 1980 *Iterated maps on the interval as dynamical systems.* Birkhäuser.

P. Collet, J.-P. Eckmann and H. Koch, 1981a, *On universality for area-preserving maps of the plane.* Physica **3D**, 457-467.

P. Collet, J.-P. Eckmann and H. Koch, 1981b, *Period-Doubling Bifurcations for Families of Maps on* $\mathbf{R}^n$. *J. Stat. Phys.* **25**, *1*.

P. Collet, J.-P. Eckmann and O. Lanford, 1980, *Universal properties of maps on an interval* Comm. Math. Phys. **76**, 211-254.

P. Collet, J. Lebowitz and A. Porzio, 1986, *Dimension spectrum for some dynamical systems.* Preprint.

P. Coullet and C. Tresser, 1978, *Iterations d'endomorphismes et groupe de renormalisation.* J. Phys. **C5**, 25-28.

P. Cvitanovic, 1987, *Hausdorff dimension of irrational windings.* Preprint, Niels Bohr Institute, Copenhagen.

J. Dieudonné, 1960, *Foundations of Modern Analysis,* Academic Press.

A. Douady and J. Hubbard, 1985, *On the dynamics of polynomial like mappings* Ann. Scient. Ec. Norm. Sup., **18**, 287-343.

H. Epstein, 1986, *New proofs of the existence of the Feigenbaum functions.* Comm. Math. Phys. **106**, 395-426.

H. Epstein and J.-P. Eckmann, *On the existence of fixed points of the composition operator for circle maps.* Comm. Math. Phys. **107**, 213-231

H. Epstein and J.-P. Eckmann, 1987, *Fixed points of composition operators.* Proceedings of the 1986 IAMP Conference in Mathematical Physics (Editors: M. Mebkhout and R. Sénéor), World Scientific, 517-530, 1987.

J.-P. Eckmann, H. Koch and P. Wittwer, 1982, *A computer-assisted proof of universality for area-preserving maps.* Preprint, Université de Génève.

D. F. Escande and F. Doveil, 1981a, *Renormalization method for computing the threshold of the large-scale stochastic instability in two degree of freedom Hamiltonian systems.* J. Stat. Phys. **26**, 257-284.

D. F. Escande and F. Doveil, 1981b, *Renormalization method for the onset of stochasticity in Hamiltonian systems.* Phys. Lett. A **83**, 307-310.

D. F. Escande, 1987, *Renormalisation strange set for KAM tori.* In preparation.

K. J. Falconer, 1985, *The geometry of fractal sets.* Cambridge University Press.

J. D. Farmer and I. I. Satija, 1984, *Renormalisation of the quasi-periodic transition to chaos for arbitrary rotation numbers.* Phys. Rev. **31A**, 3520-3522.

M. Feigenbaum, 1978, *Quantitative universality for a class of non-linear transformations.* J. Stat. Phys. **19**, 25-52.

M. Feigenbaum, 1979, *The universal metric properties of a non-linear transformation.* J. Stat. Phys. **21**, 669-706.

M. Feigenbaum, 1982, *The transition to aperiodic behaviour in turbulent systems.* Comm. Math. Phys. **77**, 65-86

M. Feigenbaum, L. Kadanoff and S. Shenker, 1982, *Quasi-periodicity in Dissipative Systems: A Renormalisation Group Analysis.* Physica **5D**, 370-386.

J.-M. Gambaudo, J. E. Los and C. Tresser, 1987, *A horseshoe for the doubling operator: topological dynamics for metric universality.* Preprint, Université de Nice.

J.-M. Gambaudo, I. Procaccia, S. Thomae and C. Tresser, 1986, *New universal scenarios for the onset of chaos in Lorenz-like flows.* Preprint.

J. M. Greene, 1968, *Two-dimensional area-preserving mappings*. J. Math. Phys. **9**, 760-768.

J. M. Greene, 1979, *A method for determining a stochastic transition.* J. Math. Phys. **20**, 1183

J. M. Greene, R. S. MacKay, F. Vivaldi and M. Feigenbaum, 1981, *Universal behaviour in families of area-preserving maps*. Physica **21D**, 468-486.

J.M. Greene, 1980, Annals of the N.Y. Acad. of Sci., **357**, 80.

J. M. Greene, R. S. MacKay and J. Stark, 1986, *Boundary circles for area-preserving maps.* Warwick Preprint.

J. Guckenheimer, 1977, *On the bifurcation of maps of the interval* Invent Math. **39**, 165-178.

J. Guckenheimer, 1979, *Sensitive dependence on initial conditions for one-dimensional maps.* Comm. Math Phys. **70**, 133-160.

M. Gundlach, 1986, *Large fluctuations of pointwise dimension, characteristic exponents and pointwise entropy in Axiom A attractors.* Warwick University M.Sc. Dissertation.

T. Halsey, M. Jensen, L. Kadanoff, I Procaccia and B.Shraiman, *Fractal measures and their singularities: the characterisation of strange sets.* Preprint 1985.

M. Herman, 1979, *Sur la Conjugaison Differentiable des Diffeomorphismes du Cercle a des Rotations.* Publ. Math. I.H.E.S. **49**, 5-234.

M. Herman, 1983, *Sur les courbes invariantes par les difféomorphismes de l'anneau.* Astérisque **103-104**.

N. Hoidn, 1985, *On Invariant Curves Under Renormalisation.* Warwick Preprint, 1985 and Warwick Ph.D. Thesis, 1986.

M. H. Jensen, P. Bak and T. Bohr, 1983, *Complete devil's staircase, fractal dimension and universality of the mode-locking structure in the circle map.* Phys. Rev. Lett. **50**, 1637-1639.

L. Jonker and D. A. Rand, 1981, *Bifurcations in one dimension. I The nonwandering set.* Invent math. **62**, 347-365.

L. Jonker and D. A. Rand, 1983, *Universal Properties of Maps of the Circle with ε-Singularities.* Comm. Math. Phys. **90**, 273-292.

L.P. Kadanoff, 1981, *Scaling for a critical KAM trajectory,* Phys. Rev. Letts. **47**, 1641.

T. Kato, 1966, *Perturbation theory for linear operators.* Springer-Verlag.

S. Kim and S. Ostlund, 1985, *Renormalization of mappings of the two-torus.* Phys. Rev. Lett. **55**, 1165-1168.

S. Kim and S. Ostlund, 1987, *Universal scalings in critical circle maps.* In preparation.

A. N. Kolmogorov, 1954, *The general theory of dynamical systems and classical mechanics,* Address to 1954 Int. Congress of Mathematicians reprinted in R. Abraham and J. Marsden, Foundations of Mechanics, Benjamin.

M. Kohmoto, L. P. Kadanoff and C. Tang, 1983, *Localisation problem in one dimension: Mapping and escape.* Phys. Rev. Lett. **50**, 1870-1872.

O. Lanford, 1982, *A Computer-Assisted Proof of the Feigenbaum Conjectures.* Bull. AMS, **6**, 427-434.

O. Lanford, 1985a, *A numerical study of the likelihood of phase locking.* Physica **14D**, 403-408.

O. E. Lanford, 1985b, *Renormalisation group methods for circle mappings.* to appear in the proceedings of the 1985 Gröningen conference on Statistical mechanics and field theory: mathematical aspects, and *Renormalisation analysis for critical circle mappings with general rotation number.* Proceedings of the 1986 IAMP Conference in Mathematical Physics (Editors: M. Mebkhout and R. Sénéor), World Scientific, 1987, 532-536.

O. E. Lanford, 1987, *Computer assisted proofs in analysis.* I.H.E.S Preprintof the text of a talk delivered at the International Congress of Mathematicians, 1986.

R.S. MacKay, 1982, *Renormalisation in area-preserving mappings,* Princeton University Ph.D.Thesis.

R.S. MacKay, 1983, *A renormalisation approach to invariant circles in area-preserving maps,* Physica **7D**, 283-300.

R. S. MacKay and I. C. Percival, 1986, *Universal small-scale structure near the boundary of Siegel disks of arbitrary rotation number.* Warwick Preprint.

R. S. MacKay, 1986, *Exact results for an approximate renormalisation scheme* In preparation.

R. S. MacKay, 1987, *A new proof of Denjoy's Theorem.* Warwick Preprint.

N. S. Manton and M. Nauenberg, 1983, *Universal scaling behaviour for iterated maps in the complex plane.* Comm. Math. Phys. **89**, 555-570.

J. N. Mather, 1982, *Existence of quasiperiodic orbits for twist homeomorphisms of the annulus,* Topology **21**, 457.

J. N. Mather, 1984, *Non-existence of invariant circles.* Ergodic Theory and Dynamical Systems, **4**, 301-311.

J.N. Mather, 1986, *A criterion for the non-existence of invariant circles.* Math. Publ. IHES, **63**, 153-204

B. D. Mestel, 1985, *A Computer Assisted Proof of Universality for Cubic Critical Maps of the Circle with Golden Mean Rotation Number.* Warwick University Ph.D. Thesis.

J. Moser, 1973, *Stable and random motions in dynamical systems with special emphasis on celestial mechanics,* Annals of Math. Studies No.77, Princeton University Press.

S. Ostlund, R. Pandit, D. A. Rand, H. Schellnhuber and E. Siggia, 1982, *The 1-dimensional Shrödinger equation with an almost-periodic potential.* Phys. Rev. Lett. **50**, 1873

S. Ostlund, D. A. Rand, J. Sethna and E. Siggia, 1983, *Universal Properties of the Transition from Quasi-Periodicity to Chaos in Dissipative Systems.* Physica **8D**, 303. [Also see Phys. Rev. Lett. 49, (1982), 132.]

W. Paluba, 1987, *The Lipschitz condition for the conjugacies of Feigenbaum-like mappings.* Preprint, Warsaw.

I. Procaccia, S. Thomae and C. Tresser, 1987, *First-return maps as a unified renormalization scheme for dynamical systems.* Phys. Rev. A, **35**, 1884-1900.

D. A. Rand, 1984, *Universality for the breakdown of dissipative golden invariant tori.* Proceedings of the 1986 IAMP Conference in Mathematical Physics (Editors: M. Mebkhout and R. Sénéor), World Scientific, 1987, 537-547, and *Universality for golden critical circle maps and the breakdown of golden invariant tori.* Cornell Preprint submitted to Comm. Math. Phys. 1984.

D. A. Rand, 1986, *The singularity spectrum for hyperbolic Cantor sets and attractors.* Warwick Preprint.

D. A. Rand, 1987, *Fractal bifurcation sets, renormalisation strange sets and their universal invariants.* To appear in the Proceedings of the Royal Society Discussion Meeting on Dynamical Chaos.

I. I. Satija, 1985, *Universal strange attractor underlying Hamiltonian stochasticity.* Preprint.

G. Schmidt and J. Bialek, 1982, *Fractal diagrams for Hamiltonian stochasticity.* Physica **5D**, 397-404.

C.L. Siegel and J.K. Moser, 1971, *Lectures on Celestial Mechanics*, Springer-Verlag.

S. J. Shenker, 1982, *Scaling behaviour of a map of a circle onto itself : empirical results.* Physica **5D**, 405.

S. J. Shenker and L.D. Kadanoff, 1982, *Critical behaviour of a KAM surface: I Empirical results,* J. Stat. Phys. **27**, 631.

D. Sullivan, 1987, *Quasiconformal homeomorphisms in dynamics, topology and geometry.* Preprint.

D. Umberger, J. D. Farmer and I. I. Satija, 1984, *A universal attractor underlying the quasi-periodic transition to chaos.* Phys. Lett. **114A**, 1986, 341.

D. Whitley, 1983, *Discrete dynamical systems in dimensions one and two.* Bull. London Math. Soc. **15**, 177-217.

M. Widom, 1983, Renormalization group analysis of quasi-periodicity in analytic maps. Comm. Math. Phys. **92**, 121-136.

J. Wilbrink, 1987, *Erratic behaviour of invariant circles fot standard-like mappings.* Preprint.

K. Wilson, 1971 Phys. Rev. **3D**, 1818.

K. Wilson, 1975, *The renormalisation group : Critical phenomena and the Kondo problem.* Rev. Mod. Phys. **47**, 773.

J.-C. Yoccoz, 1984, *Conjugaison differentiable des diffeomorphismes du cercle dont le nombre de rotation vérifié une condition Diophantienne.* Ann. Sci. de l'Ec. Norm. Sup., **17**, 333-361.

SMOOTH DYNAMICS ON THE INTERVAL
(with an emphasis on quadratic-like maps)

Sebastian van Strien
Mathematics Department,
University of Delft,
Delft, The Netherlands.

There are many motivations for studying maps of the interval. Take for example maps from the quadratic family $f_\mu(x)=(\mu-1)-\mu x^2$. Mathematically these maps are very interesting because they can have such complicated, and yet quite well understood, dynamics. (The same can be said for the family of tent-maps $x \to 1-\mu|x|$, but these are much simpler because of their linearity.) Already in the beginning of this century, Fatou and Julia observed that quadratic maps can have an infinite number of periodic points. Later, in the 1940's, Ulam and von Neumann observed that for $\mu=2$ the map f_μ has an absolutely continuous invariant measure. This implies that typical orbits of this map have stochastic behaviour. In the late 1950's P.J. Myrberg began to study quadratic maps, partly numerically. And in the 1960's, W.Parry and A.N. Sarkovskii began to describe the symbolic dynamics of these maps. Moreover A. Schwartz (1963) developed a tool which can be used to get metric results (bounded distortion results) for maps without "critical points". For a more physical motivation for studying these maps see the book of P. Collet and J.P. Eckmann (1980).

In the 1970's the theory of these maps became much more mature. A much more complete understanding of the symbolic dynamics became possible through the work of M. Metropolis & M.L. Stein & P.R. Stein (1973) and especially through J. Milnor & W. Thurston's (1977) preprint. In this last paper the kneading theory was developed. L. Jonker & D. Rand and B. Derrida & A. Gervois & Y. Pomeau used this to give a better topological description of unimodal maps. Independently, M. Feigenbaum (1978) and P. Coullet & C. Tresser (1978) discovered that certain universality properties could be understood by a renormalisation procedure. These observations were made rigorous in the 1980's by M. Campanino, P. Collet, J.P. Eckmann, H. Epstein, O.E. Lanford and D. Ruelle.

I would like to thank T. Bedford, M. Martens, W. de Melo, J. Swift and C. Tresser for many useful suggestions and remarks on a previous version of this paper.

Recently D. Sullivan (1986) announced a result which states that all maps from a very large class have these universality properties (see Theorem (8.1.2))

A new metric tool, the Schwarzian derivative, was introduced by D. Singer. The first to notice and use the power of this tool were J. Guckenheimer, M. Misiurewicz. Since this tool was introduced, most metric results for interval maps from the 70's assumed that the Schwarzian derivative of a map is negative. (This condition is satisfied for quadratic interval maps.) In particular the following metric results were proved for either expanding maps, or quadratic maps or for maps with negative Schwarzian derivative:

A. Lasota & J. Yorke, M.V. Jacobson, R. Bowen, D. Ruelle, M. Misiurewicz, W. Szlenk, P. Collet & J.P. Eckmann and others proved the existence of absolutely continuous invariant measures. In particular the result of M.V. Jacobson is remarkable. He proved that a large number of maps from the family of quadratic maps have absolutely continuous invariant measures.

J. Guckenheimer proved that maps with just one critical point cannot have wandering intervals. This implies that the topological dynamics is completely determined by the "kneading invariant" from Milnor & Thurston's theory.

Of course, there are many other extremely interesting contributions to this theory (especially of a more combinatorial type). However, rather than giving a superficial overview, we just want to emphasise certain problems on maps of the interval. Other surveys about one-dimensional maps are for example the book by P. Collet & J.P. Eckmann (1980), and the paper by Z. Nitecki (1981). We have tried to minimise the overlap with those surveys.

Our survey has two purposes:

1) First of all we will give a short summary (including proofs) of (what we consider) the most essential results on the topological dynamics of unimodal maps. Many of these results are contained in preprints by J. Milnor and W. Thurston, (1977), which were never published. Quite a few of their results cannot be found anywhere in the literature, not even in the beautiful monograph on unimodal mappings by P. Collet & J.P. Eckmann,

(1980) or in the papers by L. Jonker and D. Rand.
2) All the results on hyperbolicity or existence of certain measures from the 70's rely on the Schwarzian derivative of a map being negative. This condition is extremely strong (it is like a convexity assumption), and not very natural (for example, this condition is not coordinate invariant). In this survey we will discuss some analytic tools which are useful in getting rid of this condition. In particular we want to discuss the use of the following two tools:

In order to get rid of the Schwarzian derivative condition, R. Mañé (1985) used the techniques of A. Schwartz extensively and proved a very general hyperbolicity result "away from the critical points". (This result was proved before by M. Misiurewicz under an assumption on the Schwarzian derivative ($S(f)<0$)). Some of the main ingredients used in Mañé's result will be explained.

In the same spirit, W. de Melo & S. van Strien (1986) have generalised two Lemma's of A. Denjoy and A. Schwartz in such a way that they can also be used near critical points. Using this they have proved the non-existence of wandering intervals for general smooth unimodal maps without a flat critical point. (This result was proved before by J.Guckenheimer for unimodal maps with $S(f)<0$.)

More recently it was shown (S. van Strien (1987)) that if a C^2 map has only repelling periodic points and satisfies the so-called Misiurewicz condition (see section (7.3.)) then it has an absolutely continuous invariant measure (no assumption on $S(f)$ is necessary).

It seems likely that, with those techniques, the condition on the Schwarzian derivative can be replaced everywhere by more natural conditions.

We will go into some recent developments on the following topics: (i) non-existence of wandering intervals, (ii) hyperbolicity results, (iii) existence of invariant measures and (iv) renormalisation results. We will ask some questions which are in the same spirit as in J. Milnor (1985).

Many other interesting topics will not be discussed here. (It seems that in particular the idea of using results on iterations of rational mappings on the Riemann-sphere for quadratic-like mappings (Douady-Hubbard, Thurston, Sullivan) may well turn out to be one of the most important recent developments.)

The organisation of this paper is as follows:

1. SOME BASIC DEFINITIONS

Consider the class $\mathcal{U}$ of continuous functions $f: [-1,1] \to [-1,1]$ such that $f(-1) = f(1) = -1$ and such that f has only one turning point c. (That is f is strictly monotone on $[-1,c]$ and on $[c,1]$.) We call these maps unimodal, because they have only one "critical point". Many of the results we shall mention also hold for arbitrary maps $f: [-1,1] \to [-1,1]$ (with more critical points). However for simplicity we shall restrict our attention mostly to unimodal maps.

Often we will make additional smoothness assumptions on our maps. In this case we will call the maps *quadratic like* in order to express the similarity with the quadratic maps $f_\mu(x)=(\mu-1)-\mu x^2$.

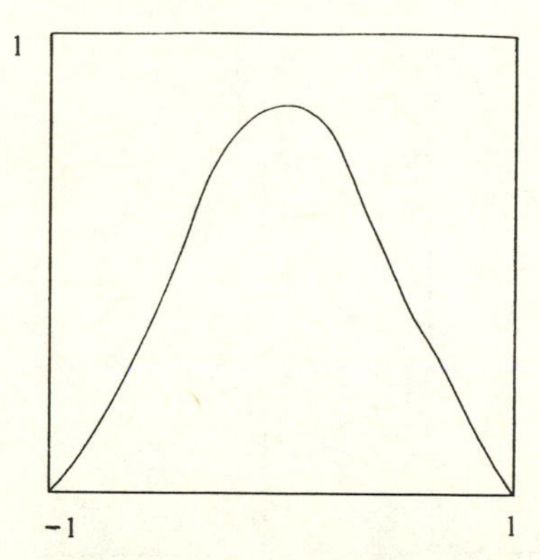

Figure 1: A typical map in $\mathcal{U}$

Let us give some general definitions: We say that p is a periodic point of f if $f^n(p)=p$ for some $n \geq 1$. The smallest $n>0$ for which $f^n(p)=p$ is called the period of p. A point x is called non-wandering if and only if there exist sequences $x_n \to x$ and integers $k(n) \to \infty$ such that a subsequence of $f^{k(n)}(x_n)$ converges again to x. The set of non-wandering points is denoted by $\Omega(f)$.

This set is closed and contains the set of periodic points of f.

Let K be an invariant set of f, i.e., f(K)=K. We define the basin of K to be the set $B_K=\{y;\ f^i(y) \to K \text{ as } i \to \infty\}$. Let O be a periodic orbit $O=\{x,f(x),\ldots, f^n(x)=x\}$. The union of those components of B_O which contain the points of O is called the immediate basin. If the immediate basin of O consists of non-empty intervals then O is called an attractor. If these intervals contain O in their interior then O is called a two-sided attractor. Otherwise O is called a one-sided attractor.

Now suppose f is C^1. Then one can prove that a periodic point p of period n can only be an attractor if $|(f^n)'(p)| \le 1$. If $|(f^n)'(p)|<1$ then p is a hyperbolic attractor. In this case all maps sufficiently C^1 close to f also have a periodic hyperbolic attractor near the orbit of p. If $|(f^n)'(p)|>1$ then p is called a hyperbolic periodic repellor.

If the immediate basin of an attracting periodic point p contains a critical point of f^n then p is called an essential attractor. Later on we shall see that for quadratic maps a periodic attractor is always essential (and since a quadratic map has just one critical point, it can have at most one attractor).

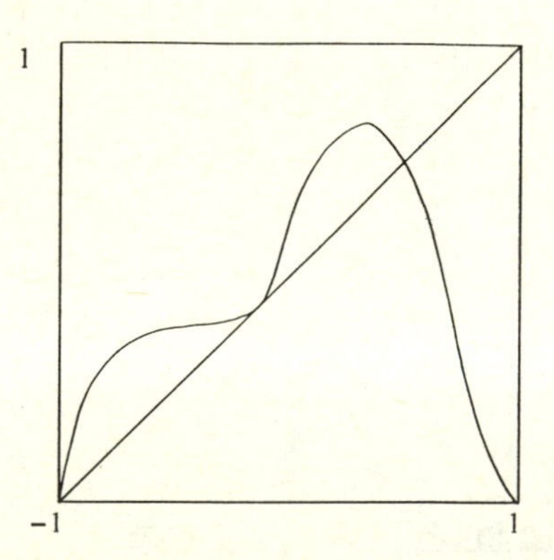

Figure 2: a one-sided attractor

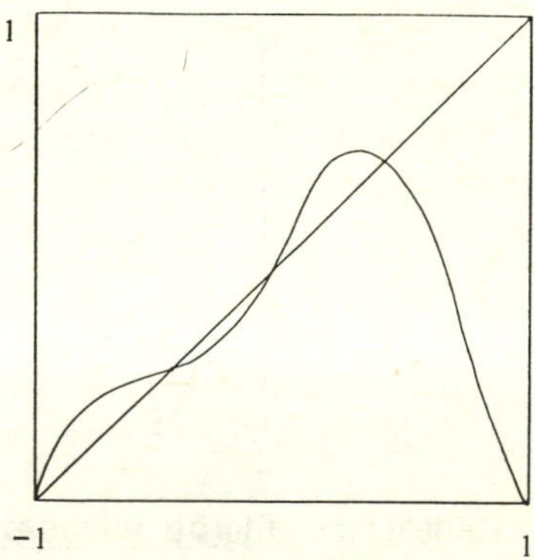

Figure 3: A two-sided attractor

Later, in section (2.1), we shall show that if a point x does not belong to the non-wandering set, then there are two possibilities:

i) x is contained in the basin of a (possibly one-sided) periodic attractor, or

ii) x is contained in a wandering interval, i.e., an interval J such that $J, f(J), f^2(J), \ldots$ are disjoint and no point of J is eventually periodic or contained in the basin of a periodic attractor.

Also we shall see that if f is unimodal, sufficiently smooth and satisfies certain non-flatness conditions (see section (4.2)) then the second possibility cannot occur.

One of the problems we are concerned with is the topological classification of maps. We say that f and g are conjugate if there is a homeomorphism h such that $h \circ f = g \circ h$.

Sometimes we shall use the following notation. Take x and y in $[-1,1]$. In order not to have to distinguish between $x<y$ and $x>y$, we let $\langle x,y \rangle$ denote the interval connecting x and y.

2. SYMBOLIC AND TOPOLOGICAL DYNAMICS OF UNIMODAL MAPS

In this section we will give an overview of some of the main results from J.Milnor & W.Thurston's (1976) preprint.

(2.1) Itineraries and Kneading Coordinates

In this subsection we shall show that the dynamics of unimodal maps, up to certain periodic or wandering intervals, is completely determined by a formal power series (called the kneading-invariant). The existence and the ordering of orbits turns out be determined by combinatorial properties of this kneading-invariant. Many of these properties date back to the paper of M. Metropolis, M.L. Stein and P.R. Stein (1973).

To describe an orbit we associate to a point x in $[-1,1]$ a sequence of symbols $A_i(x)$ which give the position of $f^i(x)$ with respect to the critical point c. More precisely, define

$$A_i(x;f) = \begin{cases} L & \text{if } f^i(x) \text{ is in } [-1,c), \\ C & \text{if } f^i(x)=c, \\ R & \text{if } f^i(x) \text{ is in } (c,1]. \end{cases}$$

In this way we get the itinerary $A_f(x)=(A_i(x))_{i\geq 0} \in \{L,C,R\}^{\mathbb{N}}$. As usual define the shift map $\sigma:\{L,C,R\}^{\mathbb{N}} \to \{L,C,R\}^{\mathbb{N}}$, by $\sigma(A_i)_{i\geq 0} = (A_{i+1})_{i\geq 0}$, for $(A_i)_{i\geq 0} \in \{L,C,R\}^{\mathbb{N}}$. Clearly the shift map commutes with the map f, i.e., $\sigma(A_f(x))= A_f(f(x))$.

Another, alternative way to describe the symbolic dynamics of the map f is to use the kneading coordinate. If f is C^1 define

$$\theta_i(x)=\begin{cases} 0 & \text{if } (f^{i+1})'(x) = 0, \quad \text{i.e., if } f^j(x)=c \text{ for some } 0\leq j\leq i, \\ -1 & \text{if } (f^{i+1})'(x) < 0, \\ 1 & \text{if } (f^{i+1})'(x) > 0. \end{cases}$$

If f is only continuous we define $\theta_i(x)$ similarly depending on whether f^{i+1} near x has an extremum, is orientation reversing or orientation preserving.

Let $\theta_f(x)$ denote the formal power series $\sum_{i\geq 0} \theta_i(x)\, t^i$. This power series is called the kneading coordinate of x. Since f is orientation preserving on $[-1,c)$ and orientation reversing on $(c,1]$, it is not difficult to express the itinerary $A_f(x)$ into the kneading coordinate $\theta_f(x)$, and vice versa. In fact $\theta_i(x)=1$ if and only if an even number of points of $\{x,f(x),\ldots, f^i(x)\}$ are contained in the interval $(c,1]$. Define the shift map on the space of formal power series similarly as before: $\sigma(\sum_{i\geq 0} \theta_i\, t^i)= \sum_{i\geq 0} \theta_{i+1}\, t^i$. We remark that the shift map does not commute with f, but that

$$\sigma\circ\theta_f(x) = \theta_0(x)\,.\,\theta_f\circ f(x).$$

For some purposes it is easier to work with the itinerary, for other purposes the kneading coordinate is more useful.

Endow the space of power series with the usual lexicographical ordering. That is $\sum_{i\geq 0}\theta_i\, t^i < \sum_{i\geq 0}\eta_i\, t^i$ if for some $k\geq 0$, $\theta_i = \eta_i$ for $i<k$ and $\theta_k < \eta_k$. This ordering defines a topology on the space of formal power series. The following Lemma is fundamental in the theory of Milnor-Thurston.

Lemma (2.1.1). The map $x \to \theta_f(x)$ is monotonically decreasing.

Proof: Take two points y,z in [0,1] with y<z. Let $n\geq 0$ be the smallest integer such that $\theta_n(y)$ and $\theta_n(z)$ are distinct. If n=0 then $y\leq c\leq z$ and $\theta_0(y) \geq \theta_0(z)$. If $n>0$ then f^n maps $\langle y,z\rangle$ homeomorphically onto $\langle f^n(y),f^n(z)\rangle$ and the interval $\langle f^n(y), f^n(z)\rangle$ contains the critical point c. There are three cases: (i) $\theta_{n-1}(y)=\theta_{n-1}(z)=-1$; (ii) $\theta_{n-1}(y) = \theta_{n-1}(z)= 1$; (iii) $\theta_{n-1}(y) = \theta_{n-1}(z)= 0$. Let us only treat the first case, (ii) is similar and (iii) is impossible since $\theta_0(y) \neq \theta_0(z)$. In this case the homeomorphism $f^n:\langle y,z\rangle \to \langle f^n(y),f^n(z)\rangle$ is orientation reversing and therefore $f^n(y)$ is in [c,1] and $f^n(z)$ in [-1,c]. Hence $(f^n)'(y)\geq 0$ and therefore $\theta_n(y)\geq 0$. Similarly $\theta_n(z)\leq 0$. Hence $\theta_n(z)\leq \theta_n(y)$.

□

Remark. In L.Jonker (1981) and J. Milnor & W. Thurston (1977) the same ordering is used but their unimodal maps have a minimum. So in this case $x\to\theta_f(x)$ is monotonically increasing. In P. Collet & J.P. Eckmann (1981) unimodal maps have a maximum, but another convention for the ordering is chosen.

From this Lemma it follows that the limits $\theta_f(x^-)$ (respectively $\theta_f(x^+)$) of $\theta_f(y)$ as y tends from the left (respectively the right) to x are well defined.

Remark. The only discontinuities of $x \to \theta_f(x)$ are at points where $f^n(x)=c$ for some $n\geq 0$.

Define the <u>kneading invariant</u> ν_f of f to be the limit $\theta_f(c^-)$ of $\theta_f(x)$ as x tends from the left to c. (This limit exists because of the previous Lemma.)

Remark. The reason for taking a limit is that $\theta_f(c)=0$. However, if c is not periodic, then $x\to\theta_f(x)$ is continuous at f(c), and we have

$$\theta_f(c^-)= 1 + t\,.\,\theta_f(f(c)).$$

If c is periodic (say of period n) then the power series $\theta_f(c^-)$ has period n or 2n (depending on whether $f^n(c^-)$ tends to c from the left or from the right). In either case

$$\theta_f(c^-)= 1 + t\,.\,\theta_f(f(c)) \pmod{t^n}.$$

From this it is clear that $\theta_f(f(c))$ contains the same information as $\theta_f(c^-)$. In the book of P. Collet & J.P. Eckmann (1980) the power series $\theta_f(f(c))$ is used. Also notice that $\theta_f(c^-)= -\theta_f(c^+)$.

The following Lemma implies that all the information about the order of orbits of f is already contained in this kneading invariant of f.

We say that a formal power series θ is ν_f-admissible, if for every $n \geq 0$

$$-\nu_f < \sigma^n(\theta) < \nu_f \text{ implies } \sigma^n(\theta)=0.$$

(with respect to the lexicograpic ordering). Observe that for each x in [-1,1] one has either $|\theta_f(f^n(x))| \geq \nu_f$ or $\theta_f(f^n(x)) = 0$. Therefore for all $n \geq 0$ one has either $|\sigma^n(\theta_f(x))| \geq \nu_f$ or $\sigma^n(\theta_f(x)) = 0$. It follows that for every point x the power series $\theta_f(x)$ is ν_f-admissible.

Lemma (2.1.2). For every ν_f-admissible series θ there exists $x \in [-1,1]$ such that θ is equal to $\theta_f(x)$, $\theta_f(x^-)$ or $\theta_f(x^+)$.

Proof: Let $x = \inf\{y; \theta_f(y) \leq \theta\}$. Then $\theta_f(x^-) \geq \theta \geq \theta_f(x^+)$. If $\theta_f(y)$ is continuous at x then $\theta_f(x)=\theta$. If $\theta_f(y)$ is not continuous at x, then for some $n \geq 0$ one has $f^n(x)=c$, and $\sigma^n(\theta_f(x^-)) = -\sigma^n(\theta_f(x^+))$ is equal to $\pm\nu_f$. Since $\sigma^n(\theta)$ is ν_f-admissible and between $\sigma^n(\theta_f(x^-))$ and $\sigma^n(\theta_f(x^+))$, one has $\sigma^n(\theta) = \pm \nu_f$ or $\sigma^n(\theta)=0$. The result follows.

□

Corollary (2.1.3). Take two maps f,g in $\mathcal{U}$, with critical points c_f and c_g, respectively. If $\nu_f = \nu_g$ then there exists an order preserving map h: $\{f^{-i}(c_f); i \geq 0\} \to \{g^{-i}(c_g); i \geq 0\}$, such that hf=gh.

Proof: For x in $\{f^{-i}(c_f); i \geq 0\}$, take $h(x) = \inf\{y; \theta_g(y) \leq \theta_f(x)\}$. As in the previous Lemma one proves that h(x) is contained in $\{g^{-i}(c_g); i \geq 0\}$. The equation hf=gh is obvious.

□

From Lemma (2.1.2) and the Corollary (2.1.3) it is clear that all information about orbits is contained in ν_f except the information as to whether or not the map $x \to \theta_f(x)$ is constant on an interval. (So the kneading invariant ν_f plays the same role in the theory of unimodal maps as the rotation number in the Poincaré-Denjoy theory for diffeomorphisms of the circle. Indeed the rotation number can be replaced by a symbolic sequence of the same kind as the itineraries from above.)

Let us show when $x \to \theta_f(x)$ is constant on some interval.

Corollary (2.1.4): If the map $x \to \theta_f(x)$ is constant on some interval J, then either

(i) every point of J is contained in the basin of a periodic attractor;

or

(ii) the interval J is <u>wandering</u>, i.e., the intervals J, f(J), $f^2(J)$, ..., are all disjoint, and no point of J is eventually periodic or contained in the basin of a periodic attractor.

Proof: The fact that the map $x \to \theta_f(x)$ is constant on some interval J, implies that there is no $x \in int(J)$ with $f^n(x)=c$. In particular $f^n|J$ is a homeomorphism for all $n \geq 0$. Assume that case (ii) is not satisfied. Then there is $n \geq 0$ and $k>0$ such that $f^n(J)$ and $f^{n+k}(J)$ are not disjoint. Then for every $p \geq 0$, $f^{n+pk}(J)$ and $f^{n+(p+1)k}(J)$ overlap. Thus $L=\bigcup_{p\geq 0} f^{n+pk}(J)$ is an interval and f^k maps this interval L injectively into itself. So points of J are eventually mapped onto fixed points of $f^k|L$ or iterate to some attracting fixed point of $f^k|L$.

□

Remark. Let K be an interval such that $f^k|K$ is a homeomorphism, for all $k \geq 0$. Such an interval is usually called a <u>homterval</u>. The previous result states that a homterval is contained either in the basin of periodic attractors or in a wandering interval. Later on we shall show that typically wandering intervals cannot exist (for unimodal maps).

From Lemma (2.1.2) one can get a wealth of information. For example one may use this to get a proof in the unimodal case of the well-known result of A.N. Sarkovskii. Define the following ordering on $\mathbb{N}$:

$$3 \succ 5 \succ 7 \succ 9 \succ \ldots \quad \ldots \succ 2.3 \succ 2.5 \succ 2.7 \succ 2.9 \succ \ldots$$
$$\succ 2^2.3 \succ 2^2.5 \succ 2^2.7 \succ 2^2.9 \ldots \succ 2^n.3 \succ 2^n.5 \succ 2^n.7 \succ 2^n.9 \ldots$$
$$\succ \ldots \succ 2^n \succ 2^{n-1} \succ \ldots \succ 8 \succ 4 \succ 2 .$$

Theorem (2.1.5). (Sarkovskii (1964), see also P. Stefan (1976) and L. Block et al (1979)).

Let $f:[-1,1] \to [-1,1]$ be a continuous map. Assume that f has a periodic point of period k, and $k \succ l$. Then f also has at least one periodic point of period l.

□

A proof of Theorem (2.1.5) in the unimodal case was given by L. Jonker (1980) using kneading theory. However this result is true for arbitrary maps whereas the power of kneading theory is based on the map being piecewise monotone.

There are many very nice results extending these ideas due to L. Alseda, L. Block, E.M. Coven, W.A. Coppel, J.M. Gambaudo, J. Guckenheimer, Y. Kan, J. Llibre, I. Malta, Z. Nitecki, M. Misiurewicz, C. Simó, C. Tresser, L.S. Young and others. We will not go into these results here.

(2.2) The number of critical points and laps for unimodal maps

In this section it will be shown how the kneading sequence is related to the number of critical points γ_n and the number of laps lap_n of f^n. In the next section it will turn out that the growth rate of these number γ_n and lap_n determines certain qualitative features of f.

Let J be an open interval in [-1,1]. We will count the number of critical points of $f^n|J$ in two ways.

Define $\gamma_n(J)$ to be the cardinality of $\{x \in f^{-n}(c);\ x \in J$, and $f^k(x) \neq c$ for $k=0,\ldots,n-1\}$. Furthermore define the power series $\gamma_f(J) = \sum_{n\geq 0} \gamma_n(J)t^n$. Let γ_n denote the cardinality of $f^{-n}(c)$, and $\gamma_f = \sum_{n\geq 0} \gamma_n t^n$. There is a very neat relation between $\gamma_f(J)$ and ν_f.

Lemma (2.2.2).
(i) $\nu_f\, \gamma_f = (1-t)^{-1} = 1+t+t^2 \ldots$;
(ii) Let $J=(a,b)$, then $\nu_f\, \gamma_f(J) = (1/2)\,(\,\theta_f(a^+) - \theta_f(b^-)\,)$.

Proof: Since $f(-1)=f(1)=-1$, one has $\theta_f(1^-) = (-1-t-t^2 \ldots)$ and $\theta_f(-1^+) = (1+t+t^2 \ldots)$. Therefore statement (i) follows from statement (ii). Let us prove (ii). Write $\nu_f = \sum_{i\geq 0} \nu_i\, t^i$. Let us compute

$$\nu_f\, \gamma_f = \sum_{n\geq 0} \left(\sum_{0\leq i\leq n} \gamma_{n-i}\, \nu_i \right) t^n.$$

Take a preimage x of $f^{-(n-i)}(c)$ which is in J. Then $f^{(n-i)}(x)=c$ and f^{n+1} has a local maximum (resp. minimum) at x if and only if $\nu_i = 1$ (resp. $\nu_i = -1$). From this it follows that $(\sum_{0\leq i\leq n} \gamma_{n-i}\, \nu_i)$ is equal to the number of maxima of f^{n+1} (restricted to int(J)) minus the number of minima of $f^{n+1}|\mathrm{int}(J)$. This number is either +1, 0 or -1; for example it is 0 iff $f^{n+1}|J$ is increasing near one of the boundary points of J and decreasing near the other boundary point. Since $f^{n+1}: J \to \mathbb{R}$ is increasing near b (resp. near a) iff $\theta_n(b^-)=1$ (resp

$\theta_n(a^+)=1$), the difference of the number of maxima and minima of $f^{n+1}|J$ is equal to $(1/2)(\theta_n(a^+) - \theta_n(b^-))$. This proves the Lemma.

□

Similarly we can define $lap_n(J)$ to be the number of laps of $f^n|J$, i.e., the number of components in J on which $f^n|J$ is monotone. Also define $lap_f(J) = \sum_{n\geq 0} lap_n(J)t^n$. Similarly define lap_n to be the number of laps of f^n and $lap_f = \sum_{n\geq 0} lap_n t^n$.

Since each critical point of f is also a turning point, the number of laps $lap_n(J)$ is equal to the number of critical points of $f^n|J$ plus one, i.e.,

$$lap_n(J) = \sum_{0\leq k\leq n-1} \gamma_k(J) + 1.$$

From this one easily gets

$$lap_f(J) = \sum_{n\geq 0} lap_n(J)t^n = (1 + t\,\gamma_f(J)) / (1-t).$$

The importance of Lemma (2.2.2) comes from the following corollary:

Corollary (2.2.3). The power series ν_f is analytic. The power series lap_f and γ_f are both meromorphic and converge for $|t|<r(f)$, where r(f) is the smallest positive real zero of $\nu(f).(1-t)$.

Proof: Since $\nu_f = \sum_{i\geq 0} \nu_i t^i$, where $\nu_i = \pm 1$, the power series ν_f converges for all t and therefore that $t\to\nu_f(t)$ is analytic. Since $\nu_f\gamma_f = (1-t)^{-1} = 1+t+t^2...$, and $lap_f = (1 + t\,\gamma_f) / (1-t)$ it follows that lap_f and γ_f are meromorphic and have the same poles in $|t|<1$. Since all coefficients of γ_f are positive it follows from Abel's theorem that if $r(f) > 0$ is the radius of convergence of $t\to\gamma_f(t)$ then r(f) is a pole of $t\to\gamma_f(t)$ (i.e. the "first" pole of γ_f is on the positive real line). Therefore if $r(f)<1$ then the power series $\nu(f)$ has the same real number r(f) as its smallest positive real zero.

□

(2.3) Topological entropy for unimodal maps

In Corollary (2.2.3) it was shown that the power series lap_f and γ_f converge for $|t|<r(f)$. In this section it will be shown that indeed the numbers lap_n and γ_n have growth-rate $(1/r(f))$. From a result of Misiurewicz and Szlenk then it follows that r(f) is related to the topological entropy of f. In this way Milnor and Thurston show that h(f) depends continuously on f.

Lemma (2.3.1). The limits of $(lap_n)^{(1/n)}$ and $(\gamma_n)^{(1/n)}$ as $n \to \infty$ both exist and are equal (call this limit s(f)). The power series $\gamma(f) = \sum_{i\geq 0} \gamma_i t^i$ and $lap_f = \sum_{n\geq 0} lap_n t^n$ both have radius of convergence $r(f)=1/s(f)$. Furthermore $r(f)\in[1/2,1]$ and r(f) is the smallest positive real solution of $\nu(f).(1-t)=0$.

Proof: Denote the number of laps of a function g by $lap(g)$. Clearly

$$lap\,(f\circ g) \leq lap(f)\,.\,lap(g).$$

Hence, using a well-known trick for subadditive sequences, it follows that $\lim_{n\to\infty}(1/n).\log(lap_n)$ and therefore $\lim_{n\to\infty}(lap_n)^{(1/n)}$ exists. Because $lap_n = \sum_{0\leq k\leq n-1} \gamma_k + 1$, $\lim_{n\to\infty}\gamma_n)^{(1/n)} = \lim_{n\to\infty}(lap_n)^{(1/n)}$. Clearly this number is equal to the radius of convergence of the powerseries $lap_f = (1 + t\,\gamma_f) / (1-t)$ and that of γ_f. Since for any x in $[-1,1]$ $f^{-1}(x)$ has at most two preimages, one has $1 \leq lap_n \leq 2^n$, and therefore $r(f)\in[1/2,1]$. Using Corollary (2.2.3) the result follows.

□

Let h(f) be the topological entropy of f. From the next Lemma it follows that $s(f)=\exp(h(f))$. (Notice that from Lemma (2.3.1) one has for unimodal maps $h(f)\in[0,\log(2)]$.)

Lemma (2.3.2). [Misiurewicz and Szlenk, (1977)]
$h(f) = \lim_{n\to\infty}(1/n) \log \gamma_n$.

□

We will not prove this here. Using this result and the results from the previous section J. Milnor and W. Thurston proved the following result which had been unanswered for quite some time.

Corollary (2.3.3.). [Milnor & Thurston (1977)]
Consider the space $\mathcal{U}^1$ of C^1 unimodal mappings with the C^1 topology. Then for $f\in\mathcal{U}^1$, $f\to h(f)$ is a continuous map.

Proof: Notice that all maps in $\mathcal{U}^1$ have only one critical point. We distinguish two cases:

(i) The critical point is non-periodic. Since $\nu(f)= \sum_{i\geq 0} \nu_i t^i = \theta_f(c^-)$, the power series $\nu(f)$ depends continuously on f in this case. Moreover the Cauchy integral theorem implies that the smallest zero of $\nu(f)$ depends continuously on f. Using the remark above, the Corollary follows.

(ii) The critical point c is periodic, say of period n. Now we need that f is C^1 and that perturbations are C^1 close to f. So take g C^1 close to f. Since c is hyperbolic for f, there is an attracting periodic orbit of g near the orbit of c such that the critical point c is in the immediate basin of this periodic point. Let $\nu(g)=\sum_{i\geq 0}\nu_i t^i$. The kneading sequence of $\nu(g)$ is periodic. More precisely if the periodic attractor of g near c is orientation preserving then $\nu_{i+n}=\nu_i$, for all $i\geq 0$; if it is orientation reversing then $\nu_{i+n}=-\nu_i$, for all $i\geq 0$. (If the periodic attractor has eigenvalue 0, then the critical point c of g is periodic and we are in the first case if g^n has a maximum at c, and in the second case otherwise.)
So one can write $\nu(g)$ in the first case as $(\sum_{0\leq i\leq n-1}\nu_i t^i).(1-t^n)^{-1}$ and in the second case as $(\sum_{0\leq i\leq n-1}\nu_i t^i).(1+t^n)^{-1}$. The poles in $|t|<1$ of these two power expansions coincide and therefore the topological entropy of all sufficiently small C^1 perturbations of f is the same as h(f). This proves the Corollary.

□

In a more general setting the map $f\to h(f)$ need not be continuous: see the following examples.

Examples (2.3.4)
(1) If we allow the perturbation of the map f to have more critical points than f, then the result above does not hold. (This can be done by C^0 small perturbations.) An example of this can be found in several places, e.g. M.Misiurewicz and W.Szlenk (1979).

(2) Also, if we take the C^0 topology on the space of C^0-unimodal mappings then the result above is not true either. For example consider the non-symmetric piecewise linear map F on [-1,1] (with slope 1 and $s_2<-2$ to the left and respectively to the right of the turning point). It is easy to check that $F^n=F$ for all $n\geq 0$ and therefore $h(F)=0$. However, perturb F to a piecewise linear map F^* with slopes $s>1$ and s_2 as in the figure below. Let p be the fixed point for F^* with $|(F^*)'(p)|=|s_2|>2$. Take q so that $F^*(q)=p$. Consider the second iterate $G=(F^*)^2$. One has $G(q)=p$, $G'(q)=s.s_2>2$, $G(p)=p$,

$G'(p)=(s_2)^2>4$. Moreover G has precisely one critical point in $\langle q,p \rangle$. From this and $(1/s_2)+(1/s_2)^2<1$, one can see that $G(\langle q,p \rangle)$ strictly contains $\langle q,p \rangle$, as in figure four. But then the number of laps of G^n is at least 2^n. Hence from Lemma (2.3.2), $h(G)\geq\log(2)$. Hence $h(F^*)\geq(1/2).\log(2)$.

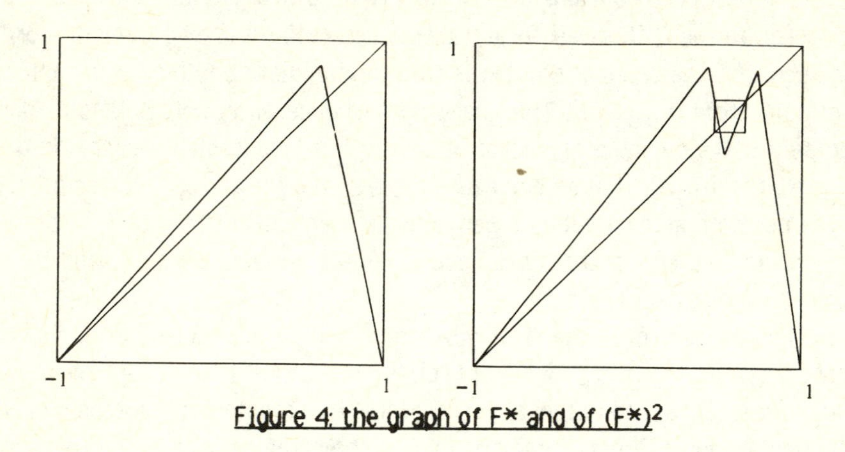

Figure 4: the graph of F^* and of $(F^*)^2$

This is a good point to remark that the idea of the last example dates back to the following result:

Theorem (2.3.4). M. Misiurewicz (1979).
Let $f : [-1,1] \to [-1,1]$ be a continuous map. Let $h(f)$ be the topological entropy of f. Then $h(f)>0$ if and only if f has a horseshoe.

□

Theorem (2.3.4) is the one-dimensional analogue of Katok's theorem for $C^{1+\epsilon}$ diffeomorphisms on surfaces.

Remark. (2.3.5) M. Misiurewicz (1986) has recently proved the following. Let $\mathcal{U}^0$ be the space of unimodal mappings with the C^0 topology. Then $\mathcal{U}^0 \ni f \to h(f)$ is continuous at a map f_0 provided $h(f_0)>0$. The proof of this is heavily based on the fact that only perturbations of f_0 are allowed with one turning point. (In fact his result gives a formula for $\inf\{h(f_n) \mid f_n \to f$, f_n has at most k critical points$\}$.)

(2.4) Maps with zero entropy: the idea of renormalising

In this section we assume that $h(f)=0$. Let us explain the dynamics of such a unimodal map using the procedure of "renormalising".

For any f in the space of C^1 unimodal mappings $\mathcal{U}$ we shall distinguish two cases (see figure 5):

case 1) there is an orientation reversing fixed point p (i.e. $f(p)=p$ and $f'(p)<0$);

case 2) there is no such fixed point.

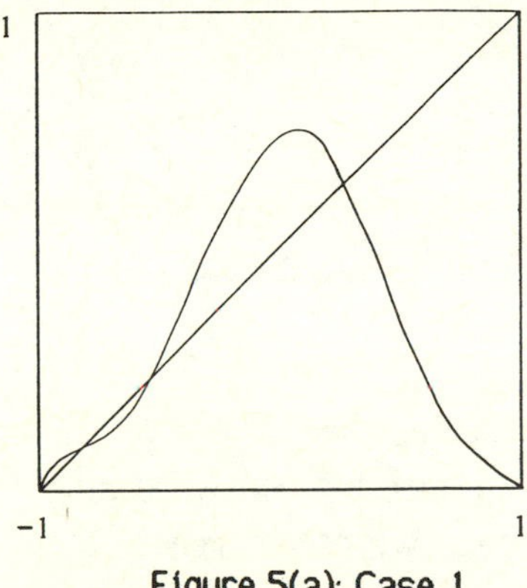

Figure 5(a): Case 1.

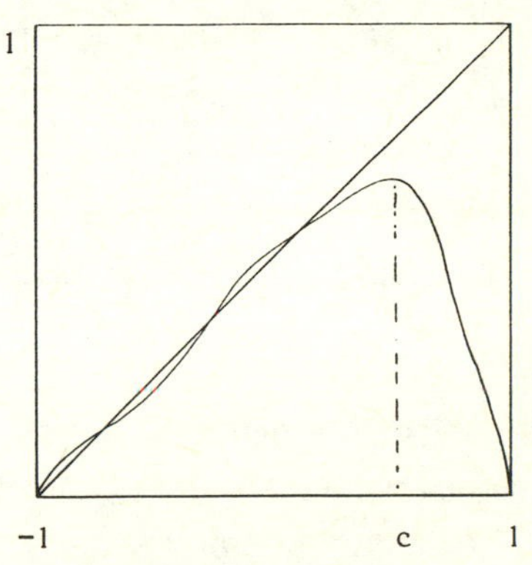

Figure 5(b): Case 2.

In case 1:

We will renormalise: Let $p'\neq p$ be the point such that $f(p')=f(p)$. Since f is unimodal there is precisely one such point p'.

Clearly $f[p',p]=[p,f(c)]$ and therefore $f^2|[p',p]$ has precisely one critical point. Also $f^2(p')=f^2(p)=p$.

Claim. f^2 maps (p',p) into itself.

Proof: If not, then f^2 maps $[p',p]$ twice over itself, as in Figure 6(a) below. But then the number of laps of $(f^2)^n$ is at least 2^n. Hence from Lemma (2.3.2), $h(f^2)\geq\log(2)$. This contradicts $h(f)=0$. (Compare this with the second example in (2.3.4).) □

Hence we see that f^2 maps $J_1=[p',p]$ into itself with one critical point, and moreover f^2 maps the boundary points of $[p',p]$ into p. So after scaling and reversing orientation on the interval we can consider f^2 restricted to J_1 as a map f_2 in $\mathcal{U}$ with $h(f_2)=0$. So in this case we can repeat the procedure. Remark that J_1 and $f(J_1)$ have precisely one point in common and are contained in $\mathrm{int}(J_0)=(-1,1)$.

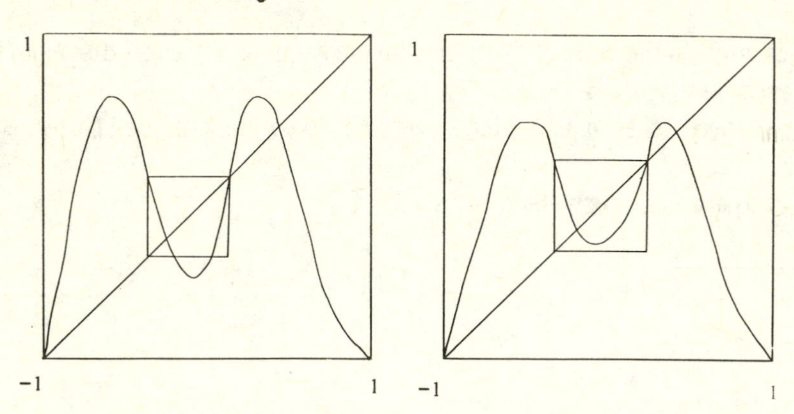

Figure 6(a): In this case $h(f)>0$ Figure 6(b): $f^2: [p',p] \mapsto [p,p']$

In case 2:
The situation is simple: all points in $[-1,1]$ iterate towards one of the fixed points (which are all contained in $(-1,c]$). Hence the non-wandering set of f consists just of fixed points. In this case we cannot repeat the procedure.

Suppose we can repeat this procedure n-times. By induction we get intervals J_n with $[-1,1]=J_0 \supset \ldots \supset J_{n-1} \supset J_n$ around the critical point c such that for $k(n)=2^n$,

i) $f^{k(n)}(J_n)$ is contained in J_n and is, up to scaling, a map in $\mathcal{U}$; (Similarly $f^{k(n)+i}(J_n)$ is contained in $f^i(J_n)$, for $i<k(n)$).

ii) $J_n, f(J_n), f^2(J_n), \ldots, f^{k(n)-1}(J_n)$ have no interior points in common;

iii) $f^i(J_n)$, $f^{i+k(n)/2}(J_n)$ are strictly contained in the interior of the previous interval $f^i(J_{n-1})$, for $i<(k(n)/2)=k(n-1)$.

It follows that $K_n = J_n \cup f(J_n) \cup f^2(J_n) \cup \ldots \cup f^{k(n)-1}(J_n)$ consists of $k(n)/2= 2^{n-1}$ disjoint intervals.

There are two possibilities:

A) The process terminates (after r steps we are in case 2), and the non-wandering set consists only of periodic points of periods $1,2,4,\ldots, 2^r$.

B) The process repeats infinitely many times. In this case the non-wandering set of f consists of periodic points of periods $1,2,2^2,2^3,\ldots$ and a subset of an invariant Cantor-like set (it is a countable intersection of a nested sequence of intervals)

$$\Lambda = \bigcap_{n\geq 0} \{J_n \cup f(J_n) \cup f^2(J_n) \cup \ldots \cup f^{k(n)-1}(J_n)\}$$
$$= \bigcap_{n\geq 0} K_n$$

where $k(n)=2^n$ and K_n is a nested decreasing sequence of intervals. K_n has 2^{n-1} components, two in each component of K_{n-1}.

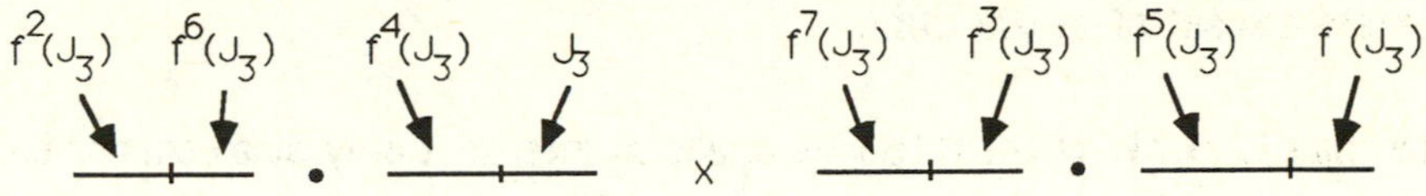

x = fixed point; • = point of period 2; + = point of period 4

Figure 7: The set K_3

If Λ does not contain an interval, then Λ is a Cantor set and since $f^{k(n)+i}(J_n)$ is mapped into $f^i(J_n)$, each point of Λ is non-wandering. If Λ does contain an interval then f^n restricted to this interval is a homeomorphism for each $n\geq 0$ and f has a homterval (i.e., a wandering interval). The intersection of the non-wandering set with Λ is equal to the ω-limit set of the critical point c. Later we shall see that a large class of maps cannot have any wandering intervals.

Remark. It is this renormalising that led M. Feigenbaum and P. Coullet & C. Tresser to an explanation of the universality of some numbers. (See section 8.) In the infinite decomposition, when Λ is a Cantor-set, one also speaks of the map "at the accumulation of period doubling bifurcations".

(2.5) Maps with positive entropy: Semi-conjugacies with tent-maps.

Everywhere in this section we assume that $h(f)>0$, i.e., that $r(f)=1/s(f)<1$. Following J.Milnor & W.Thurston's preprint, we shall show how to semi-conjugate f with a tent-map. (Presently we will define what we mean by "semi-conjugacy" and by "tent-map".) This result is the analogue of Poincaré's result that any homeomorphism of S^1 with an irrational rotation number is semi-conjugate to a rigid rotation. So tent-maps play the same role of "model maps" in the space of unimodal maps as rigid rotations in the space of homeomorphisms of the circle.

Take an interval J in [-1,1]. Recall that the powerseries $lap_f(t)$ converges for $|t|<r=r(f)<1$. Since $0 \le lap_f(J) \le lap_f$, it follows that

$$\Lambda(J) = \lim_{t \to r} lap_f(J)(t) \,/\, lap_f(t)$$

exists and satisfies $0 \le \Lambda(J) \le 1$.

Lemma (2.5.1). If the intervals J_1 and J_2 intersect only at a common end point then $\Lambda(J_1 \cup J_2) = \Lambda(J_1) + \Lambda(J_2)$.

Proof: Clearly $lap_n(J_1) + lap_n(J_2)$ differs by at most 1 from $lap_n(J_1 \cup J_2)$. Hence the difference $lap_f(J_1,t) + lap_f(J_2,t) - lap_f(J_1 \cup J_2,t)$ remains bounded as $t \to r$ (recall that $r=r(f)<1$). Dividing by $lap_f(t)$ and passing to the limit as $t \to r$ we get $\Lambda(J_1) + \Lambda(J_2) - \Lambda(J_1 \cup J_2) = 0$.

□

Lemma (2.5.2). Let $s=s(f)$. If J does not contain the critical point c, then $\Lambda(f(J)) = s.\Lambda(J)$.

Proof: In this case $f|J$ is a homeomorphism, and $lap_{n+1}(J) = lap_n(f(J))$. Thus implies that $lap_f(J,t) = 1 + t.\, lap_f(f(J),t)$. This implies the result.

□

Lemma (2.5.3). The number $\Lambda(J)$ depends continuously on the end points of the interval J.

Proof: By the Lemma (2.5.1) above it suffices to show that $\Lambda(J) \to 0$ as J shrinks to a point. But if the interval J is small enough to be contained in one lap of f^n, then it follows inductively from the previous Lemma (2.5.2), that

$$\Lambda(J) = s^{-n}\, \Lambda(f^n(J)) \le s^{-n}.$$

Let $n(J)$ be the maximal integer such that J is contained in a lap of $f^{n(J)}$. Clearly $n(J) \to \infty$ as $|J| \to 0$. Since $s>1$, and $\Lambda(J) \le s^{-n(J)}$ one has that $\Lambda(J) \to 0$

as $|J| \to 0$.

□

Define F_s: $[-1,1] \to [-1,1]$ to be the (symmetric) piecewise linear map (or tent-map) in $\mathcal{U}$ with slopes equal to s and $-s$.

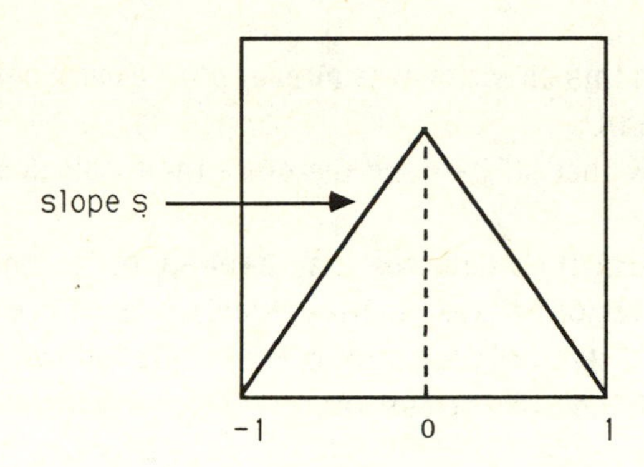

Figure 8: a tent-map F_s

Now we will show that f is semi-conjugate to some tent-map F_s. That is, there exists a monotone continuous surjective map $h:[-1,1] \to [-1,1]$ such that $h \circ f = F_s \circ h$. In general h need not be injective.

Theorem (2.5.4). [Milnor &Thurston, (1976)] Let $s=s(f)>1$, then there exists a semi-conjugacy between f and F_s. More precisely there exists a continuous surjective monotone map $h:[-1,1] \to [-1,1]$ which maps the critical point c of f onto the critical point 0 of F_s such that

$$h \circ f = F_s \circ h.$$

Proof: First define a map $k:[-1,1] \to [0,1]$ and a map $F:[0,1] \to [0,1]$ as follows. Define $k(x)=\Lambda[-1,x]$. Remark that each lap of f^n in $[-1,c)$ corresponds to precisely one lap of f^n in $(c,1]$. Therefore $k(c)= \Lambda[-1,c]=1/2$. Let us show that $k(x)=k(y)$ implies $k(f(x))=k(f(y))$. If $k(x)=k(y)$ and $x<y$ then it follows that $\Lambda[x,y]=0$ and using Lemma (2.5.2), $\Lambda(f[x,y])=0$. Hence $k(f(x))=k(f(y))$. This shows that there exists a map F with $k \circ f = F \circ k$. We want to check that F is the tent map on $[0,1]$ with slopes $\pm s$. For $x<c$, we have

$F(k(x)) = k(f(x)) = \Lambda[f(-1),f(x)] = \Lambda(f[-1,x]) = s.\Lambda[-1,x] = s.k(x)$.

It follows that $x \to F(k(x))$ has slope s for $k(x)<(1/2)$. Similarly one proves that $F(k(x))$ has slope $-s$ for $k(x)>(1/2)$. Using an affine change of coordinates F is conjugate to the piecewise linear map $F_s:[-1,1] \to [-1,1]$. Therefore f is conjugate to F_s, $h \circ f = F_s \circ h$, where h is the composition of k and the affine coordinate change.

□

Remarks.
(1) A partial result in this direction was already proved much before by W. Parry, see [Parry (1964)].
(2) It is easy to show that $s(F_s)=s$, and therefore the topological entropy of f and F_s is the same.
(3) Let $F=F_s$. In general it is not true that $\theta_f(x) = \theta_F(h(x))$. Since h is a semi-conjugacy and $h(c)=0$, $f^i(x)<c$ implies $F^i(h(x))\leq 0$, and $f^i(x)>c$ implies $F^i(h(x))\geq 0$. (So $|\theta_f(x)| \geq |\theta_F(h(x))|$, for all x in $[-1,1]$.) It follows that $\theta_f(x) = \theta_F(h(x))$, precisely if $F^i(h(x))\neq 0$ for all $i\geq 0$.

Let us describe the non-wandering set $\Omega(F_s)$.

Proposition (2.5.5). If $\sqrt{2} < s^m \leq 2$ then $\Omega(F_s)$ consists of m-disjoint intervals $I_0, I_1, \ldots, I_{m-1}$ and a finite number of periodic points. Furthermore
i) F_s maps I_i linearly onto $I_{(i+1) \bmod (m)}$, for $0\leq i<m$;
ii) F_s maps I_0 onto I_1 with precisely one fold;
iii) if $(F_s)^n(0)=0$ then $s^n>2$.

Proof: If $\sqrt{2}<s\leq 2$ then $\Omega(F_s)=I_0\cup\{-1\}$, where $I_0=[(F_s)^2(0),F_s(0)]$. Indeed take an interval U. If for all $k\geq 0$, $0\notin(F_s)^k(U)\cap(F_s)^{k+1}(U)$ then $|(F_s)^{2k}(U)|\geq(s^2/2)^k\,|U|$, where $|U|$ denotes the length of U. So for some $k\geq 0$, $(F_s)^k(U)=[-1,1]$ which is a contradiction with $0\notin(F_s)^k(U)\cap(F_s)^{k+1}(U)$. So for some k one has $0\in(F_s)^k(U)\cap(F_s)^{k+1}(U)$ and therefore $(F_s)^{k+2}(U)\supset I_0= [(F_s)^2(0),F_s(0)]$. From this one gets $\Omega(F_s)= I_0\cup\{-1\}$. If $(F_s)^n(0)=0$ then since $F_s(0)=s-1\neq 0$ and $(F_s)^2(0)=-(s-1)^2\neq 0$ one has $n\geq 3$ and $s^n\geq s.s^2>2$. This proves i),ii) and iii) for $s>\sqrt{2}$.

If $\sqrt{2}<s^2\leq 2$, then F_s has two fixed points -1 and $x_0=(s-1)/(s+1)$. $(F_s)^2$ maps $[-x_0,x_0]$ into itself and has slope $\pm s^2$ where $s^2 > \sqrt{2}$. So using the previous argument $\Omega((F_s)^2|\,[-x_0,x_0]\,)$ consists of one interval and the point x_0. Therefore $\Omega(F_s)$ consists of two intervals I_0 and I_1 (which are mapped onto each other) and the two fixed points -1 and x_0. (Alternatively, notice that F_s folds $I_0= [(F_s)^2(0),(F_s)^4(0)]$ onto $I_1=[(F_s)^3(0),F_s(0)]$ and maps I_1 linearly onto I_0. So $(F_s)^2:I_0\to I_0$ is a surjective expanding map and therefore $\Omega((F_s)^2 \mid I_0)=I_0$.) If $(F_s)^n(0)=0$ then 0 is a periodic point of $(F_s)^2$: $[-x_0,x_0]\to[-x_0,x_0]$ and therefore $(F_s{}^2)^k(0)=0$. Since the slope of $(F_s)^2$ is $\pm s^2$ and $\sqrt{2}<s^2\leq 2$ one can use the previous argument and one gets $s^n=s^{2k}>2$.

By induction one can treat the general case that $\sqrt{2}<s^m\leq 2$.

□

Corollary (2.5.6). If f in $\mathcal{U}$ is C^1 and $F^n_s(0)=0$ (where $s=s(f)>1$), then the semi-conjugacy h collapses an interval J around the critical point c to a point. Furthermore

i) f^n maps J into J, f^n: $J\to J$ is a map with precisely one critical point and f^n maps the boundary of J into one of the boundary points of J. (So f^n restricted to $J=h^{-1}(0)$ can be considered as a map in $\mathcal{U}$.)

ii) $h(f^n)>h(f^n|J)$; in otherwords the entropy decreases:

$$h(f)>h(f|K_n),$$

where $K_n=J \cup f(J) \cup \ldots \cup f^{n-1}(J)$.

Proof: Since h is a semi-conjugacy $J=h^{-1}(0)$ is either an interval or a point and f^n maps J into itself. Let us assume by contradiction that J is just a point. Since J contains the critical point c this implies that $f^n(c)=c$. Since f is C^1 and $(f^n)'(c)=0$, there is an interval neighbourhood J of c such that $f^n(J)$ is contained in J, and $f^{kn}(J)\to c$ as $k\to\infty$. Since h(J) would a neighbourhood of 0, one would have $F_s{}^{kn}(h(J))\to 0$. This is impossible since F_s is expanding. It follows that J is a non-trivial interval. Since 0 has (minimal) period n, $J,f(J),\ldots f^{n-1}(J)$ are disjoint, and therefore $f^n|$ J has precisely one critical point and maps the boundaries of J into itself. Let us now prove (ii). From Corollary (2.5.5) we have that $(F_s)^n(0)=0$ implies that $s^n>2$. So $h(f)=\log(s)>(1/n).\log(2)$. Since f^n is a unimodal map of J into itself, $h(f^n|J)\leq\log(2)$, and therefore $h(f|K_n)\leq(1/n)\log(2)<h(f)$.

□

Compare the situation from Corollary (2.5.5) with Figure 6(b). The map f can be " decomposed into a map F_s and a unimodal map $f^n:J\to J$ ".

Corollary (2.5.7). If $f\in\mathcal{U}$ is semi-conjugate to F_s (where $s=s(f)>1$) and $F^n_s(0)\neq 0$, then the semi-conjugacy h only collapses the basins of periodic attractors and wandering intervals.

Proof: Write $F=F_s$ and assume that $h^{-1}(y)$ is some non-trivial interval J. There are two cases.

Case i): For some $i \geq 0$, $F^i(y)$ is a periodic point y', i.e., $F^n(y')=y'$ for some $n \geq 0$. But then $J'=f^i(J)$ is a periodic interval of f, i.e., $f^n(J')$ is contained in J'. From $F^n(0) \neq 0$ for all $n \geq 0$ it follows that $y' \neq 0$. From this and from $h(c)=0$ one gets that $J', f(J'), f^2(J'), \ldots$ do not contain c. So f^n maps J' homeomorphically into J'. Hence all points of J' (and therefore of J) are in the basin of periodic attractors.

Case ii): All the points $y, F(y), F^2(y), \ldots$ are distinct. But then $J, f(J), f^2(J), \ldots$ are all mutually disjoint and J is a wandering interval.

□

(2.6) Decomposition of the non-wandering set: Using renormalising.

From the semi-conjugacy result of the previous section we can get a good understanding of the dynamics of unimodal maps. One obtains a decomposition of the map in simpler maps (using "renormalising"). This decomposition was first proved by L. Jonker and D. Rand (1981).

Theorem (2.6.1). Let f be in $\mathcal{U}$. Then there are exist $0 \leq r \leq \infty$, intervals $J_0 \supset J_1 \supset J_2 \supset \ldots$ containing c in their interior, and integers k(n) such that for each finite n with $n \leq r$, $f^{k(n)}|J_n : J_n \to J_n$ is a map in $\mathcal{U}$ (up to scaling).

For $n<r$ there are two possibilities:

(A1) $f^{k(n)}|J_n : J_n \to J_n$ has positive topological entropy and is semi-conjugate to a piecewise linear map F_s with 0 a periodic point of F_s;

(B1) $f^{k(n)}|J_n : J_n \to J_n$ has zero topological entropy but has a fixed point which is not orientation preserving.

For $n=r$ (and $r<\infty$) there are two possibilities:

(A2) $f^{k(n)}|J_n : J_n \to J_n$ is semi-conjugate to a piecewise linear map F_s with 0 not a periodic point of F_s;

(B2) $f^{k(n)}|J_n : J_n \to J_n$ is a map which has only periodic points of period one and all these fixed points are orientation preserving.

Furthermore

i) $J_n, f(J_n), f^2(J_n), \ldots, f^{k(n)-1}(J_n)$ have no interior points in common;

ii) the intervals $f^{j.k(n-1)}(f^i(J_n))$, $j=0,1,\ldots,k(n)/k(n-1)$, are strictly contained in the interior of the previous interval $f^i(J_{n-1})$, for $0 \le i < k(n-1)$.

Proof: Roughly speaking the procedure is as follows. One finds a nested sequence J_n of intervals containing the critical point c in its interior, such that for some $k(n) \ge 0$ the map $f^{k(n)}$ maps J_n into itself and has one critical point. So after scaling, the map $f^{k(n)}|J_n : J_n \to J_n$ can be considered as a unimodal mapping. So once a point x is mapped into the interval J_n, its forward orbit is "trapped" in the union of the intervals $J_n, \ldots, f^{k(n)-1}(J_n)$. The dynamics of the point x can then be described by the dynamics of $f^{k(n)}|J_n : J_n \to J_n$ and the relative position of the intervals $J_n, \ldots, f^{k(n)-1}(J_n)$. After some time the orbit may get mapped into J_{n+1} and then the dynamics can be better understood from the map $f^{k(n+1)}|J_{n+1} : J_{n+1} \to J_{n+1}$. In this way the procedure is repeated.

Let us now be more precise. First define $k(0)=1$ and $J_0=[0,1]$. We define $k(i)$ and J_i inductively as follows. Assume that we have defined $k(n)$ and J_n. The map $f^{k(n)}|J_n : J_n \to J_n$ can be considered as map in $\mathcal{U}$. Denote this map by f_n. Now distinguish the case that the map $f_n = f^{k(n)}|J_n : J_n \to J_n$ has zero entropy and the case that this map has positive entropy.

(B) If f_n has <u>zero entropy</u> then, as in section (2.4), there are two cases

(B1) there is an orientation reversing fixed point p. In this case the interval $J_{n+1}=[p',p]$ (from section (2.4)) is mapped by $(f_n)^2$ into itself. We set $k(n+1)=2.k(n)$. Note that

*) J_{n+1} and $f_n(J_{n+1})$ are both contained in the interior of J_n and have only the point p in common.

(B2) there is no orientation reversing fixed point p of f_n, and all points are in the basin of the fixed points of f_n (see section (2.4 case 2). In this case we do not define J_{n+1} and $k(n+1)$; $r=n$.

(A) If f_n has <u>positive entropy</u>, then there exists a semi-conjugacy h with the piecewise linear map F_s, $h \circ f_n = F_s \circ h$ (where $\log(s)=h_{top}(f_n)$ is the topological entropy of f_n). There are again two cases:

(A1) There exists a (minimal) k such that $F_s^k(0)=0$. As we saw in the previous section, in this case $h^{-1}(0)$ is a non-trivial interval J_{n+1}, and $(f_n)^k$: $J_{n+1} \to J_{n+1}$ is a map with precisely one critical point and $(f_n)^k$ maps the boundary of J_{n+1} into one of the boundary points. So $(f_n)^k: J_{n+1} \to J_{n+1}$ can be considered as a map in $\mathcal{U}$. Furthermore

**) $\qquad J_{n+1}, f(J_{n+1}), \ldots, (f_n)^{k-1}(J_{n+1})$ are disjoint.

Set $k(n+1)= k. k(n)$.

(A2) $F_s^n(0) \neq 0$ for all $n \geq 0$. Since f_n is semi-conjugate to F_s, the critical point of f_n and therefore of f also cannot be periodic. In this case we do not define J_{n+1} and $k(n+1)$.

In the cases where we do not define J_{n+1} and $k(n+1)$ we set r=n. In the other cases we repeat the constructions inductively. From *) and **) the statements i) and ii) from the Theorem follow.

□

Now write

$$K_n = J_n \cup f(J_n) \cup \ldots \cup f^{k(n)-1}(J_n).$$

Clearly f maps K_n into K_n. It follows from the Theorem above that K_n consists of at least $k(n-1)$ disjoint intervals (and if f_n is as in case (A1) or (A2) then it consists of at least $k(n)$ disjoint intervals).

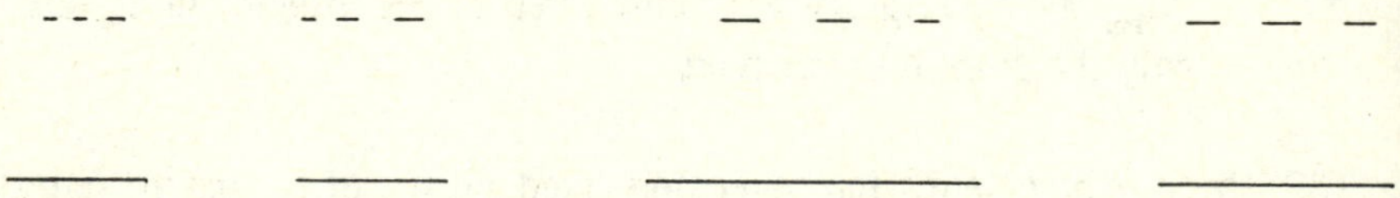

Figure 9: Examples of the sets K_1 and K_2

If $r=\infty$ then we say that we are in the "infinite decomposition case". Notice that if f_n is as in case (A1) then f_k is also as in case (A1) for $k\leq n$. (Because otherwise f_k would be as in case (B1) and then $h(f_n)\leq h(f_k)=0$. This would imply $h(f_n)=0$, a contradiction.) By analogy to section (2.4) we call this the map at the accumulation of period multiplying bifurcations.

Corollary (2.6.2). (Decomposition of the non-wandering set)
Write $\Omega_n=\Omega(\, f \mid Cl(K_n\backslash K_{n+1}))$. If $r=\infty$ then define

$$\Lambda = \bigcap_{n\geq 0} \{J_n \cup f(J_n) \cup f^2(J_n) \cup \dots \cup f^{k(n)-1}(J_n)\}$$
$$= \bigcap_{n\geq 0} K_n.$$

If Λ does not contain intervals then set $\Omega_\infty=\Lambda$. If Λ does contain intervals then define $\Omega_\infty = \Lambda \cap \Omega(f)$. Then

$$\Omega(f) = \bigcup_{0\leq n\leq r} \Omega_n .$$

If $f_n: J_n \to J_n$ is as in:
Case (A1); then $f\colon \Omega_n \to \Omega_n$ is semi-conjugate to a subshifts of finite type.
Case (B1); then $f\colon \Omega_n \to \Omega_n$ has just periodic points of period $k(n)$ and Ω_n does not contain any other points.
Case (A2); then $f\colon \Omega_n \to \Omega_n$ is as in case (A1) but all point of J_n are attracted to the fixed points of $f\colon \Omega_n \to \Omega_n$.
Case (B2); then $f\colon \Omega_n \to \Omega_n$ is semi-conjugate to a piece-wise linear map with a non-periodic critical point.

Proof: The proof follows quite easily from the previous Theorem. For details see the paper of L. Jonker and D. Rand (1981) and also van Strien (1979 & 1981).

□

From Corollary (2.5.6) $h(\, f \mid \Omega_n) < h(\, f \mid \Omega_{n-1})$ for $n\leq r$.

In van Strien (1979 &1981) it is shown that if f is C^3 and also satisfies the condition $S(f)<0$ (see section (3.2)) then:

i) $f\colon \Omega_n \to \Omega_n$ is hyperbolic for $n<r$.
ii) $\Lambda = \Omega_\infty$ is a Cantor-set.
iii) the semi-conjugacies from the corollary above are in fact conjugacies.

3. ANALYTIC TOOLS: BOUNDED DISTORTION OF LENGTHS AND CROSS-RATIOS.

In order to get metric results on the asymptotic dynamics of f, one needs to have tools to estimate the non-linearity of f^n, for high $n \geq 0$. The easiest case is when the derivative of f is "away from zero".

(3.1) Bounded distortion techniques: Denjoy's & Schwartz's Lemma

Denjoy and Koksma developed a technique which is related to the idea of bounded variation. We say that g has <u>bounded variation</u> if there exists a constant $V(f) < \infty$, such that for any subdivision $-1 \leq x_0 \leq x_1 \ldots \leq x_j \leq 1$,

$$\sum_{0 \leq i \leq j-1} |f(x_{i+1}) - f(x_i)| \leq V(f).$$

We say that intervals $I_1, \ldots, I_n$ <u>cover</u> $[-1,1]$ <u>at most N-times</u> if each point in $[-1,1]$ is contained in at most N of the intervals $I_1, \ldots, I_n$.

Theorem (3.1.1). [Denjoy, (1932)] Assume that $\log |f'|$ has bounded variation. Let $N < \infty$. There exists $C < \infty$ such that given any interval J if for some $n \geq 0$ the intervals $J, f(J), f^2(J), \ldots, f^n(J)$ cover $[-1,1]$ at most N times, then for each y,z in J one has

$$1/C < (|(f^n)'(y)| \;/\; |(f^n)'(z)|) < C.$$

Proof: Using the chain rule one gets

$$| \log|(f^n)'(y)| - \log|(f^n)'(z)| | \leq \sum_{0 \leq i \leq n-1} | \log|f'(f^i y)| - \log|f'(f^i z)| |$$

Since the intervals $J, f(J), \ldots f^n(J)$ cover $[-1,1]$ at most N times, the last expression is less or equal to $N.\mathrm{var}(\log|f'|)$.

□

Denjoy uses this tool for analysing diffeomorphisms of the circles. Since the ordering of the circle is preserved by such maps, one can get good intervals J with the required disjointness properties. For maps which do not preserve ordering on the interval, this disjointness is often too restrictive.

However if $\sum_{0 \leq i \leq n-1} |f^i(J)|$ is not too big then we can do something similar, provided $\log|f'|$ is Lipschitz. ($|f'|(x)$ is said to be well defined if f' exists at x or if the left and right derivatives at x both exist and have opposite signs.) Moreover we say that a function h is Lipschitz (with Lipschitz constant K), if there exists $K < \infty$ such that $|h(x)-h(y)| \leq K|x-y|$ for all x,y. (In this case h also has bounded variation and $var(h) \leq K$.)

Lemma (3.1.2). (Lipschitz version of Denjoy's Lemma). Assume that $\log|f'|$ has Lipschitz constant K. Take an interval J and $x \in J$. Let $S = \sum_{0 \leq i \leq n-1} |f^i(J)|$. Then

$$e^{-K.S} . |f^n(J)| / |J| \leq |(f^n)'(x)| \leq e^{+K.S} . |f^n(J)| / |J|.$$

Proof: Exactly as the bounded variation case of Denjoy's Lemma.

□

Using this, one can get an extremely powerful tool, introduced first by A. Schwartz, and later reinvented by Z. Nitecki.

Denote the length of an interval J by |J|. For later use we introduce the following conventions: for intervals $T \supset J$ let L and R be the two components of $T \setminus J$.

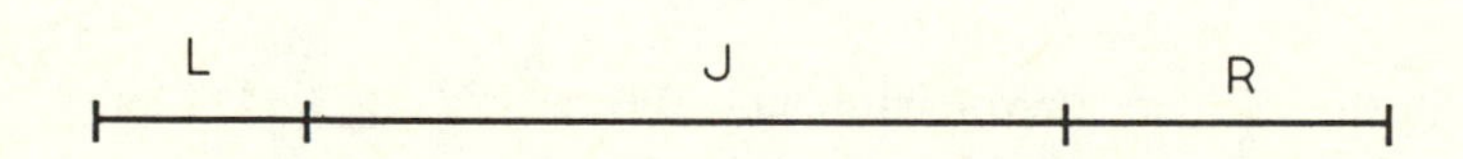

Figure 10: the intervals L, J and R.

Schwartz Lemma (3.1.2). [A. Schwartz (1963)]. Assume that $\log|f'|$ is Lipschitz. Then there exists $\delta > 0$, such that for any pair of intervals $T \supset J$ as above with

$$\sum_{0 \leq i \leq n-1} |f^i(J)| \leq 1, \quad |L|/|J| \text{ and } |R|/|J| < \delta,$$

one has for all $m \leq n$,

$(*_m)$ $\qquad |f^i(T)| \leq 2. |f^i(J)|$, for $i=1,\ldots,m-1$,

$(**_m)$ $\qquad \{\exp(-2.K)\} . |f^m(J)| / |J| \leq |(f^m)'(x)| \leq \{\exp(2.K)\} . |f^m(J)| / |J|.$

Proof: Take $\delta < (1/2)\exp(-2K)$. Let us prove $(*_m)$ and $(**_m)$ by induction. $(*_1)$ is obvious and $(**_2)$ follows from the fact that $\log|f'|$ has Lipschitz constant $\leq K$. Assume that $(*_m)$ and $(**_m)$ are satisfied. $(**_m)$ implies that $|(f^m)'(x)| \leq \exp(2.K).|f^m(J)|/|J|$ for all $x \in T$. Since $|T| \leq (1+2\delta).|J|$ this gives

$$|f^m(T)| \le |f^m(J)| + 2\delta |J| \exp(2K).|f^m(J)|/|J|.$$

By the choice of $\delta>0$ this implies $(*_{m+1})$. Now summing this over i , one has $\sum_{0\le i\le k-1} |f^i(T)| \le 2$. Using this, the result from the previous Lemma (3.1.2) and the mean-value theorem one gets $(**_{m+1})$.

□

(3.2) Schwarzian derivative

As we saw, we can use the technique of A. Schwartz to get analytic estimates on f^n for high n, provided the function log|f'| is Lipschitz for a map f. However for smooth maps with turning points, log|f'| is definitely not Lipschitz. So we cannot use the previous techniques automatically.

(3.2.1) Schwarzian derivative: Some immediate observations

In order to get results similar to those of A. Schwartz from the previous section one can use the so-called Schwarzian derivative (introduced by H.A. Schwarz, (1868)). (One begins to wonder whether any other mathematician with a name of the form ?.A.Schwar?z will get involved with maps of the interval.)

Define the Schwarzian derivative for a C^3 map $g:\mathbb{R}\to\mathbb{R}$, as the following function

$$S(g)= g'''/g' - (3/2)\,(g''/g')^2.$$

The first to observe the use of the Schwarzian derivative in the study of one-dimensional maps are D. Singer (1978) and D. Allwright (1978). It was also used in M. Herman (1979).

<u>Lemma</u> (3.2.1) (Elementary properties of the Schwarzian derivative S(f))
1. Assume that $S(f)<0$. If f is monotone on [x,z] then |f'| has no positive local minima on [x,z]. In particular if $y\in(x,z)$ and $u\in(x,y)$ and $v\in(y,z)$ then

$$|f'(y)| > \min\{ |f'(u)|, |f'(v)| \}.$$

2. Let f and g be two C^3 functions $S(f\circ g)= S(f)\circ g.(g')^2 + S(g)$. In particular $S(f)<0$ implies $S(f^n)<0$.

<u>Proof</u>: We remark that if $f''(y)=0$ then $f'(y).f'''(y)<0$. Statement 1 follows.

Statement 2 follows from a direct calculation.

□

Corollary (3.2.2) [D.Singer, (1978)] Assume f is C^3 and satisfies $S(f)<0$. Then each periodic attractor of f contains a critical point or a point on the boundary of $[-1,1]$ in its basin. In particular a unimodal map f in $\mathcal{U}$ has at most one periodic attractor (apart from possibly the fixed point $\{-1\}$).

Proof: Let x be a point on the periodic attractor and assume x is not on the boundary of $[-1,1]$. Let n be the period of the attractor. Then $f^n(x)=x$ and $|(f^n)'(x)|\leq 1$. By taking $2n$ instead of n, we have $0\leq(f^{2n})'(x)\leq 1$. Let J be the maximal interval containing x for which f^{2n} is monotone. Since $S(f)<0$, and therefore $S(f^{2n})<0$, Lemma (3.2.1) implies that $|(f^{2n})'|$ has no positive local minima on J. Since $0<(f^{2n})'(x)\leq 1$ this implies that one has $0\leq(f^{2n})'<1$ on one of the components of $J-\{x\}$. Hence the endpoint of this component of $J-\{x\}$ iterates towards the orbit of x. Therefore this endpoint is a critical value of f^{2n} or a boundary point of $[-1,1]$ and one of these points is contained in the basin of the attractor. If $f\in\mathcal{U}$ then the result follows from $f(-1)=f(1)=-1$.

□

Remark: It is not hard to show that for unimodal maps the critical point is contained in the immediate basin of the periodic attractor.

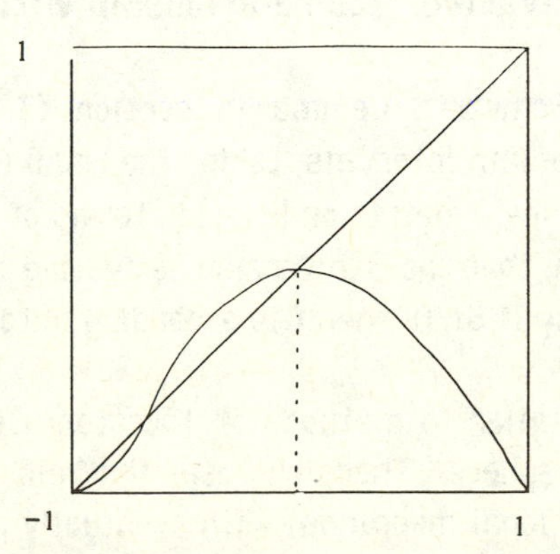

Figure 11: A unimodal map with $S(f)<0$ and an attractor in $\{-1\}$ and in $\{0\}$

The condition that $S(f)<0$ is much more powerful than one might think at first. Let us quote a few properties. We will not prove these properties here because they can be proved in a general context using the operators A and B from section (3.2.b).

Lemma (3.2.3) If $S(f)<0$ and $T=[a,b]$ is an interval such that $f'\neq 0$ on T, then
1. $$|f'(a)| \,.\, |f'(b)| \geq (|f(T)|/|T|)^2.$$
2. For any x in (a,b), let L and R be the two components of $(a,b)\backslash\{x\}$. Then one has
$$f'(x)|.(|f(T)|/|T|) \geq (|f(L)|/|L|).\,(|f(R)|/|R|).$$
3. For each $\delta>0$ there exists $C>0$, such that for any two points $x<y$ in T, one has
$$1/C < |f'(x)|/|f'(y)| < C,$$
provided $|f([a,x])| / |f(T)| > \delta$, $|f([y,b])| / |f(T)| > \delta$.

□

Clearly this last property is almost as good as bounded distortion: it gives bounded distortion "as long as one stays away from the critical values". (The Koebe inequality is the analogue for conformal mappings.) Properties (1) & (2) were discovered by several people independently. In fact property (1) is one of the three main ingredients in the proof of the existence of absolutely continuous measures. Property (3) was first used in van Strien, (1979 &1981) and later rediscovered by Guckenheimer (1986).

(3.2.b) Schwarzian derivative: Its relationship with cross-ratios

The idea of the Schwartz Lemma in section (3.1) is based on comparing the length of certain intervals, taking the usual Euclidean metric on $[-1,1]$. One can also define a metric on $[-1,1]$ in terms of cross-ratios. In this section we shall show that the Schwarzian derivative is related to the cross-ratio. More precisely if $Sf<0$, then f is expanding in terms of this new metric.

This approach is similar to methods of Thurston used for rational mappings on the Riemann-sphere. There one uses the Teichmüller space of conformal structures. Rational mappings (with eventually periodic critical points) act as expansions on this Teichmüller space. We shall see that something similar is the case for maps $S(f)<0$. So in some sense they can be considered as the analogue of "conformal mappings on the Riemann-sphere".

Let us first show that the Schwarzian derivative is related to cross-ratio's. Let J be a subinterval of an interval T such that $T\setminus J$ consists of two connected components L and R. Define the following two cross-ratios:

$$C(T,J) = \{ |J| . |T| \} \ / \ \{ |J \cup L| . |J \cup R| \},$$

$$D(T,J) = \{ |J| . |T| \} \ / \ \{ |L| . |R| \}.$$

For g monotone define

$$A(g,T,J) = C(g(T),g(J)) \ / \ C(T,J),$$

$$B(g,T,J) = D(g(T),g(J)) \ / \ D(T,J).$$

Moreover, let M be the group of Moebius transformation, i.e., maps of the form $t \to (at+b)/(ct+d)$. One can show that $Sf=0$ is equivalent to f being a Moebius transformation.

Lemma (3.2.4). The Schwarzian derivative of a function f is identically zero ($S(f)=0$) if and only if f is a Moebius transformation.

Proof: It is useful to observe that Sf can be written as $Sf=(f''/f')' -(1/2)(f''/f')^2$. By integrating $(f''/f')' -(1/2)(f''/f')^2 =0$ the result follows.

□

Lemma (3.2.5). Let f be C^3 and $S(f)<0$. Consider an interval T such that f' is non-zero on T and let J be some subinterval of T as above. Then

$$A(f,T,J) > 1, \text{ and } B(f,T,J) > 1.$$

Proof: Compose f with Moebius transformations ϕ and Ψ on the right and on the left. Using property (2) from Lemma (3.2.1) and Lemma (3.2.4), it follows that this composition $\phi \circ f \circ \Psi$ still has the property $S(\phi \circ f \circ \Psi)<0$. So we can assume that $T=[0,1]$ and $J=[a,b]$, $0<a<b<1$, and $f(0)=0$, $f(a)=a$ and $f(1)=1$. Since $|f'|$ cannot have any local minima, f cannot have solutions of $f(x)=x$ on $[0,1]$ apart from $0,a,1$, and $f(x)<x$ for $x\in(0,a)$ and $f(x)>x$ for $x\in(a,1)$. Therefore

$$B(f,T,J) = \{(|f(J)|/|J|).(f(T)|/|T|) \}/\{(|f(L)|/|L|).(|f(R)|/|R|)\}$$

$$= \{(|f(J)|/|J|)\}.\{|R|/|f(R)|\} > 1.1 = 1.$$

Similarly one proves $A(f,T,J)>1$. □

(3.3) Beyond the Schwarzian derivative

As we saw the Schwarzian derivative is an extremely useful tool. However the assumption that $S(f)<0$, is not very natural. For example, the property $S(f)<0$ is not invariant under smooth coordinate changes. Moreover the property is not "generic". So we would like to relax the condition that $S(f)<0$. We will find a coordinate independent statement which often seems strong enough to replace the condition $S(f)<0$.

One way to think about this is to try to replace the class of maps corresponding to the conformal maps (with $S(f)<0$), by a class which is "quasi-conformal". It is not clear at present whether this analogy can be made precise. However let us see how badly behaved general smooth maps are.

(3.3.a) How does a smooth map distort the cross-ratio?

Let us estimate the distortion of the cross-ratio operators A and B by a smooth map.

Let us say that f is <u>non-flat</u> at a critical point c, if there exists $k \geq 3$ such that f is C^k at c and $f^{k-1}(c) \neq 0$. This condition is certainly satisfied if f is analytic. Remark that if f is non-flat at c, then $Sf(x)<0$ for x near c. It is this that we need.

Lemma (3.3.1). [de Melo & van Strien (1986)]. Assume that f is C^3 and is non-flat at the critical point. Then there exists a constant $C_0=C_0(f)$ such that if $T \supset J$ as before and f' is non-zero on T, then

$$A(f,T,J) \geq 1-C_0|L||R| \text{ , and } B(f,T,J) \geq 1-C_0|T|^2.$$

Sketch of Proof: First we need to remark that if a critical point of f is non-flat then an elementary calculation shows that $Sf < 0$ on a neighbourhood of this critical point. Now distinguish three cases.
(i) T is contained in a small neighbourhood U of the critical point of f, where $Sf<0$. In this case from the previous Lemma one has $A(f,T,J)>1$ and $B(f,T,J)>1$.
(ii) T is completely contained in the complement of U. Then using the fact that $|f'|$ is bounded away form zero by $(|f'(x)|;x \notin U)>0$, and the fact that f'' is Lipschitz, one can show the required inequalities.

(iii) T is partly contained in U. In this case one has to use a mixture of the two previous arguments.
For a complete proof see de Melo & van Strien (1986).

□

(3.3.b) Distortion of the cross-ratios under iterations: Analogues of Denjoy's and Schwartz's Lemma's

In this section we come to a version of the theorems of Denjoy and Schwartz for maps which do not have the required bounded distortion properties.

We assume in the rest of this section that F has non-flat critical points, and that $T \supset J$ and that $T \setminus J$ consists of two components L and R. Assume that $f' \neq 0$ on T.

The first result resembles the Theorem of Denjoy: it assumes that the total length of the orbit of the bigger interval (i.e. $\sum |f^k(T)|$) is bounded in some way.

Lemma (3.3.2). [de Melo & van Strien (1986)] Assume that f is C^3 and has no flat critical points. Then there exists a constant C, such that if f^m is a diffeomorphism on T, then

$$B(f^m,T,J) \geq 1 - C.\{\sum_{0 \leq k \leq m-1} |f^k(T)|^2\}.$$

□

The proof of this Theorem is quite easy. A weaker version of this (based on $B(f,T,J) \geq 1 - C_0|T|$) is used in papers on smooth homeomorphisms of the circle see J.C. Yoccoz (1984) and G. Swiatek (1985).

The second Theorem is more reminiscent of the results of Schwartz: It only requires that the total length of the orbit of the smaller interval J (i.e. $\sum |f^k(J)|$) is small.

Lemma (3.3.3). [de Melo & van Strien, (1986)]
Assume that f is C^3 and has no flat critical points. Then there exist constants δ and ε>0 such that if
(i) f^m is a diffeomorphism on T,

(ii) $\sum_{0\le k\le m}|f^k(J)| < \delta$, and
(iii) $|L||R| < \varepsilon|J|^2$, then

$$A(f^m,T,J) \ge 1 - 8.\{\, |L|\,|R| \,/\, |J|^2 \,\}.$$

□

The proof of this theorem is technical, and is similar in spirit to Schwartz's proof. It can be found in de Melo & van Strien (1986).

4. WANDERING INTERVALS AND HOMTERVALS

From section 1 we saw that everything about the topological aspects of the dynamics can be deduced from the kneading invariant of f, except the information of whether $x \to \theta_f(x)$ is constant on an interval. Suppose $\theta_f(x)$ is constant on an interval I. Then we saw that there were two possibilities: either

1) every point of I is contained in the basin of a periodic attractor; or
2) I is a wandering interval.

As before we say that J is a wandering interval, if J, f(J), $f^2(J)$,... are disjoint and J is not contained in the basin of a periodic attractor.

However, if f is unimodal and one poses some smoothness and "non-flatness" conditions on f, then one can show that f cannot have any wandering intervals. Let us first deal with the case in which we can apply the "Schwartz lemma" , i.e., we first assume that $\log|f'|$ is Lipschitz.

(4.1) The bounded distortion case

Let $\log|f'|$ be Lipschitz as in section 3.

Theorem (4.1.1). [based on Schwartz (1963)] If $\log|f'|$ is Lipschitz, then f cannot have any wandering intervals.

Proof: First we make the following observation. Take a component U of the immediate basin of a periodic attractor which is, say, of period n. Clearly f^n maps U onto itself and maps ∂U into itself. So the two boundary points of U are fixed points of f^{2n}. Moreover these fixed points (on ∂U) of f^{2n} are not

locally attracting on U. (That is, for any neighbourhood V in U of these points one has that $|f^n(V)|$ stays bounded away from 0 as $n\to\infty$.)
Now we are ready to give the proof of the Theorem. Assume by contradiction that f did have a wandering interval J_0. Choose that biggest interval $J \supset J_0$ which is non-wandering. We want to show that for any interval $T \supset J$ with $T\neq J$ one has that $|f^i(T)|$ stays bounded away from zero. We will distinguish two cases.
(i) T contains a point which is in the basin of a periodic attractor. Since J is wandering, no point of J is contained in the basin of this periodic attractor and therefore T contains in its interior a point of the boundary of the basin of the periodic attractor. But the observation above implies that the boundary point is eventually periodic and not attracting on U (in the same sense as above). Therefore we have that $|f^i(T)|$ stays away from zero as $i\to\infty$
(ii) T does not contain points which are in the basin of periodic attractors. But then by the maximality of J there exists $n\geq 0$ and $k>0$ such that $f^{n+k}(T)\cap f^n(T)\neq\emptyset$. It follows that, for every $j\geq 0$,

$$f^{n+(j+1)k}(T) \cap f^{n+jk}(T)\neq\emptyset.$$

Thus $I=\bigcup_{j\geq 0} f^{n+jk}(T)$ is an interval and f^k maps I into itself. Since no point of T is contained in the basin of a periodic attractor, $f^k:I\to I$ will have a fixed point in the interior of I which is not attracting (in the sense as above). But then for some $i(0)\geq 0$, $f^{i(0)}(T)$ contains this fixed point and $|f^i(T)|$ stays away from zero as $i\to\infty$.

Now we will use Schwartz's Lemma. One has that $J,f(J),f^2(J),\ldots$ are all disjoint and therefore $\sum_{i\geq 0}|f^i(J)| \leq 1$. Using the Lemma of Schwartz, one can find an interval $T \supset J$ such that $T\setminus J$ consists of two components and such that

$$|f^i(T)| \leq 2.|f^i(J)|, \text{ for all } i\geq 0,$$

and therefore $|f^i(T)|\to 0$. This contradicts the statement in the first part of the proof. Thus we have proved the Theorem.

□

Remark. In the proof from above we go "over the critical point", i.e., the images of $f^n|T$ were allowed to contain the critical point c. In the first part of the proof this is not serious. But in order to be able to use the Lemma of Schwartz in the second part we need that $\log|f'|$ is also Lipschitz at the

turning points of f (and therefore that the left and right derivatives of f' at turning points of f are the same). Let us show that this condition is not necessary.
It suffices to assume the following: (i) f is continuous, (ii) f is piecewise monotone (i.e., there exists a finite number of points $-1=x_0<x_1<\ldots<x_n=1$ such that $f|[x_i,x_{i+1}]$ is monotone and at each point x_i f has either a local maximum or a local minimum), (iii) $|f'|>0$ on $[-1,1]$, (iv) $\log|f'|$ is Lipschitz on each interval on which f is monotone.

This follows since at a turning points of such a map, f is almost symmetric in the following sense. One can find $1<k<2$ such that if T and J are intervals with (i) $T \supset J$, (ii) $|T|<k.|J|$, and (iii) J does not contain a critical point of f, then there exists $T \supset T' \supset J$ such that T' does not contain critical points in its interior and such that $f(T')=f(T)$. So one can shrink T to an interval $T' \supset J$ without changing the f-image of T. Using a slightly modified version of the Lemma of Schwartz (with the constant 2 in the Lemma replaced by $1<k<2$), we can shrink T (at each stage when $f^i(T)$ hits a turning point), to an interval $T_i \supset J$ such that $f^i|T_i$ is a homeomorphism, while still $f^i(T_i)=f^i(T)$. So as before we would get $|f^i(T)| = |f^i(T_i)| \leq k.|f^i(J)|$, for all $i\geq 0$, and therefore $|f^i(T)|\to 0$. So again we would reach a contradiction.

This remark may also be of use to obtain a proof in the case where $|f'|$ is allowed to become zero.

(4.2) The smooth case with critical points

Assume that f is C^3. As before we say that a critical point c of f is non-flat, if for some $k\geq 2$ the map f is C^{k+1} in a neighbourhood of c and $f^k(c)\neq 0$. This implies $Sf(x)<0$ for x near c. (For our purposes it is sufficient to assume that $Sf(x)<0$ for x near c.)

Unimodal maps with no flat critical points cannot have any wandering intervals.

Theorem (4.2.1). [de Melo & van Strien (1986)]
Let f be a C^3 unimodal map, such that the critical point of f is non-flat. Then f has no wandering intervals.

□

This result was previously proved for unimodal maps with $Sf<0$ by J. Guckenheimer (1979). The topological part of the proof in our case is identical to Guckenheimer's proof. That the non-flatness condition is necessary follows from an example of C.R. Hall (1981).

This Theorem is the analogue of the Denjoy result for C^2 diffeomorphisms of the circle (for diffeomorphisms with irrational rotation number). Also there is a corresponding theorem in the complex case. D. Sullivan (1985) proved that there are no wandering components in the complement of the Julia's set for iterations of holomorphic maps of the Riemann sphere.

Open Question: It is not known whether the result from above holds in the case with more critical points (even if $Sf<0$).

However if f satisfies the condition of Misiurewicz (which we state below), then f has no wandering points (even if f has more than one critical point).

Let C(f) be the set of critical points $C(f)=\{x;f'(x)=0\}$. If there exists a neighbourhood V of C(f) such that for each $c\in C(f)$, $f^n(c)$ is outside V for all $n\geq 1$ then f is said to satisfy the "Misiurewicz" condition (we will return to this condition in section (7.3)).

Theorem (4.2.2). [de Melo & van Strien, (1986)]
Let N be either $[-1,1]$ or S^1. Assume that $f:N\to N$ is a C^3 endomorphism, with no flat critical points and which satisfies the Misiurewicz condition. Then f has no wandering interval.

Let us give the proof of Theorem (4.2.2). The proof of Theorem (4.2.1) uses the topological ideas of J. Guckenheimer combined with Lemmas (3.3.2), (3.3.3) and can be found in de Melo & van Strien (1986).

Proof of Theorem (4.2.2): Assume by contradiction that f has a wandering interval J. By possibly replacing J with some iterate of J we can assume that $f^n(J)$ does not contain a critical point for all $n\geq 0$, and that $\sum_{n\geq 0}|f^n(J)| \leq \delta$, where δ is the number from Lemma (3.3.3) above. Also, by possibly increasing J, we can assume that J is not contained in a strictly bigger wandering interval.

Let T_n be the maximal interval containing J such that $f^n|T_n$ is a homeomorphism. Since $f^n(J)$ does not contain a critical point, $T_n \setminus J$ consists of two intervals L_n and R_n. From the maximality of J one can easily see that $|L_n|, |R_n| \to 0$. So for n sufficiently big, we have that $|L_n||R_n| < \varepsilon |J|^2$, where $\varepsilon > 0$ is the number from Lemma (3.3.3). Therefore from this Lemma we get that for all $n \geq 0$,

$$(*) \qquad A(f^n, T_n, J_n) \geq 1 - 8.\{ |L_n|\,|R_n| / |J|^2 \}.$$

Let us show that this gives a contradiction with the fact that the forward orbit of the critical values of f stay away from the critical points. For this we choose a subsequence n(i) as follows:

Claim: There exists a sequence of numbers $n(i) \to \infty$, such that $f^{n(i)}(J)$ converges to some critical point.

Proof: Assume $f^n(J)$ is outside some neighbourhood U of the set of critical points. If one modifies f inside U then this will still be true. Change f into a map F (by a modification which is only inside U), such that $\log|F'|$ is Lipschitz on [-1,1]. F still has J as a wandering interval. Clearly this contradicts Theorem (4.1.1) and so the claim is proved.

Therefore, by the maximality of the intervals T_n, the intervals $f^{n(i)}(J \cup L_{n(i)})$ and $f^{n(i)}(J \cup R_{n(i)})$ contain critical values of $f^{n(i)}$ (in their boundary) and also points converging to a critical point of f. Since the forward orbit of the critical values of f stay away from the critical points of f, one has for this choice of n(i) that

$$|f^{n(i)}(J \cup L_{n(i)})|, \quad |f^{n(i)}(J \cup R_{n(i)})|, \text{ and } |f^{n(i)}(T_{n(i)})|$$

are bounded away from 0 as $i \to \infty$. Since the iterates of J are disjoint,

$$|f^{n(i)}(J)| \to 0,$$

as $i \to \infty$. But then $A(f^{n(i)}, T_{n(i)}, J) =$

$$\frac{\{|f_{n(i)}(J)|/|J|\} \times \{|f_{n(i)}(T_{n(i)})|/|T_{n(i)}|\}}{\{|f^{n(i)}(J \cup L_{n(i)})|/|J \cup L_{n(i)}|\} \times \{|f^{n(i)}(J \cup R_{n(i)})|/|J \cup R_{n(i)}|\}}$$

goes to zero. This contradicts (*) above, and proves Theorem (4.2.2).

□

5. AXIOM A MAPS AND STRUCTURAL STABILITY.

Let us now consider perturbations of maps of the interval. We say that a C^r map $f:[-1,1]\to[-1,1]$ is C^r-structurally stable if every C^r map $g:[-1,1]\to[-1,1]$ which is sufficiently C^r close to f is conjugate to f. Structural stability is closely related to hyperbolicity.

(5.1) Hyperbolicity

As usual we say that a compact invariant set K is hyperbolic for a map f in $\mathcal{U}$ if there exists a $\lambda>1$ and a constant $C>0$, such that for y in K either,

$$|(f^n)'(y)| \geq C.\,\lambda^n \qquad \text{for } n\geq 0 \qquad \text{or}$$

$$|(f^n)'(y)| \leq C^{-1}.\,\lambda^{-n} \quad \text{for } n\geq 0.$$

It is quite amazing but maps f in $\mathcal{U}$ are hyperbolic on very large sets. Let us quote the most general result in this direction:

Theorem (5.1.1). [Mañé (1985)]
Let f be a C^2 map. If K is a compact invariant set which does not contain critical points of f, and which also does not contain any non-hyperbolic periodic points, then K is hyperbolic.

□

Mañé's result also holds for interval or circle maps with more critical points. Previously a proof of this fact was given for maps with negative Schwarzian derivative by van Strien (1979 &1981) in the unimodal case, and simultaneously in 1979 by Misiurewicz (1981) in the case with more critical points. H. Nusse (1983) has an alternative proof. Misiurewicz states his theorem in the following form. By a compactness argument his result follows from the previous Theorem.

Corollary (5.1.2). [Misiurewicz, (1981)] Assume f is C^3, has no periodic attractors or non-hyperbolic periodic points and that $S(f)<0$. Take a neighbourhood U of the critical points of f. Then there exists $n>0$ and $\lambda>1$, such that if $x,f(x),f^2(x),\ldots,f^n(x) \notin U$, then $|(f^n)'(x)|>\lambda$.

□

Mañé's proof is very much based on the use of Schwartz's Lemma. Therefore he does not get estimates for points that come close to the critical points. However, there is the following result for maps satisfying the "Misiurewicz condition" from sections (4.2) and (7.3):

Theorem (5.1.3). [van Strien (1987)]
Assume that $f:[-1,1]\to[-1,1]$ is a C^3 endomorphism, with no flat critical points such that the forward orbit of each critical point does not accumulate to the set of critical points $C(f)$. Furthermore assume that f has only periodic orbits which are hyperbolic and repelling. Then there exists a constant $K<\infty$ such that the following holds. Let I_n be a maximal interval on which f^n is monotone. Then

$$\sum_{0\le k\le n}|f^k(I_n)| \le K.$$

□

This result was already proved by T. Nowicki (1986 (b)), for unimodal maps with $S(f)<0$. It is not hard to show that Theorem (5.1.3) implies Theorem (5.1.1) (the assumption that f is C^3 instead of C^2 is only used "near" the critical point).

(5.2) Axiom A maps and structural stability

Let us consider the space M^r of C^r maps $f:[-1,1]\to[-1,1]$ satisfying $f(-1)=f(1)=-1$ with the C^r topology. As before for f in M^1 let $C(f)=\{x;f'(x)=0\}$ be the set of critical points.

We say that a map f in M^1 is an <u>Axiom A</u> map if:

i) The closure of the set of periodic points $Per(f)$ is equal to $\Omega(f)$; and

ii) $\Omega(f)$ is hyperbolic.

Furthermore we say that f satisfies the <u>no-cycle</u> condition if for each $x,y\in C(f)$, $f^n(x)\ne f^m(y)$ for all $n,m\ge 0$. We use this terminology since if $c\in C(f)$ is

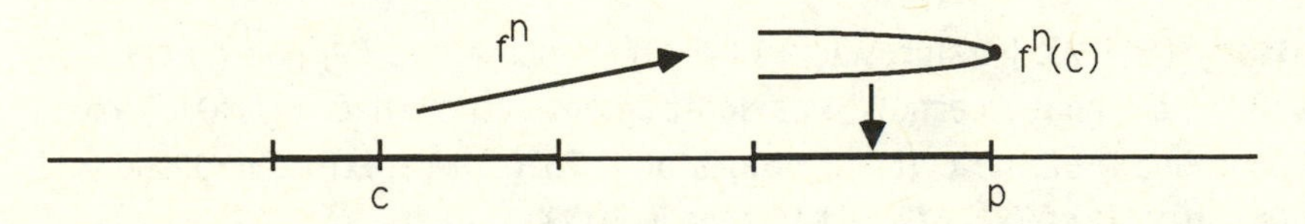

Figure 12: A critical point is mapped onto a periodic point p

eventually mapped onto a periodic repelling point p then one gets a "homoclinic tangency". (If c is periodic then one gets a "one-cycle".) A neighbourhood of c "folds over" and the turning point is mapped to p (see Figure 12).

Using techniques from theory of hyperbolic sets it is not hard to prove the following:

Theorem (5.2.1). [see Jacobson (1971), Nitecki (1973) or Mañé (1985)]
Let $f \in M^2$ be an Axiom A map satisfying the no-cycle condition. Furthermore let
i) for all $c \in C(f)$ one has $f''(c) \neq 0$;
ii) $-1, 1$ are not in $C(f)$.
Then f is C^2 structurally stable.

□

If $\Omega(f)$ is hyperbolic and f has no cycles then $Cl(Per(f)) = \Omega(f)$. This follows from the proof of the following Closing Lemma:

Theorem (5.2.2). [L.S. Young (1970)]
Let $0 \leq r \leq \infty$ and let f be in M^r. Then arbitrarily C^r close to f there exists a map g such that (i) holds for g, i.e., $Cl(Per(g)) = \Omega(g)$.

□

This is the analogue of C. Pugh's Closing Lemma for diffeomorphisms on compact manifolds. However for diffeomorphisms this result is extremely hard to prove and only known for r=1. For interval maps the result is comparatively easy to prove.

The following conjectures are among the deepest in the theory of smooth maps of the interval:

Conjecture 1: Let f be in M^r. Then C^r arbitrarily close to f there exists a map g which satisfies the Axiom A and the no-cycle condition.

Conjecture 2: Let f be a structural stable map in M^r. Then f satisfies the Axiom A and the no-cycle condition.

Conjectures 1 and 2 would the analogue of Peixoto's results for vectorfields on compact orientable surfaces.

Conjecture 3: Let f be in M^r. Then C^r arbitrarily close to f there exists a map g which is structurally stable.

Conjecture 3 is much weaker than the first two conjectures. For unimodal maps the last conjecture can be reduced to the following conjecture (see de Melo (1984)):

Conjecture 4: Let f be in M^r and assume all the critical points of f are non-flat. Then the period of attracting periodic points and non-hyperbolic periodic points of f is uniformly bounded.

(Remark: de Melo and I think that we are close to proving conjecture 4 for unimodal maps.)

6. SENSITIVE DEPENDENCE ON INITIAL CONDITIONS

In this section we describe a result of Guckenheimer. We say that h has <u>sensitive dependence on initial conditions</u>, if there is a subset Y of $[-1,1]$ of positive Lebesgue measure and $\varepsilon>0$ such that for every $x \in Y$ there exists y arbitrarily near x and an $n\geq 0$ such that $|f^n(x)-f^n(y)|>\varepsilon$.

Theorem (6.1). [Guckenheimer, (1979)] Let f be a C^3 unimodal map with $S(f)<0$ and no periodic attractors. Then f has sensitive dependence on initial condition if and only if we are in the finite decomposition case (see section (2.6)).

Proof: Remark that from the result of section 4, f has no wandering intervals.

If we are in the infinite decomposition case, then it is not hard to show from the hyperbolicity result of the previous section that almost all points (in Lebesgue measure) converge to the set Ω_∞ from section (2.6). (Since f is C^2 this follows from the hyperbolicity results of the previous section using methods which are now well-known.) Since $S(f)<0$ the map f has no wandering intervals and therefore Ω_∞ is a Cantor set. It follows that f has no sensitive dependence.

Assume now that f is as in the finite decomposition case. Again from the hyperbolicity result Theorem (5.1.1) or Corollary (5.1.2) it follows that Lebesgue almost all points iterate towards the orbit of the last interval in the decomposition. Take this last interval and call it I. Let n be the period of I. The iterate $f^n|I$ is up to scaling a map in $\mathcal{U}$. Since f has no periodic attractors, $f^n|I$ is necessarily semi-conjugate to a tent-map F_s. Since f has no periodic attractors and no wandering intervals, this semi-conjugacy is in fact a conjugacy (see Corollary (2.5.6). Since F_s is expanding F_s and therefore f has sensitive dependence on initial conditions. This proves the result.

□

So in this case we have sensitive dependence on a set of full measure.

Question: We should remark that it is not clear at all whether the result above is true if $S(f)>0$ in some points and therefore f may have a periodic attractor (which does not contain the critical point in its basin). In this case one would have to estimate the measure of the union of the basins of the attractors.

7. ERGODIC PROPERTIES

We are interested in the asymptotic properties of orbits. One of the ways to describe these asymptotic properties is using an invariant probability measure μ. That is, require that $\mu(f^{-1}(A))= \mu(A)$ for every measurable set A. This property implies that the support of the measure μ is contained in the non-wandering set of f.

In order to avoid trivial measures we require that μ does not give positive measure to the set of periodic points of f. Ideally, one would like an even stronger property: the measure μ should be <u>absolutely continuous</u> with respect to the Lebesgue measure. This means that if the Lebesgue measure of A is zero then $\mu(A)=0$.

(7.1) General results on the existence of absolutely continuous invariant measures

There are some very general conditions for when absolutely continuous invariant measures do exist. Let us first give a result which also holds in a much more abstract setting:

We say that f is <u>non-singular</u> (w.r.t. to the Lebesgue measure λ) if $\lambda(A)=0$ implies that $\lambda(f^{-1}(A))=0$. Then the following are equivalent statements:

Theorem (7.1.1). [See e.g. S.R. Foguel (1969)]
Let f be non-singular. Then the following are equivalent:
1) there exists an invariant measure for f which is absolutely continuous with respect to λ;
2) There exists $\alpha<1$, such that $\lambda(A) < \alpha$ implies $\limsup \lambda(f^{-n}(A))<1$;
3) There exists $\alpha<1$, such that $\lambda(A) < \alpha$ implies

$$\limsup \left((1/n) \sum_{0\le k\le n-1} \lambda(f^{-k}(A)) \right) < 1.$$

□

(This result dates from the 50's and is due to Y.N. Dowker, A. Calder and others.) For references see S.R. Foguel (1969). Later E. Straube (1981), reinvented part of this Theorem.

Another, completely different way to determine whether some measure is absolutely continuous is by using the Radon-Nikodym Theorem. A probability measure μ is absolutely continuous with respect to the Lebesgue measure λ if and only if there exists an L^1 "density"-function $p(t)$, such that

$$\mu(A)= \int_A p(t).d\lambda(t) \text{, for every measurable } A \text{ in } [-1,1].$$

Using these densities, A. Lasota and J. Yorke proved the existence of a measure which is absolutely continuous with respect to the Lebesgue measure for maps which are piecewise C^2 and expanding:

Theorem (7.1.2). [A. Lasota & J. Yorke (1973)].
If $f:[-1,1]\mapsto[-1,1]$ is piecewise C^2 and expanding then f has an invariant measure which is absolutely continuous with respect to the Lebesgue measure.

□

Here we say that $f:[-1,1]\to[-1,1]$ is piecewise C^2 and expanding if:
i). There exist a finite number of points $x_0=0, x_1, \ldots, x_n=1$ such that f can be extended to a C^2 map on $[x_i, x_{i+1}]$ (f need not be continuous at x_i); and,
ii). $|f'|>1$ on $[-1,1]$.

The proof of Lasota and Yorke uses the Frobenius-Perron operator: This operator is a transformation on densities of measures (corresponding to pushing forward measures under f). Applying the Frobenius-Perron operator n-times one gets a measure whose density p_n has bounded variation. Making some estimates one proves that p_n has a limit p which has also bounded variation (here one needs C^2). Since maps with bounded variation are also L_1 the result follows. The result of Lasota & Yorke can also be proved for $C^{1+\epsilon}$ maps.

Question: It would be interesting to know whether there exists a proof of Lasota-Yorke's result using Theorem (7.1.1). One motivation for this is that the Frobenius-Perron operator becomes badly behaved if the map f has critical points.

Example (7.1.3). Let f be a map as in Figure 13 below. This map is conjugate to the tent-map F_2 with slope ± 2. The map F_2 has the Lebesgue measure as an invariant measure. However f does not have an absolutely continuous probability measure. One can prove this using Theorem (7.1.1) from above and the Lemma of Schwartz.

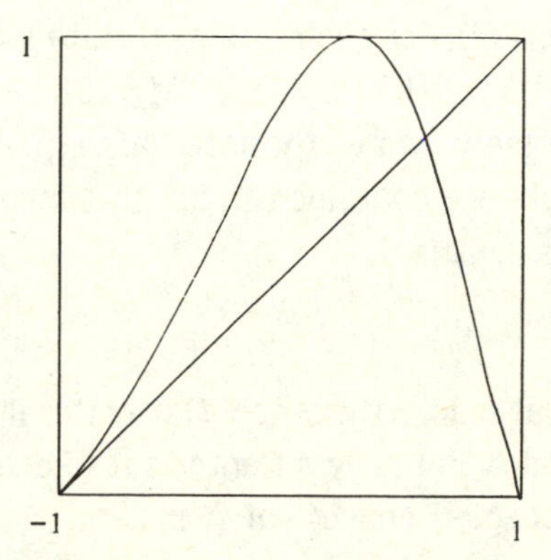

Figure 13: f does not have an absolutely continuous invariant measure

(7.2) Maps with zero entropy

If the map f has zero entropy, then either

(i) The decomposition in section (2.4) is finite. In this case the non-wandering set of f consists only of periodic points (with period uniformly bounded). So there is no interesting measure in this case.

(ii) The decomposition in section (2.4) is infinite. In this case the non-wandering set of f consists of periodic orbits of periods 2^n, and a set of the following form:

$$\Lambda = \bigcap_{n\geq 0} \{J_n \cup f(J_n) \cup f^2(J_n) \cup \ldots \cup f^{k(n)-1}(J_n)\}$$

where $k(n)=2^n$ and J_n is an interval containing the critical point. The intervals J_n, $f(J_n)$, $f^2(J_n)$, ..., $f^{k(n)-1}(J_n)$ are mutually disjoint and $f^{k(n)}$ maps J_n into itself. Furthermore J_n has a non-empty intersection with precisely two of the intervals J_{n+1}, $f(J_{n+1})$, ..., $f^{k(n+1)-1}(J_{n+1})$.

Theorem (7.2.1). [M.Misiurewicz (1981)], see also [P. Collet & J.P.Eckmann (1980)]

(i) There exists a unique invariant probability measure on Λ which does not associate positive measure to any subset of the periodic points.

(ii) The measure μ is Bowen-Ruelle: for any continuous function g on [-1,1] and any point x in the basin $B\Lambda$ of the set Λ one has

$$\lim_{N\to\infty} (1/N)\sum_{0\leq n\leq N-1} g(f^n(x)) = \int g\,d\mu.$$

(iii) the measure μ is ergodic.

Proof: Let us only show (i). For each $n\geq 0$, associate measure $(1/2^n)$ to each of the intervals I_n, $f(I_n)$, $f^2(I_n)$, ..., $f^{k(n)-1}(I_n)$. Since I_n has a non-empty intersection with precisely two of the intervals I_{n+1}, $f(I_{n+1})$, ..., $f^{k(n+1)-1}(I_{n+1})$ (and those two intervals are contained in I_n), this measure is well defined. Clearly this measure is invariant.

□

If f is a unimodal map with a non-flat critical point, then f has no wandering intervals and Λ is really a Cantor set. Recently J. Guckenheimer has proved that the Lebesgue measure of Λ is zero.

Theorem (7.2.2). [Guckenheimer (1986)] If $f \in \mathcal{U}$ satisfies $S(f)<0$, $h(f)=0$ and f has an infinite decomposition, then the measure of the Cantor set in the non-wandering set is zero.

□

Question: Does this result hold for maps which are C^2 and have no flat critical points? (A counterexample was constructed by M. Misiurewicz (1982) for a map with a flat critical point.) This question is related to the question of which C^2 maps (with infinite decomposition and $h(f)=0$) are contained in the stable manifold of the period doubling operator from section 8, see Theorem (8.1.2).

(7.3) Maps f absolutely continuous invariant measures do exist

Let us quote a result about the existence of absolutely continuous measures. This result is a combination of results of M. Misiurewicz (1981), P. Collet & J.P. Eckmann (1984) and T. Nowicki (1985). Let $C(f)=\{x;f'(x)=0\}$.

Theorem (7.3.1). Let f be a C^3 map such that $S(f)<0$ and f has no periodic attractors or non-hyperbolic periodic points. Assume either that

(*) $\qquad C(f) \cap \text{closure}\{f(C(f)),f^2(C(f)), f^3(C(f)),....\} = \emptyset$

or that f is unimodal and that closure $\{f(c),f^2(c),... \}$ is uniformly hyperbolic in the sense that the exist $C_0>0$, $\lambda_0>1$ such that

(**) $\qquad |(f^n)'(f(c))| \geq C_0\lambda_0{}^n, \quad$ for $n\geq 0$

Then f has an absolutely continuous invariant measure.

□

We encountered condition (*) before. It is frequently called the <u>Misiurewicz condition</u>.

Remarks: 1) R. Mañé's result from section 6 shows that if f is C^2 and if f has no periodic attractors or no non-hyperbolic points, then (*) implies (**).
2) T. Nowicki has shown that if f is unimodal and $S(f)<0$, then condition (**) implies that closure$\{\bigcup_{i\geq 1} f^{-i}(c)\}$ is uniformly hyperbolic in the sense that there exist $C_1>0$ and $\lambda_1>1$ such that

(***) $\qquad n>0$ and $f^n(z)=c$ implies $|(f^n)'(z)| \geq C_1\lambda_1{}^n$.

P. Collet and J.P. Eckmann (1981), prove that (**), (***) and $S(f)<0$, together imply the existence of absolutely continuous invariant measures.
3) Recently A.M. Blokh and M.Ju. Ljubich (1987) have shown that if f is unimodal and has an absolutely continuous invariant measure then f is ergodic and therefore has at most one such invariant measure.

More recently there are some results which show that the condition on the Schwarzian derivative can be dispensed with.

Theorem (7.3.2). [van Strien (1987)]
Suppose that f is C^2, has no flat critical points and no periodic attractors or non-hyperbolic periodic points. Then (*) implies the existence of an absolutely continuous invariant measure. □

Theorem (7.3.3). [Nowicki & van Strien (1987)]
Suppose that f is C^2, has no flat critical points and no non-hyperbolic periodic points. Then (**) and (***) imply the existence of an absolutely continuous invariant measure. □

There are also some results which show that the exponential growth of $|Df^n(f(c))|$ as $n\to\infty$ can be replaced by subexponential growth, see [Keller (1987a), (1987b)].

(7.4) Maps f which are conjugate to piecewise linear maps but where absolutely continuous invariant measures do not exist.

For a number of years it was an open question whether maps which are conjugate to piecewise linear maps have absolutely continuous invariant measures. Recently S.Johnson (1985) has proved that this is not the case:

Theorem (7.3.1). [S. Johnson, 1985].
There exists an example of a quadratic map f of the form $x\to(\mu-1)-\mu x^2$, such that f is conjugate to a piecewise linear map and has no absolutely continuous invariant measure.

□

Question: Assume $S(f)<0$. Does the existence of an absolutely continuous invariant measure only depend on the kneading invariant of f? A similar question can be asked for smooth maps which do not satisfy $S(f)<0$.

8. ANOTHER ANALYTIC TOOL: RENORMALISATION TECHNIQUES

(8.1) Period doubling

Let $\mathcal{M}$ be the space of C^2 unimodal mappings $g:[-1,1]\to[-1,1]$ in $\mathcal{U}$ such that the critical point is at 0. For g in $\mathcal{M}$ define $\lambda=g(1)$ and let

$$Tg(x)= \lambda^{-1}g(g(\lambda x)).$$

$T:\mathcal{M}\to \mathcal{M}$ is called the renormalisation operator for period doubling in one dimension.

Independently, M. Feigenbaum and P. Coullet & C. Tresser, conjectured that this operator has a fixed point and that the spectrum is related to the universality of certain numbers. These results were proved by several people, some by hand and others relying on computer assisted estimates.

Theorem (8.1.1). [M. Campanino, H.Epstein, O.E. Lanford and D. Ruelle in papers appearing between 1979 and 1986]

(i) There exists an analytic fixed point Ψ of $T: \mathcal{M}\to \mathcal{M}$. This map Ψ is symmetric.

(ii) The map $T: \mathcal{M}\to \mathcal{M}$ is C^1, $dT(\Psi)$ is a compact operator and the spectrum of $dT(\Psi)$ consists of one simple eigenvalue $\delta>1$ (whose corresponding eigenspace is one-dimensional) and the remainder of the spectrum is contained in a disc with radius less that one centred at 0.

□

Remark: The expanding eigenvalue of the operator is related to the fact that for a family $x\to 1-\mu x^2$ period doubling happens "at an exponential rate" (corresponding to the eigenvalue δ). One can find more about this in [P. Collet & J.P. Eckmann, (1980)].

It follows from stable manifold theory, see for example M. Hirsch & C.

Pugh (1970), that there exists a unique local codimension-one stable invariant manifold $W^s_\varepsilon(\Psi)$ in $\mathcal{M}$. More precisely this stable manifold theory shows that there exists $\varepsilon>0$ such that $W^s_\varepsilon(\Psi)=\{g \in \mathcal{M}, T^n(g)$ is ε-close to Ψ for all $n\geq 0\}$ is a C^2 manifold and that this set is contained in the set $W^s(\Psi)=\{g;\ \ T^n(g)\to \Psi$ as $n\to\infty\}$. More recently, D.Sullivan has announced a result which says that this stable manifold $W^s(\Psi)$ is quite big.

Theorem (8.1.2). [D.Sullivan, (1986)]
Assume that f in $\mathcal{U}$ satisfies the following two conditions
1) f is conjugate to the fixed point of the doubling operator Ψ;
2) there is a disk in the domain of the complex analytic extension F of f which (a) contains the interval [-1,1], (b) contains exactly one critical point of F, and (c) whose boundary is carried outside itself by F. (In other words F is quadratic-like mapping in the sense of Douady and Hubbard.)
Then f is contained in $W^s(\Psi)$, i.e., $T^n(f)\to \Psi$, as $n\to \infty$.

□

D. Sullivan's result also includes some statement about the asymptotic metric properties of the Cantor set $\Lambda(f)$ in the non-wandering set of f. (Part of those results consists of joint-work of D. Sullivan and M. Feigenbaum.) Here $\Lambda(f)$ is the closure of the orbit of the critical point.

That the renormalising results are related to some smoothness results follows from this and from the following result.

Theorem (8.1.3). [W.Paluba, (1986)]
Assume that f and g are both contained in $W^s(\Psi)$. Then $\Lambda(f)$ and $\Lambda(g)$ are Lipschitz conjugate.

□

By this the following is meant: there is conjugacy h between f and g such that there is $K<\infty$ so that for $x \in \Lambda(f)$ and $y \in [-1,1]$, one has $|h(x)-h(y)| \leq K.\ |x-y|$ and similarly for h^{-1}. Of course one cannot hope that this will be true for x and y both arbitrary in [-1,1], because the eigenvalues of corresponding periodic points of f and g would then all have to be the same. (W. Paluba has even stronger results in this direction.)

This last result shows that the idea of renormalising may well give smoothness results in the same spirit as in M. Hermann's results for diffeomorphisms of the circle. We will go into this more in the next section.

(8.2) More generalised renormalising

As we saw in the result on decomposition in section (2.6), one can have two types of maps: Either

1) f is in the finite decomposition case; or
2) f is in the infinite decomposition case.

In the first case f either has periodic attractors or it has sensitive dependence on initial conditions. In the second case, these maps are much more similar to diffeomorphisms of the circle. Let us explain this now. In this case there exists a strictly decreasing sequence of intervals J_0, J_1, J_2, ...,J_n containing the critical point c, such that for some (minimal) k(n) the iterate $f^{k(n)}$ maps J_n into itself (and is up to scaling a unimodal map), and

$$K_n = J_n \cup f(J_n) \cup f^2(J_n) \dots \cup f^{k(n)-1}(J_n)$$

consists of a disjoint union of at least k(n-1) intervals. K_{n+1} consists of k(n+1) intervals with k(n+1)/k(n) in each of the connected components of K_n. In the case of period doubling k(n+1)=k(n)+k(n)=2.k(n).

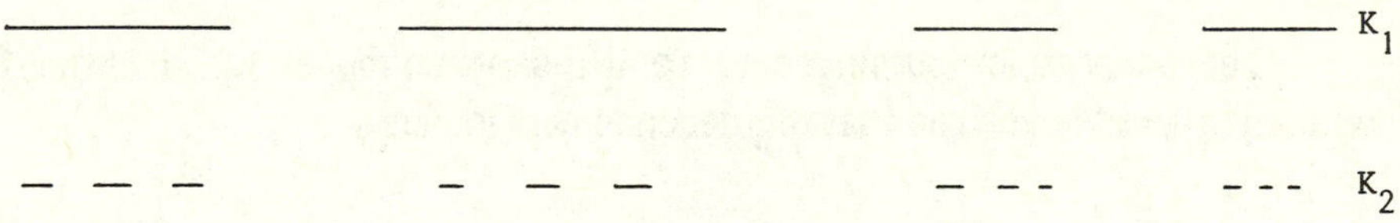

K_1

K_2

Figure 14: the sets K_n.

But more generally there exist unimodal maps for which $k(n+1)=(i_n+1).k(n)$, where the sequence i_n is completely arbitrary. The sequence i_n can be considered as an analogue of the continued fraction expansion of the rotation number for diffeomorphisms of the circle.

In the period doubling case one has $i_n=1$ for all n=1,2,... . In other words this case corresponds to a diffeomorphism of the circle with continued fraction expansion consisting of 1's, i.e. the golden mean. For other one-dimensional maps of this type with i_n equal to some constant i, it seems there is a result similar to the previous Theorem (9.1). For such maps again an analytic fixed of the period-multiplying operator (defined by $T(f)(x)=(1/\lambda) f^i (\lambda x)$) has been found (at least numerically).

For diffeomorphisms of the circle one has the classical problem, of whether a diffeomorphism is smoothly conjugated to a pure rotation. This turns out to be a "small denominator" K.A.M. problem and to depend on how badly the rotation number is approximated by rational numbers. Arnold and Moser deal with the case where the diffeomorphisms of the circle are close to pure rotations of the circle. M. Hermann treats the case where a diffeomorphism is not necessarily close to a pure rotation.

For interval maps one may hope to have a similar situation: One could hope that for maps where the sequence i_n is bounded (or corresponds to the continued fraction of a number which is badly approximated by rational numbers, one can get smoothness results for maps close to fixed points (as in the case of Arnold-Moser's result) or even better for arbitrary maps not necessarily close to the fixed points (as in M. Hermann's result)

An even more general renormalisation scheme also involving circle diffeomorphisms and using the language of substitutions on symbols has been proposed by I. Procaccia, S. Thomae and C. Tresser (1985).

9. ONE-PARAMETER FAMILIES OF UNIMODAL MAPS

Let us restrict ourselves to families of unimodal $f_\mu:[-1,1]\to[-1,1]$, where $\mu\in[0,1]$. We assume that df_μ depends continuously on μ.

(9.1) An intermediate value theorem for families of unimodal maps.

Let us first show that if we know the kneading sequences $\nu(f_0)$ and $\nu(f_1)$ that $\mu \to \nu(f_\mu)$ has to "pass all kneading sequences in between". We call a power series ν admissible if $|\sigma^n(\nu)| \geq \nu$, for all $n\geq 0$. (Here σ is the shift-map from section (2.1).)

Theorem (9.1.1). [M. Metropolis, M.L. Stein and P.R. Stein (1973)]
Let ν be an admissible power series with $\nu(f_0) < \nu < \nu(f_1)$. Then there exists $\mu(0)\in(0,1)$ such that $\nu(f_{\mu(0)}) = \nu$.

Proof: Let $\mu(0)=\sup\{ \mu \mid \nu(f_\mu) < \nu \}$. It follows that $\nu(f_{\mu(0)}) \geq \nu$. There are two cases:

i) The map $\mu \to \nu(f_\mu)$ is continuous from the left at $\mu= \mu(0)$ and therefore $\nu(f_{\mu(0)}) = \nu$;

ii) The map $\mu \to \nu(f_\mu)$ is discontinuous from the left at $\mu= \mu(0)$. Then

(*) $\qquad \nu(f_{\mu(0)}) \geq \nu > \nu(f_\mu) \quad$ for $\mu < \mu(0)$.

and the critical point of $f_{\mu(0)}$ is periodic. As in Corollary (2.3.3) we need that f is C^1, that for every μ the map f_μ is unimodal and that f_μ depends C^1 on the parameter μ. As in Corollary (2.3.3) we get that for μ near $\mu(0)$, $\nu(f_\mu)$ is either of the form

(**) $\quad (\sum_{0\leq i\leq n-1} \nu_i t^i).(1+t^n)^{-1}$ or $(\sum_{0\leq i\leq n-1} \nu_i t^i).(1-t^n)^{-1}$.

Clearly the first power series is smaller than the second one. From (*) it follows that $\nu(f_{\mu(0)})$ is equal to the second power series and that for μ near $\mu(0)$, $\nu(f_\mu)$ is equal to the first power series. Now it not hard to check that there is no admissible power series between the two power series in (**). This and (*) implies that $\nu(f_{\mu(0)}) = \nu$.

□

(9.2) Monotonicity of the kneading sequence

For the quadratic family one can prove more:

Theorem (9.2.1). [Douady (1982), Douady & Hubbard (1984) and Thurston (1984)]
Let $f_\mu=(\mu-1)-\mu x^2$ be the usual one-parameter family of quadratic maps in $\mathcal{U}$. Then $\mu \to \nu(f_\mu)$ is monotone.

□

The proofs of this result rely on considering f_μ as a rational map on the Riemann-sphere. The basic step in the proof of Douady and Hubbard is to show the following. Let D be a connected domain in the complex plane of parameter values where f_μ has a hyperbolic periodic attractor z_μ of period n. Using the quasi-conformal mapping theorem they show that the map $\mu \to |df^n(z_\mu)|$ is injective on D. It easy to show that this implies the Theorem. More about rational maps on the Riemann-sphere can be found in P. Blanchard (1984) and P. Sad (1983).

Question: Is $\mu \to \nu(g_\mu)$ also monotone for families of unimodal maps g_μ sufficiently near the quadratic family f_μ?

(9.3) Bifurcation diagram for unimodal mappings

In order to clarify the bifurcations of the quadratic family $\mu \to f_\mu$ let us draw a bifurcation diagram. For each μ we draw the set of periodic points $Per(f_\mu)$ of f_μ. From the proof of Douady & Hubbard for Theorem (9.2.1) one can show that the bifurcations of the periodic points are generic codimension-one saddle-node and flip (pitchfork) bifurcations. More about these bifurcations can be found in Guckenheimer, (1977). See also the survey paper of Palis (1977).

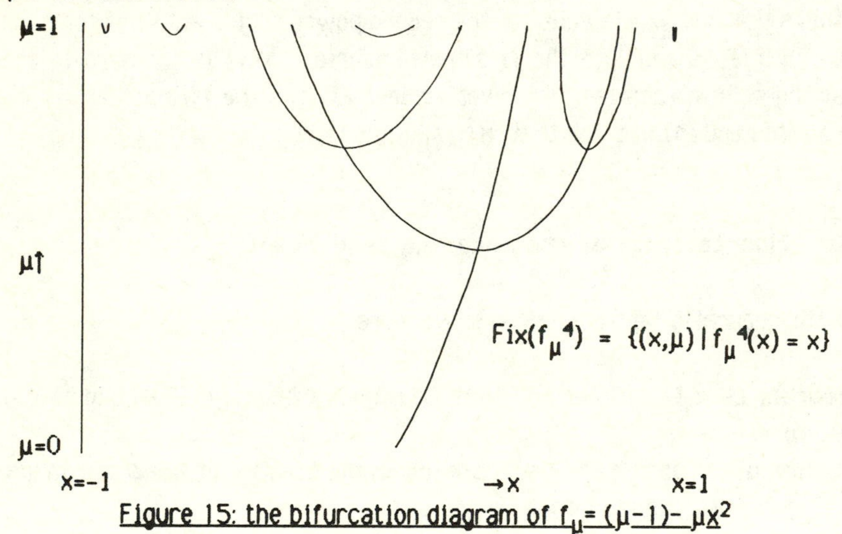

Figure 15: the bifurcation diagram of $f_\mu = (\mu-1) - \mu x^2$

Of course other families of unimodal mappings will undergo the same bifurcations (see section (9.1)) but possibly in a different order. For two-parameter families of maps with more than one critical point the bifurcations are much less universal, see R. Mackay & C. Tresser (1985).

(9.4) Invariant measures for one-parameter families of maps

M.V. Jacobson proved the following remarkable Theorem. Let f_μ be the quadratic family from above. Let $A = \{ \mu;\ f_\mu$ has an absolutely continuous invariant measure).

Theorem (9.3.1). [M.V. Jacobson (1981), see also M. Benedick & L. Carleson (1985), Johnson (1985), M. Rees (1984), (1986) and M.R. Rychlik(1986)]
The Lebesgue measure λ of A is positive. In particular $\mu=2$ is a density point of A, i.e. $\lim_{\varepsilon\to 0} (1/\varepsilon) . \lambda\{ [2-\varepsilon,2] \cap A\} = 1$.

□

For $\mu \in A$, f_μ has certainly no attracting periodic points. Therefore, if the conjectures of section (5.2) are true then A would certainly not contain intervals. So A would be a Cantor-set with positive measure.

10. LITERATURE

(No attempt has been made to be complete. Nevertheless, I would like to apologise if some references have been omitted which should have been included. For a more complete list of references see Z. Nitecki's survey paper, and P. Collet & J.P. Eckmann's book.)

Allwright, D.J., (1978): Hypergraphic functions in recurrence relations. *SIAM J. Appl. Math.*, **34**, 687-691.

Benedicks, M. and Carleson, L., (1985): On iterations of $1-ax^2$ on $(-1,1)$, *Ann. of Math. (2)*, **122**, 1-25.

Blanchard, P., (1984): Complex dynamics on the Riemann sphere, *Bull. Amer. Math. Soc.*, 85-141.

Block, L., Guckenheimer, J., Misiurewicz,M., Young, L.S., (1980): Periodic points and topological entropy of one-dimensional maps, Springer Lecture Notes in Math., Vol. 819, 18-34.

Blokh, A.M. & Ljubich, M. Ju., (1987): Attractors of maps of the interval, preprint.

Bowen, R., (1979): Invariant measures for Markov maps of the interval, *Comm. Math. Phys.*, **69**, 1-17.

Campanino, M. and Epstein, H., (1981): On the existence of Feigenbaum's fixed point. *Comm. Math. Phys.*, **79**, 261-302.

Campanino, M., Epstein, H. and Ruelle, D., (1982), On the existence of

Feigenbaum's fixed point, *Topology*, **21**, 125-129.

Collet, P., (1984): Ergodic properties of some unimodal mappings of the interval, preprint, Institut Mittag-Leffler, report no 11.

Collet, P. and Eckmann, J.-P., (1980): *Iterated maps on the intervals as dynamical systems*, Birkhauser: Basel.

Collet, P. and Eckmann, J.-P.,, (1983): Positive Liapunov exponents and absolute continuity for maps of the interval, *Ergod. Th. & Dynam. Syst.*, **3**, 13-46.

Coullet, P. & Tresser,C., (1978a): Itérations d'endomorphismes et groupe de renormalisation, *C. R. Acad. Sci. Paris*, **287**, 577.

Coullet, P. & Tresser,C., (1978b): Itérations d'endomorphismes et groupe de renormalisation. *J. Physique* **C5**, 25.

Denjoy, A., (1932): Sur les courbes definie pas les equations differentielles a la surface du tore, *J. Math. Pures Appl.(9)*, 11.

Derrida, B., Gervois, A. and Pomeau, Y. , (1978): Iterations of endomorphisms of the real line and representations of numbers. *Ann. Inst. H. Poincaré*, **29**, 305.

Derrida, B., Gervois, A. and Pomeau, Y., (1979): Universal metric properties of bifurcations of endomorphisms. *J. Phys.* **A12**, 269.

Douady, A., (1982): Systemes dynamiques holomorphes, in Séminaire Bourbaki, 599.

Douady, A. and Hubbard, J.H., (1985): Etude dynamiques des polynomes complexes, I & II, preprint.

Epstein, H., (1986): New proofs of the existence of the Feigenbaum function, *Comm. Math. Phys.*, **106**, 395-426.

Fatou, P., (1919 &1920): Sur les équations fonctionelles, *Bull. Soc. Math. France*, **47**, 161-270, (1919); **48**, 33-95, 208-314 (1920).

Feigenbaum, M., (1978,1979a): Quantitative universality for a class of non-linear transformations, *J. Statist. Phys.*, **19**, 25-52; 21, 669-709.

Feigenbaum, M., (1979b): The onset of turbulence, *Phys. Lett.*, **74A**, 375.

Feigenbaum, M., (1980): The transition to aperiodic behaviour in turbulent systems, *Comm. Math. Phys.*, 65-86.

Foguel, S.R., (1969): *The Ergodic Theory of Markov Processes.* Van Nostrand Math. Studies, No. 21, New York.

Guckenheimer, J., (1977): Bifurcations of maps of the interval, *Invent.*

Math., **39**, 165-178.

Guckenheimer, J., (1979): Sensitive dependence on initial conditions for one-dimensional maps, *Comm. Math. Phys.* **70**, 133-160.

Guckenheimer, J., (1986): Limit sets of S-unimodal maps with zero entropy, preprint.

Hall, C. R. , (1981): A C^∞ Denjoy counterexample, *Ergod. Th. and Dynam. Sys.*, **1**, 261-272.

Henry, B., (1973), Escape from the unit interval under the transformation $x \to \lambda x(1-x)$, *Proc. Amer. Math. Soc.*, **41**, 146-150.

Herman, M., (1979): Sur la conjugaison différentiable des diffeomorphismes du circle a des rotations, *Inst. Hautes Etudes Sci. Publ. Math.*, **49**, 5-234.

Hirsch, M. and Pugh, C., (1970): Stable manifolds and hyperbolic sets, in *Proc. Symp. in Pure Math.*, Vol.14, AMS, Providence, R.I., 122-164.,

Hofbauer, F., (1980): Maximal measures for simple piecewise monotonic transformations, *Z. Warsch. Verw. Gebiete* **52**, 289-300.

Jacobson, M.V., (1971): On smooth mappings of the circle into itself, *Math. U.S.S.R.-Sb* **16**, 161-185.

Jacobson, M.V., (1981): Absolutely continuous invariant measures for one-parameter families of one-dimensional maps, *Comm. Math. Phys.*, **81**, 39-88.

Johnson, S., (1985): Continuous measures and strange attractors in one dimension, Thesis Cornell University.

Jonker, L., (1981): Periodic orbits and kneading invariants, *Proc. London Math. Soc.*, **3**, 428-450.

Jonker, L. and Rand, D., (1981): Bifurcations in one-dimension.
I. The non-wandering set, *Invent. Math.*, **62**, 347-365.
II. A versal model for bifurcations, *Invent. Math.*, **63**, 1-15.

Julia, G., (1918): Memoire sur l'iteration des fonctions rationelles, *J. de Math.*, **8**, 47-245.

Keller, G., (1987a): Invariant measures and Lyapounov exponents for S-unimodal maps, preprint.

Keller, G., (1987b): Invariant measures and expansion along the critical orbit for S-unimodal maps, preprint.

Kopwalski, Z.S., (1976): *Invariant measures for piecewise monotonic transformations.* Lecture Notes in Mathematics, Vol. 472, 77-94,

Springer Verlag, Berlin.

Lanford III, O. E., (1979): Remarks on the accumulation of period doubling bifurcations, in Lecture Notes in Physics, Vol. 74, Springer Verlag, New York.

Lanford III, O. E., (1982): A computer assisted proof of the Feigenbaum conjecture, *Bull. Amer. Math. Soc.*, **6**, 427-434.

Lanford III, O. E., (1984): A shorter proof of the existence of the Feigenbaum fixed point, *Comm. Math. Phys.*, **96**, 521-538.

Lasota, A. & Yorke, J., (1973): On the existence of invariant measures for piecewise monotonic transformations, *Trans. Amer. Math. Soc.* **186**, 481-488.

Ledrappier, F., (1981): Some properties of absolutely continuous invariant measures of an interval, *Ergod. Th. and Dynam. Sys.*, **1**, 77-93.

Li, T. & Yorke, J.A. , (1975): Period three implies chaos, *Amer. Math Monthly*, **82**, 985-992.

Mackay, R.S. & Tresser, C., (1985), Transitions to chaos for circle maps,preprint.

Malta, I. , (1986): On Denjoy's Theorem for Endomorphisms, *Ergod. Th. and Dynam. Sys.*, **6**, 259-264.

Mañé, R., (1985): Hyperbolicity, sinks and measure in one dimensional dynamics, *Comm. Math. Phys.* **100**, 495-524.

May, R.B., (1975): Biological populations obeying difference equations equations, *J. Theoret. Biol.*, **51**, 511-524.

May, R.B., (1976): Simple mathematical models with very complicated dynamics, *Nature*, **261**, 459-467.

de Melo, W., (1984): A finiteness problem for one-dimensional maps, preprint.

de Melo, W. & van Strien, S.J., (1986): A structure theorem in one-dimensional dynamics, preprint.

Metropolis, M. & Stein M.L. and Stein, P.R., (1973): On finite limit sets for transformations of the unit interval, *J. Combin. Theory*, **15**, 25-44.

Milnor, J. and Thurston, W., (1976): On iterated maps of the interval: I,II, preprint, Princeton.

Milnor, J., (1985): On the concept of attractor, *Comm. Math. Phys.*, **99**, 177-195, and **102**, 517-519.

Misiurewicz, M., (1979): Horseshoes for mappings of the interval, *Bull.*

Acad. Polon. Sci. Sér. Sci. Math., **27**, 167-169.

Misiurewicz, M., (1981): Structure of mappings of an interval with zero entropy, *Inst. Hautes Etudes Sci. Publ. Math.*, **53**, 1-16.

Misiurewicz, M., (1981): Absolutely continuous measures, *Inst. Hautes Etudes Sci. Publ. Math.*, **53**, 17-52.

Misiurewicz, M., (1982): Attracting Cantor set of positive measure: for a C^∞ map of an interval, *Ergod. Theory and Dynam. Syst.*, **2**, 405-415.

Misiurewicz, M., (1986): private communication.

Misiurewicz, M. and Szlenk, W. (1977): Entropy of piecewise monotone mappings, *Astérisque* **50**, 299-310 .

Myrberg, P.J., (1959 &1963): Iteration der reellen Polynom zweites Grades, *Ann. Acad. Sci. Fenn.*, **256A**, 1-10, (1959), and **336A**, 1-18, (1963).

Nitecki, Z., (1972): Smooth, Non - singular Circle Endomorphisms, Preliminary Report, unpublished.

Nitecki, Z., (1973): Factorization of nonsingular circle endomorphisms, *Salvador Symposium on Dynamical Systems*, ed. by M. Peixoto, Academic Press.

Nitecki, Z., (1981): Topological dynamics on the interval, *Ergodic Theory and Dynamical Systems, vol II, Proceedings of the Special Year, Maryland, 1979-1980*, Progress in Math. Birkhäuser, Boston.

Nowicki, T., (1985(a)): Symmetric S-unimodal mappings and positive Liapounov exponents, *Ergod. Th. and Dynam. Sys.*, (1985), **5**, 611-616.

Nowicki, T., (1985(b)): On some dynamical properties of S-unimodal mappings, *Fund. Math.*, **126**, 27-43.

Nowicki, T., (1986(a)): Positive Liapounov exponent of critical value of S-unimodal mappings imply uniform hyperbolicity, preprint, Warsaw Agricultural University.

Nowicki, T., (1986(b)): On hyperbolic properties of some maps of the interval, preprint, Warsaw Agricultural University.

Nowicki, T. & Przytycki, F., (1986): Collet-Eckmann's maps of the interval are Hölder conjugates to piecewise linear maps, preprint, Warsaw Agricultural University.

Nowicki, T. & van Strien S.J., (1987):Absolutely continuous invariant measures for C^2 unimodal maps satisfying the Collet-Eckmann conditions, preprint (1987).

Nowicki, T. & van Strien S.J., (1987): Hyperbolicity properties of C^2 multi-modal Collet-Eckmann maps without Schwarzian derivative conditions, preprint (1987).

Nusse, L., (1983): Chaos, yet no chance to get lost, Thesis Utrecht University.

Palis, J., (1977): Some developments on stability and bifurcations of dynamical systems, In: *Geometry and Topology*, Springer Lecture Notes in Mathematics, Vol. 597, 495-509.

Parry, W., (1964): Symbolic dynamics and transformations of the unit interval, *Trans. Amer. Math. Soc.*, **122**, 368-378.

Paluba, W., (1986): private communication and preprint Warsaw University (1987).

Peixoto, M., (1962): Structural stability on two dimensional manifolds, *Topology* **1**, 101-120.

Pianigiani, G., (1979): Absolutely continuous invariant measures for the process $x_{n+1}=Ax_n(1-x_n)$, *Boll. Un. Mat. Ital.*, **16**, 374-378.

Preston, C., (1983): *Iterates of maps on the interval*, Springer Lecture Notes in Math., Vol. 999, Berlin.

Procaccia, I., Thomae, S., and Tresser, C., (1985): First return maps as a unified renormalization scheme for dynamical systems, *Phys. Rev. A* **35** (1987) 1884-1900.

Ruelle, D., (1977): Applications conservant une mesure absolument continue par rapport a dx sur [0,1], *Comm. Math. Phys.*, **55**, 47-51.

Rees, M., (1984): Ergodic rational maps with dense critical point forward orbit, *Ergod. Th. & Dynam. Sys.*, 4, 311-322.

Rees, M., (1986): Positive measure sets of ergodic rational maps, preprint.

Rychlik, M.R., (1983): Bounded variation and invariant measures, *Studia Math.*, **76**, 69-80.

Rychlik, M.R., (1986): Another proof of Jacobson's theorem and related results, preprint, (1986), University of Warsaw.

Sad,. P., (1983): Introduçao a dinamica das funçoes racionais na esfera de Riemann, 14ᵉ colóquio Brasiliero de Matemática.

Sarkovskii, A.N., (1964), Coexistence of cycles of a continuous map of a line into itself. *Ukrain. Mat. Z*, **16**, 61-71.

Schwartz, A., (1963): A generalization of a Poincaré-Bendixon theorem to closed two-dimensional manifolds, *Amer. J. Math.*, **85**.

Schwarz, H.A., (1868): Über einige Abbildungs aufgaben. *J. Reine Angew. Math.*, **70**, 105-120.

Shub, M., (1969): Endomorphisms of compact manifolds, *Amer. J. Math.*, **91**, 175-199.

Singer, D. , (1978): Stable orbits and bifurcations of maps of the interval, *SIAM J. Appl. Math.*, **35**, 260.

Stefan, P, (1977): The theory of Sarkovskii on the existence of periodic orbits of continuous endomorphisms of the real line, *Comm. Math. Phys*, **54**, 237-248.

Straube, E., (1981): On the existence of invariant absolutely continuous measures, Comm. *Math. Phys.*, **81**, 27-30.

van Strien, S.J., (1979 &1981): On the bifurcations creating horseshoes, preprint Utrecht, part of this was published in: *Dynamical Systems and Turbulence*, Warwick 1980, Lecture Notes in Mathematics, 898, 316-351.

van Strien, S.J., (1987): Absolutely continuous measures for C^3 maps of the interval, in preparation.

Sullivan, D., (1985): Quasiconformal homeomorphisms and dynamics I: a solution of Fatou-Julia problem on wandering intervals, *Ann. of Math.*, **122**, 401-418.

Sullivan, D., (1986): announcement at the International Congress of Mathematics, Berkeley 1986.

Swiatek, G., (1985): Endpoints of rotation intervals for maps of the circle, preprint, Univ. of Warsaw.

Szlenk, W., (1979): Some dynamical properties of certain differentiable mappings, *Bol. Soc. Mat. Mexicana*, vol 24, no 2, 57-87.

Thurston, W., (1985): On the dynamics of iterated rational maps, preprint.

Tresser, C., (1986): private communication.

Ulam, S.M. & von Neumann, J. , (1947): On combinations of stochastic and deterministic processes, *Bull. Amer. Math. Soc.*, **53**, 1120.

Walters, P. , (1975): Invariant measures and equilibrium states for some mappings which expand distances, *Trans. Amer. Soc.*, **236**, 121-153.

Yoccoz, J.C., (1984): Il n'y a pas de contre exemple de Denjoy analytique, *C. R. Acad. Sci. Paris*, **t.298**, serie I, no 7.

Young, L.S., (1970): A closing Lemma on the interval, *Invent. Math.*, **54**, 179-187.

GLOBAL BIFURCATIONS IN FLOWS

Paul Glendinning
Mathematics Institute
University of Warwick
Coventry CV4 7AL, U.K.

1. INTRODUCTION

Bifurcation theory is frequently used to understand the appearance of complicated motion in dynamical systems. As a parameter of the problem is varied, a family of differential equations may pass through a structurally unstable system at which the qualitative behaviour of the system changes: a bifurcation occurs. Usually, the effect of a bifurcation is to create or remove one or more periodic orbits from the system as, for example, in saddle-node or period-doubling bifurcations. These familiar bifurcations are examples of local bifurcations; the structurally unstable object is a non-hyperbolic stationary point (of the flow or an associated return map) and they can be investigated by a local analysis of the map near the non-hyperbolic stationary point. In this review we study some global bifurcations. Here, the structurally unstable object depends on global properties of the flow. In particular, we shall concentrate on bifurcations associated with homoclinic orbits (trajectories bi-asymptotic to the same stationary point of the flow in both forwards and backwards time, see Fig. 1). The goal of the work presented here is similar to the goal in local bifurcation theory: a description of systems near structurally unstable systems of low codimension. However, the approach below is geometric and depends on global properties of the flow, whilst the approach to local bifurcations (normal forms) is algebraic.

Consider autonomous ordinary differential equations in $\mathbb{R}^n$,

$$\frac{dx}{dt} = f(x) , \qquad (1)$$

where $x \in \mathbb{R}^n$ and $f: \mathbb{R}^n \to \mathbb{R}^n$ is a smooth (continuous with first derivatives Lipschitz continuous) vector field. We shall assume that $f(0)=0$ so that the origin, 0, is a stationary point of the flow. Suppose that the system (1) has a homoclinic orbit to 0; we want to describe the implications of such a trajectory in a flow and in perturbations (e.g. one parameter families containing this flow) of this flow. Although the existence of homoclinic orbits is extremely hard to prove in most practical applications, there is strong numerical evidence that homoclinic orbits appear in many one-parameter families of flows in which complicated motion has been observed. Rodriguez (1986) has adapted Melnikov's method (see e.g. Guckenheimer and Holmes, 1983) to autonomous ordinary differential equations, and this may allow the existence of homoclinic orbits to be proved in many more examples. The work of Coullet, Tresser and Arnéodo (1979), Gaspard, Kapral and Nicolis (1984) and Glendinning and Sparrow (1984) amongst others suggests that the theoretical analysis of these flows provides a very useful organising idea when studying examples.

Trajectories which remain in a neighbourhood of a homoclinic orbit spend much of their time near the stationary point. Thus the behaviour of solutions near the origin is important. Assuming that 0 is a hyperbolic stationary point, the Jacobian matrix $Df(0)=\partial f_i/\partial x_j(0)$ will have k_u (resp. k_s) eigenvalues λ_i, $1 \le i \le k_u$ (resp. γ_j, $1 \le j \le k_s$) with real part greater than (resp. less than) zero, where $k_u+k_s=n$. Order these so that $0<\mathrm{Re}\ \lambda_1 \le \mathrm{Re}\ \lambda_2 \le \ldots$ and $0>\mathrm{Re}\ \gamma_1 \ge \mathrm{Re}\ \gamma_2 \ge \ldots$; then almost all trajectories which approach 0 in backwards time do so tangent to the eigenvectors associated with (λ_i) such that $\mathrm{Re}\ \lambda_i=\mathrm{Re}\ \lambda_1$ and almost all those which approach 0 in forwards time do so tangent to the eigenvectors associated with (γ_j) such that $\mathrm{Re}\ \gamma_j=\mathrm{Re}\ \gamma_1$. Fig. 1 shows the three cases of homoclinic orbits when

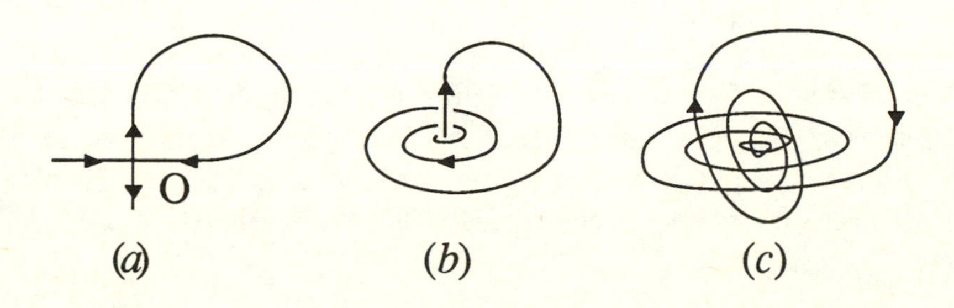

Figure 1: Homoclinic orbits bi-asymptotic to (a) a saddle; (b) a saddle-focus; (c) a bifocus.

the eigenvalues λ_1 and γ_1 are simple. These are

(a) the saddle: $\lambda_1 \in \mathbb{R}$, $\gamma_1 \in \mathbb{R}$ and $\lambda_1 < \mathrm{Re}\ \lambda_j$ for $2 \le j \le k_u$, $\gamma_1 > \mathrm{Re}\ \gamma_j$ for

$2 \leq j \leq k_s$, this can occur for systems in $\mathbb{R}^n$, $n \geq 2$;

(b) the saddle-focus: $\lambda_1 \in \mathbb{R}$, $\gamma_1 = \bar{\gamma}_2$, $\mathrm{Im}\ \gamma_1 \neq 0$ and $\lambda_1 < \mathrm{Re}\ \lambda_j$ for $2 \leq i \leq k_u$, $\mathrm{Re}\ \gamma_1 > \mathrm{Re}\ \gamma_j$ for $3 \leq j \leq k_s$, this can occur for systems in $\mathbb{R}^n$, $n \geq 3$;

(c) the bifocus: $\lambda_1 = \bar{\lambda}_2$, $\gamma_1 = \bar{\gamma}_2$, $\mathrm{Im}\ \lambda_1$, $\mathrm{Im}\ \gamma_1 \neq 0$ and $\mathrm{Re}\ \lambda_1 < \mathrm{Re}\ \lambda_j$ for $3 \leq i \leq k_u$, $\mathrm{Re}\ \gamma_1 > \mathrm{Re}\ \gamma_j$ for $3 \leq j \leq k_s$, this can occur for systems in $\mathbb{R}^n$, $n \geq 4$.

Although we shall be more interested in perturbations away from systems with homoclinic orbits, the following theorem due to Shil'nikov (1970) in a formulation by Tresser (1984a) shows how complicated the orbit structure can be in a neighbourhood of a homoclinic orbit to a saddle-focus or a bifocus.

Let $\mathbb{N}$ be the set of all strictly positive integers and $\mathbb{N}^{\mathbb{Z}}$ be the set of bi-infinite sequences $\underline{s}=(s_i)$, $i \in \mathbb{Z}$, $s_i \in \mathbb{N}$. Write $\Sigma^{*,\alpha}$ for the subset of $\mathbb{N}^{\mathbb{Z}}$ such that for each i, $s_{i+1} \geq \alpha s_i$ for some $0 < \alpha < 1$. Note that $\Sigma^{*,\alpha}$ is invariant under the usual shift operation σ, where $\sigma(\underline{s})=\underline{s}'$ with $s_i'=s_{i+1}$, so σ restricted to $\Sigma^{*,\alpha}$ is a subshift of finite type on infinitely many symbols.

Theorem 1 (Shil'nikov's Theorem)

Consider analytic ordinary differential equations in $\mathbb{R}^n$, $n \geq 3$, with a stationary point, 0, at which the characteristic equation of the linear flow has roots (γ_j, λ_i), $1 \leq j \leq k_s$, $1 \leq i \leq k_u$ and $k_s + k_u = n$, with $\mathrm{Re}\ \gamma_j < 0$ and $\mathrm{Re}\ \lambda_i > 0$. Suppose these are ordered so that $\mathrm{Re}\ \gamma_1 \geq \mathrm{Re}\ \gamma_2 > \ldots$ and $\mathrm{Re}\ \lambda_1 \leq \mathrm{Re}\ \lambda_2 < \ldots$, with $\gamma_1 = \alpha_0 + i\omega_0$, $\omega_0 \neq 0$ and $\lambda_1 = \alpha_1 + i\omega_1$. Assume that

$$\alpha_1 > -\alpha_0 > 0 \tag{2}$$

and that there is a homoclinic orbit, Γ, which tends to 0 as $t \to \pm\infty$, and is bounded away from any other stationary point and a certain matrix is non-singular. Then, in an arbitrarily small neighbourhood U of Γ, there exist trajectories in one-to-one correspondence with the shift on $\Sigma^{*,\delta}$ where $\delta = -\alpha_0/\alpha_1$.

□

Tresser (1984a) shows that this result remains true for C^1 systems with a homoclinic orbit to a saddle-focus ($\omega_1 = 0$ above). Note that these results show that there exists a countable number of periodic orbits in U (Shil'nikov, 1965, 1967).

To prove this theorem, and most of the results given here, a return map is derived on a suitable plane near the stationary point, O. This is done by splitting the flow into two parts. In a neighbourhood of O, containing the return plane, the flow is essentially linear and can be solved explicitly, giving a map from the return plane to another surface near O. Trajectories near the homoclinic orbit, Γ, are mapped by the flow from this surface back to the return plane. If Γ does not pass close to another stationary point, this global reinjection can be modelled by an affine map. The full return map is obtained by composing these two maps.

These results, and their extensions to one-parameter families, are relatively well known and we give a brief review in section 2. Recently more effort has been put into understanding codimension two bifurcations, involving a pair of homoclinic orbits, so in section 3 we describe the effect of symmetry on the codimension one bifurcations (since this introduces bifurcations involving pairs of homoclinic orbits in a natural way) and in section 4 we describe some of the codimension two bifurcations that are understood. Finally, in section 5, we argue that the techniques used to investigate homoclinic bifurcations can be used to study a variety of other problems.

2. ONE-PARAMETER FAMILIES

When studying an example derived, say, from physics on a computer there are frequently parameters in the problem. Changing a parameter may lead to qualitatively different behaviour being observed. Many such changes (bifurcations) can be explained, or at least understood, in terms of simple models such as one-dimensional maps and diffeomorphisms. So, for example, when a sequence of period-doubling bifurcations is observed numerically it is reasonable to expect a regime of complicated or chaotic behaviour. Unfortunately this is not always especially helpful, and other patterns of behaviour are often encountered. In this section we give some results which show how the existence of homoclinic orbits of various types in families of differential equations can be used as organising ideas in many problems. Like other bifurcations, a homoclinic bifurcation is a means of creating or destroying periodic orbits, and it is sometimes useful to think of a homoclinic orbit as a periodic orbit with "infinite period". To begin our account of these bifurcations we choose to ignore the complicated dynamics described in Theorem 1 and concentrate a simpler problem.

2.1 Generating a single periodic orbit.

Despite the complicated motion near a homoclinic orbit to a saddle-focus with $\delta < 1$ as described in Shil'nikov's theorem above, it is clear that nothing very untoward can happen in $\mathbb{R}^2$ and, indeed, a homoclinic bifurcation in autonomous planar systems simply creates or destroys a single periodic orbit (provided the equations are not area-preserving and there is no symmetry, see section 3). Shil'nikov (1968) gives more general conditions for this to be the case.

Recall that the origin, O, is a stationary point of the flow (1) with k_u eigenvalues (λ_i) with positive real part, and k_s eigenvalues (γ_i) with negative real part, ordered as described in the introduction to this chapter.

Theorem 2

Consider the one-parameter family of systems

$$\dot{x} = f(x;\mu), \quad x \in \mathbb{R}^n, \quad \mu \in \mathbb{R} \tag{3}$$

where $f: \mathbb{R}^n \times \mathbb{R} \to \mathbb{R}^n$ is C^1 with uniformly Lipschitz continuous partial derivatives and vanishes at the origin, O, in $\mathbb{R}^n$, which is a hyperbolic stationary point with eigenvalues as described above in some neighbourhood of $\mu=0$. Suppose that for $\mu=0$ there is a trajectory Γ bi-asymptotic to O which approaches O tangent to the eigenplanes associated with the eigenvalues of smallest absolute real parts as $t \to \pm\infty$ and that $\lambda_1 \in \mathbb{R}$ with

$$\lambda_1 < -\mathrm{Re}\ \gamma_1. \tag{4}$$

Then Γ generates a single periodic orbit in an arbitrarily small neighbourhood of Γ into either $\mu < 0$ or $\mu > 0$. This orbit is stable if $k_u = 1$.

□

The side of $\mu=0$ for which the periodic orbit exists depends upon the nature of the stationary point O and the behaviour of trajectories near Γ outside a small neighbourhood of O. Note that the theorem only describes behaviour in a small neighbourhood of the homoclinic orbit, and outside this region of phase space more complicated motion is possible (see Tresser, 1983b, for an example). Also, the period T of the periodic orbit tends to infinity as μ tends to zero from above or below like

$$T = -(c \ln|\mu|)/ \lambda_1 + O(1). \quad (5)$$

We leave the proof of these results to the reader (see Shil'nikov, 1968 or use the methods described below).

2.2 The saddle-focus

As we have seen (Theorem 1), the structure of solutions in a tubular neighbourhood of a homoclinic orbit to a saddle-focus can be exceedingly complicated. In this section we examine the way in which this complicated structure is created in typical one-parameter families, drawing on the work of Shil'nikov (1970), Arnéodo, Coullet and Tresser (1981b) and Tresser (1984a) for results on the symbolic description of orbits and Gaspard (1984a,b), Gaspard, Kapral and Nicolis (1984) and Glendinning and Sparrow (1984) for the bifurcation structure.

Consider a one-parameter family of vector fields in $\mathbb{R}^3$ for which the origin is a stationary point of saddle-focus type, and for some value of the parameter, $\mu=0$ say, there is a homoclinic orbit Γ which leaves 0 tangent to the one-dimensional unstable manifold, $W^u(0)$, and spirals in to 0 as $t \to \infty$ on the two-dimensional stable manifold, $W^s(0)$ as shown in Fig. 2b. Without loss of generality choose the parameter, μ, so that the behaviour of $W^u(0)$ is as shown in Fig. 2.

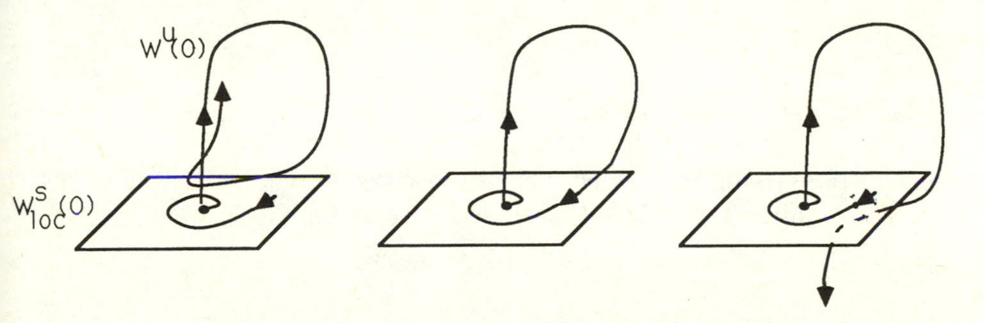

Figure 2: Behaviour of the unstable manifold of the origin as μ passes through zero. (a) $\mu > 0$; (b) $\mu=0$; (c) $\mu < 0$.

If the eigenvalues of the characteristic equation of the linearised flow at 0 are λ and $\rho \pm i\omega$ ($\lambda, -\rho, \omega > 0$) then for $\lambda < -\rho$ the conditions of Theorem 2 are satisfied and a single periodic orbit is created as μ increases through $\mu = 0$. However, if $\lambda > -\rho$, the conditions of Theorem 1 are satisfied and there are sequences of bifurcations which accumulate on $\mu = 0$. We shall concentrate on two types of bifurcation. The first concerns the bifurcations of periodic orbits which pass once through a tubular neighbourhood of Γ in each period, and the second about the existence of homoclinic orbits which pass twice through a tubular neighbourhood of Γ. These orbits are called double-pulse homoclinic orbits.

Theorem 3

Consider the one-parameter family of systems

$$\dot{x} = f(x;\mu),\ x \in \mathbb{R}^3,\ \mu \in \mathbb{R} \tag{6}$$

where $f: \mathbb{R}^3 \times \mathbb{R} \to \mathbb{R}^3$ is C^r, $r \geq 3$, and vanishes at the origin, O, in $\mathbb{R}^3$ which is a hyperbolic stationary point with eigenvalues λ and $\rho \pm i\omega$ (λ, $-\rho$, $\omega > 0$) when $\mu = 0$. Suppose that $\lambda > -\rho$ and that a homoclinic orbit Γ exists when $\mu = 0$. Then there is a natural choice of parameterisation for which the following results hold.

(i) Consider periodic orbits which pass once through a small neighbourhood of O on each period. If these orbits have period T, then there is a curve of such orbits in (μ, T) space given (asymptotically, as $|\mu| \to 0$) by

$$\mu = c\, e^{\rho T} \cos(\omega T + \Phi) \tag{7}$$

where c and Φ are constants.

(ii) There is an infinite sequence of saddle-node bifurcations, involving the periodic orbits described in (i) at parameter values (μ_n) with $\mu_{2n} > 0$, $\mu_{2n+1} < 0, \ldots$ which accumulates on $\mu = 0$ at the rate

$$\lim_{n \to \infty} (\mu_{n+1} - \mu_n)/(\mu_n - \mu_{n-1}) = \exp(\pi\rho/\omega). \tag{8}$$

(iii) There is an infinite sequence (μ_n^*) of parameter values at which there is a homoclinic orbit to O which follows a tubular neighbourhood of Γ twice before tending to O. This sequence of double-pulse homoclinic orbits accumulates on $\mu=0$ from one side at the rate

$$\lim_{n\to\infty} (\mu_{n+1}^* - \mu_n^*)/(\mu_n^* - \mu_{n-1}^*) = \exp(-\pi\lambda/\omega). \tag{9}$$

Similar results hold for systems in $\mathbb{R}^n$, with the addition of some genericity conditions. In proving this theorem, which is really just the tip of the iceberg, the role of the periodic orbit (described by (i)) which eventually forms the homoclinic orbit is crucial, and some understanding of both observable features of particular examples and the source of the recurrence properties described in Theorem 1 is obtained by considering the bifurcations which occur on branches of this orbit, defined by (7). It is also possible to determine the stability of the periodic orbits described in (i) and (ii). In the case $k_u = 1$ this leads to the existence of stable periodic orbits arbitrarily close to $\mu=0$ when $1>\delta>1/2$. For this choice of δ, one of the orbits born in a saddle-node bifurcation at a turning points of (7) is stable, whilst the other is a saddle. Suppose this bifurcation occurs in $\mu<0$. As μ increases the period of one of the orbits increases and the period of the other decreases. The stable periodic orbit loses stability by a period-doubling bifurcation becoming a saddle, then, in $\mu>0$, it regains stability by an inverse period-doubling bifurcation and is destroyed in a saddle-node bifurcation with a different saddle. The original saddle exists through $\mu=0$, and is then destroyed in a saddle-node bifurcation with another stable periodic orbit in $\mu>0$. A similar sequence of bifurcations occurs if $\delta<1/2$, except that one of the periodic orbits is unstable, whilst the other is a saddle. For more details see Glendinning and Sparrow (1984) and Gaspard, Kapral and Nicolis (1984).

2.3 Bifocal homoclinic orbits.

Due to the large numbers of examples of homoclinic orbits to a saddle-focus that have been found over the last few years and the total lack of examples of orbits bi-asymptotic to a bifocal stationary point (which can only occur in dimensions greater than or equal to four) the fact that Shil'nikov's 1970 paper applies to bifocal homoclinic orbits as well as the saddle-focus is frequently overlooked. The bifurcation results for the bifocal homoclinic orbit are complicated by resonances between ω_0 and ω_1.

Theorem 4

Consider analytic one-parameter families of ordinary differential equations in $\mathbb{R}^4$ which, after a near-identity change of coordinates and, possibly, time reversal, can be written in the form

$$\begin{aligned}\dot{x} &= \lambda_0 x - \omega_0 y + f_1(x,y,z,w;\mu)\\ \dot{y} &= \omega_0 x + \lambda_0 y + f_2(x,y,z,w;\mu)\\ \dot{z} &= \lambda_1 z - \omega_1 w + f_3(x,y,z,w;\mu)\\ \dot{w} &= \omega_1 z + \lambda_1 w + f_4(x,y,z,w;\mu)\end{aligned} \qquad (10)$$

with $\omega_i > 0$, $i=0,1$, $\lambda_1 > -\lambda_0 > 0$ and $f_i(x,y,z,w;\mu)$, $i=1,2,3,4$, vanish, together with their first derivatives, at the origin. If, for $\mu=0$, there exists a trajectory, Γ, bi-asymptotic to the origin then, letting T denote the period of a periodic orbit of the system,

(i) there is a continuous curve in (μ,T) space, $\mu=m(T)$, such that periodic orbits exist with period T at parameter values $m(T)$, this curve intersects $\mu=0$ an infinite number of times;

(ii) if $\omega_1/\omega_0 \neq 2n$, there are sequences of double-pulse homoclinic orbits at parameter values which accumulate on $\mu=0$ from both sides;

(iii) if $\omega_1/\omega_0 = 2n$, there are sequences of double-pulse homoclinic orbits at parameter values which accumulate on $\mu=0$ from one side.

□

Theorem 4(i) is proved in Fowler and Sparrow (1984) and the results for the sequences of subsidiary homoclinic bifurcations are described in Glendinning (1987). Note that by reversing the direction of time (if necessary) it is always possible to have $\lambda_1 > -\lambda_0 > 0$, so bifocal homoclinic orbits always involve complicated dynamics. As with the saddle-focus, these results almost certainly remain true for flows of this type in $\mathbb{R}^n$, $n > 4$, provided some genericity conditions are satisfied.

Thus we have seen in this chapter that the behaviour in one-parameter families of differential equations near homoclinic orbits to a stationary point which is a saddle does not generate complicated behaviour close to the homoclinic orbit, whilst if the stationary point is a saddle-focus complicated motion is generated when an inequality is satisfied between the leading eigenvalues of the characteristic equation of the linearised flow. Complicated motion is always generated in a neighbourhood of a bifocal homoclinic orbit.

3. SYMMETRY

If the system is invariant under some symmetry which maps 0 to itself, the existence of a homoclinic orbit implies the existence of other homoclinic orbits which are the image of the first under the symmetry. The symmetry

$$(x,y,z) \to (-x,-y,-z) \tag{11}$$

in $\mathbb{R}^3$ has been studied by Holmes (1980) and Glendinning (1984) for the saddle-focus and by Tresser (1983b) for the saddle. The net effect of this symmetry is that a pair of homoclinic orbits produce three periodic orbits (or curves of periodic orbits, c.f. Theorem 3): a pair of periodic orbits that each follow one of the homoclinic orbits and a symmetric periodic orbit which follows one then the other. The effect of the symmetry

$$(x,y,z) \to (-x,-y,z) \tag{12}$$

is more interesting. Note that it is impossible to have a homoclinic orbit to a saddle-focus with this symmetry for flows in $\mathbb{R}^3$, so we consider one-parameter families of the form

$$\begin{aligned} \dot{x} &= \lambda_1 x + P(x,y,z;\mu) \\ \dot{y} &= -\lambda_2 y + Q(x,y,z;\mu) \\ \dot{z} &= -\lambda_3 z + R(x,y,z;\mu) \end{aligned} \tag{13}$$

with $\lambda_i > 0$, $i=1,2,3$, $\lambda_2 > \lambda_3$, $\mu \in \mathbb{R}$ and the usual smoothness conditions on P, Q and R, which are invariant under (12). Note that this includes the Lorenz equations (Lorenz, 1963). We assume that for $\mu=0$ the system has a pair of homoclinic orbits, Γ_0 and Γ_1, which must be as shown in Fig. 3, and that

$$\lambda_3 \neq \lambda_2 - \lambda_1. \tag{14}$$

The value of $\delta = \lambda_3/\lambda_1$ is again important: both δ greater than one and δ less than one can lead to attracting chaotic motion, but in very different ways. We begin by deriving a return map T on part of the surface $S=\{(x,y,z): z=h\}$ where h is small.

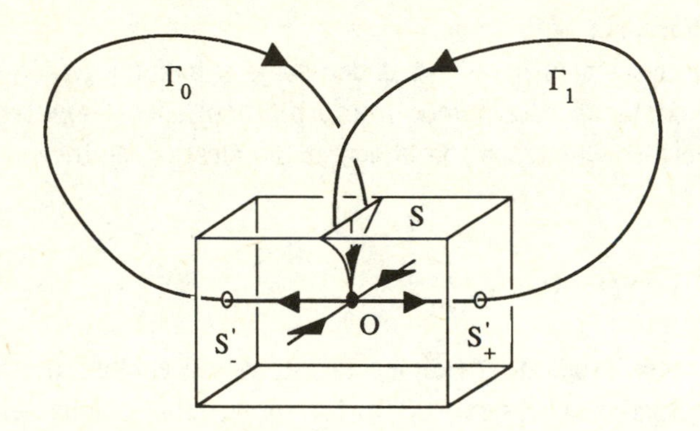

Figure 3: The homoclinic orbits and return plane for (13), invariant under (12).

By (14), a theorem of Belitskii (1978) allows us to make a smooth change of coordinates which linearises the flow in a neighbourhood, U, of O. Choose h small enough so that part of S lies in U and let $S'_\pm = \{(x,y,z): x = \pm H, H \ll 1\}$, part of which is also in U (see Fig. 3). Note that the local stable manifold of O, $W^s_{loc}(O)$, intersects S on the line x=0, and the local unstable manifold, $W^s_{loc}(O)$, intersects $S'_\pm$ at the points $(\pm H,0,0)$. The linear flow in U induces two maps $T_0^\pm: S \to S'_\pm$ where

$$T_0^+(x, y, h) = (H, y(x/h)^{\lambda_2/\lambda_1}, h(x/h)^{\lambda_3/\lambda_1}) \quad \text{for } x > 0,$$

and T_0^- for $x<0$ is given by a similar expression. We now define two maps $T_1^\pm: S'_\pm \to S$ to model the flow near the homoclinic orbits back into U. To lowest order these maps can be taken to be affine, so

$$T_1^+(H, y, z) = (p(\mu) + az + by, q(\mu) + cz + dy, h)$$
$$T_1^-(-H, y, z) = (-p(\mu) - az - by, -q(\mu) - cz - dy, h)$$

where $ad-bc \neq 0$. The positive branch of the unstable manifold of O strikes S'_+ at $(h,0,0)$ and is then mapped to $(p(\mu), q(\mu), h)$ on S, thus the condition that a homoclinic orbit exists for $\mu=0$ implies that

$$p(0) = 0, \quad q(0) \neq 0.$$

In fact, $p(\mu)$ is the natural parameterisation to use, so writing $\mu'=p(\mu)$, $\delta = \lambda_3/\lambda_1$, rescaling, dropping primes and letting $T^{\pm} = T_1^{\pm} \circ T_0^{\pm}$ gives the return map

$$\begin{aligned} T^{+}(x,y) &= (\mu + ax + byx^{\lambda_2/\lambda_1},\ q(\mu) + cx^{\delta} + byx^{\lambda_2/\lambda_1}) \qquad &\text{for } x>0 \\ T^{-}(x,y) &= -T^{+}(-x,y) \qquad &\text{for } x<0, \end{aligned}$$

which is defined for $(x, y, h) \in S \cap U$. Note that $\delta < \lambda_2/\lambda_1$, and that the map is only valid in some small neighbourhood of $(x,y)=(0,0)$, so for sufficiently small μ the x coordinate decouples and the one-dimensional map

$$f(x;\mu) = \begin{cases} \mu + ax^{\delta} & \text{if } x > 0 \\ -f(-x;\mu) & \text{if } x < 0 \end{cases} \tag{15}$$

is often used as a simple model of such flows, and we shall use this simplification below. This reduction to a one-dimensional map can be made rigorous if there is a stable foliation of the return plane. If $\delta > 1$, the existence of a stable foliation for sufficiently small $|\mu|$ is generic (Gambaudo, Glendinning, Rand and Tresser, 1986), whilst it is persistent, but not generic if $\delta < 1$ (Robinson, 1981). An example which emphasises the danger of this dimensional reduction is described in Holmes and Whitley (1984). Note also that higher order terms in x have been ignored in (15); since we are only interested in topological features of the orbits for $|x|$ and $|\mu|$ small this does not effect the results.

There are two different ways of creating complicated motion in maps such as (15) as μ varies through zero, depending upon whether δ is greater than or less than one.

3.1 Cascades of homoclinic orbits.

The case $\delta > 1$ leads to the appearance of sequences of homoclinic orbits analogous to period-doubling bifurcations in one-hump maps. Following Arnéodo, Coullet and Tresser (1981a) we shall only consider the case $a > 0$ here. A similar analysis of the other case produces similar results. Chaotic behaviour is only found outside a neighbourhood of $\mu=0$ and $x=0$ so the conclusions cannot be applied strictly to generic flows of the type being considered, but the easy observability of the predictions make them useful, and they apply rigorously to a class of semi-flows on a branched manifold (by arguments similar to those of Guckenheimer and Williams, 1979).

Assume $\delta > 1$ and let

$$g(x;\mu) = \mu + a|x|^{\delta} \tag{16}$$

which is the standard one-hump map (see e.g. Collet and Eckmann, 1980). Then

$$f(x;\mu) = \begin{cases} g(x;\mu) & \text{if } x > 0 \\ -g(x;\mu) & \text{if } x < 0 \end{cases}$$

and so, using the symmetry $x \to -x$ of the equations,

$$\begin{aligned} f^n(x;\mu) &= (-1)^k g^n(x;\mu) \\ &= [\operatorname{sgn} \partial/_{\partial x}\, g^n(x;\mu)]\, g^n(x;\mu) \end{aligned} \tag{17}$$

where k is the number of points on the orbit with $x<0$. Also note that

$$\partial/_{\partial x}\, f^n(x;\mu) = |\, \partial/_{\partial x}\, g^n(x;\mu)\, |. \tag{18}$$

These relations give a correspondence between orbits of $g(x;\mu)$ and $f(x;\mu)$. Suppose that $g(x;\mu)$ has a p-periodic orbit, then $f(x;\mu)$ has

- a pair of asymmetric p-periodic orbits which map onto each other under the symmetry if $\partial/_{\partial x}\, g^p(x;\mu) > 0$;

- a pair of homoclinic orbits if $\partial/_{\partial x}\, g^p(x;\mu) = 0$;

- a symmetric $2p$-periodic orbit if $\partial/_{\partial x}\, g^p(x;\mu) < 0$.

Thus the well-known period-doubling sequences of $g(x;\mu)$ can be reinterpreted for $f(x;\mu)$ and hence for the flows (13). Consider the first period-doubling sequence of $g(x;\mu)$, which occurs as μ decreases through zero. Let $x^*(\mu)$ be a point on the stable 2^n-cycle created by a period-doubling bifurcation when $\mu=\mu_n$. At $\mu=\mu_n$, $\partial/_{\partial x}\, g^{2^n}(x^*;\mu_n) = 1$ and decreases as μ decreases. The periodic orbit is superstable at μ'_n, where $\partial/_{\partial x}\, g^{2^n}(x^*;\mu'_n) = 0$, and at μ_{n+1} the orbit period-doubles with $\partial/_{\partial x}\, g^{2^n}(x^*;\mu_{n+1}) = -1$ creating a stable 2^{n+1}-cycle. This sequence of bifurcations is repeated, accumulating at some parameter value μ_∞ after which there are windows of stable cycles and a set of parameter values with chaotic motion.

For $f(x;\mu)$, μ_n corresponds to the birth of a pair of asymmetric 2^n-cycles. As μ decreases towards μ'_n these approach $x=0$ and at μ'_n there is a pair of homoclinic orbits. When $\mu'_n > \mu > \mu_{n+1}$ there is a stable symmetric orbit of period 2^{n+1} which loses stability at $\mu=\mu_{n+1}$ in a symmetry-breaking bifurcation ($\partial/_{\partial x}\, f^{2^{n+1}}(x^*;\mu_{n+1}) = 1$) producing two stable asymmetric 2^{n+1}-cycles. As μ decreases, more and more complicated periodic orbits are created by homoclinic bifurcations which accumulate at μ_∞. For $\mu < \mu_\infty$ it is possible to find chaotic motion at some parameter values.

These results only depend on topological properties of the maps $g(x;\mu)$. To obtain quantitative results for this cascade of bifurcations assume that δ is constant over the range of parameter values $\mu=[0,\mu_\infty]$ (this is often not true in applications). Let $\Phi(x)$ be the fixed point of the renormalisation operator $\mathfrak{R}g(x) = -{}^1/_\lambda\, g\circ g(-\lambda x)$ on $[-1,1]$, where $\lambda=-\Phi(1)$. Then (Collet, Coullet and Tresser, 1985) $\Psi(x) = -(\mathrm{sgn}\, x)\Phi(x)$ is a fixed point of the operator $\mathfrak{N}$ defined by

$$\mathfrak{N}f(x) = {}^1/_\lambda\, f\circ f(\lambda x) \tag{19}$$

for maps like $f(x;\mu)$. Thus all the results of Feigenbaum (1978) and Collet, Coullet and Tresser (1985) hold. In particular, the sequence of homoclinic orbits described above accumulate to μ_∞ geometrically at the rate

$$\lim_{n\to\infty} (\mu_{n+1} - \mu_n)/(\mu_n - \mu_{n-1}) = D(\delta) \tag{20}$$

where $D(\delta)$ is the usual positive eigenvalue of the doubling operator, $\mathfrak{R}$, $D(1+\varepsilon)=2+O(\varepsilon \ln\varepsilon)$ for ε small (Collet, Eckmann and Lanford, 1980). Collet,

Coullet and Tresser (1984) also derive a cross-over exponent for asymmetric perturbations of (15) and scaling behaviour for the topological entropy. From the standard symbolic description of orbits of one-hump maps (Collet and Eckmann, 1980), using 1 to denote points in $x > 0$ and 0 to denote points in $x < 0$, it is possible to derive the symbolic description of the homoclinic orbits. For the flow the symbol i, i=0,1, denotes a passage through a tubular neighbourhood of of the homoclinic orbit Γ_i. Table 1 shows the sequences for the first doubling (or homoclinic) cascade. Procaccia, Thomae and Tresser (1986) have generalised these sequences for the non-symmetric case.

TABLE I

Symbol sequences for the period-doubling sequence of $g(x;\mu)$ and the homoclinic cascades of $f(x;\mu)$ for the cases $a > 0$ and $a < 0$. If W is a finite sequence of 1s and 0s then $0*W=WAW$, where A=1 (resp. A=0) if the number of 0s in W is odd (resp. even). Similarly, $1\times W=W0W'$, where W' is obtained from W by exchanging 1s with 0s. C denotes the point $x=0$ for the map $g(x;\mu)$.

μ	$g(x;\mu)$	$f(x;\mu)$	
		$a>0$	$a<0$
μ_1	C0	10	11
μ_2	C010	1001	1100
μ_3	$C010^3 10$	$10^2 101^2 0$	$1^2 0^4 1^2$
μ_4	$C010^3 101010^3 10$	$10^2 101^2 0^2 1^2 010^2 1$	$1^2 0^4 1^2 0^2 1^4 0^2$
μ_n	$C(0*)^{n-1} 0$	$1(1\times)^{n-1} 0$	$1(1\times)^{n-1} 1$

Table 1 allows sequences of homoclinic orbits to be recognised as either orientable or non-orientable when found in examples (see Arnéodo, Coullet and Tresser, 1981a).

Problem: Prove the existence of a strange attractor for some flow by constructing a flow for which $f(x;\mu)$, with μ chosen so that $f(x;\mu)$ is chaotic, is a rigorous reduction of the return map.

Problem: Arnéodo, Coullet and Tresser (1981a) and Lyubimov and Zaks (1983) give explicit examples of systems which appear to behave as described above for $a>0$. Find an example with a cascade of homoclinic orbits satisfying $a<0$.

3.2 The Lorenz equations.

The second case, $\delta<1$, has been studied by many authors: a strange invariant set is created on one side of the homoclinic bifurcation. This situation has become so well-understood because of its applicability to the Lorenz equations (Lorenz, 1963), which are a set of o.d.e.s in $\mathbb{R}^3$, usually written as

$$\begin{aligned}\dot{x} &= \sigma(y-x)\\ \dot{y} &= rx-y-xz\\ \dot{z} &= -bz+xy\end{aligned} \tag{21}$$

where $\sigma=10$, $b=8/3$ and $0\leq r<\infty$. This was one of the first explicit examples of a system with complicated or chaotic behaviour. Lorenz (1963) showed that when $r=28$ trajectories are apparently chaotic. It was later shown (Kaplan and Yorke, 1979, Sparrow, 1982) that the origin of this behaviour is a pair of homoclinic orbits biasymptotic to 0 which are formed when $r\approx 13.926$. Over the range of parameters of interest the flow takes the form (13) of this section but with $\delta<1$, and the equations are invariant under the symmetry (12). Hence the return map, (15), can be used to investigate trajectories near the homoclinic orbits in a neighbourhood of $r\approx 13.926$.

Theorem 5

Let

$$\dot{x}=f(x;\mu),\ x\in\mathbb{R}^3,\ \mu\in\mathbb{R}$$

be invariant under the symmetry (12), with $f:\mathbb{R}^3\times\mathbb{R}\to\mathbb{R}^3$ C^1 with uniformly Lipschitz continuous partial derivatives and vanishing at the origin, 0, in $\mathbb{R}^3$, which is a hyperbolic stationary point with eigenvalues $(\lambda_1,\lambda_2,\lambda_3)$, $-\lambda_2>\lambda_1>-\lambda_3>0$, $\lambda_3\neq\lambda_2+\lambda_1$. Suppose that for $\mu=0$ there is a pair of homoclinic orbits, Γ_0 and Γ_1, bi-asymptotic to 0, which are mapped onto each other by the symmetry. Then, generically, for some small neighbourhood of $\mu=0$ in parameter space, on one side of the bifurcation ($\mu<0$, say) no trajectories remain in a small neighbourhood, V, of Γ_0 and Γ_1 for all time, whilst on the other side ($\mu>0$, say), the invariant set in V is homeomorphic to the full shift on two symbols.

□

The proof of this result can be found in Sparrow (1982) or Guckenheimer and Holmes (1984). For larger μ, outside a neighbourhood of $\mu=0$ for which this analysis is valid, this set can become attracting in a region of parameter space where there is a dense set of homoclinic bifurcations whose effect is to remove strange invariant sets with large numbers of consecutive 1s or 0s in their symbolic description. Lyubimov and Zaks (1983) give some scaling results for some of these bifurcations under the assumption that (15) remains a reasonable model return map for the flow. Sparrow (1982) gives an excellent account of what actually happens in the Lorenz equations and Afraimovich and Shil'nikov (1983) describe the mathematical structure of the attracting sets of such maps in a more general context.

4. CODIMENSION TWO BIFURCATIONS

In the absence of symmetry or other constraints the existence of a pair of homoclinic orbits is (at least) of codimension two. Intuitively one parameter is needed to control the reinjection of each of two branches of the unstable manifold of the stationary point. The description of codimension two homoclinic bifurcations is by no means complete, but we begin this account with a fairly general description of the gluing bifurcation, which is the codimension two result analogous to Theorem 2.

4.1 The gluing bifurcation

In the symmetric configuration with δ greater than one discussed in the previous section two periodic orbits approach a stationary point as a parameter is varied until, at some parameter value, there is a pair of homoclinic orbits which break to form a single symmetric periodic orbit. The two initial periodic orbits can be thought of as having been "glued together" at the stationary point to form the symmetric periodic orbit. The gluing bifurcation is the name used to describe a generalisation of this situation (Coullet, Gambaudo and Tresser, 1984, Gambaudo, Glendinning and Tresser, 1985a,b). Consider systems of the form

$$\dot{x}=f(x), \quad x\in\mathbb{R}^n$$

where $f:\mathbb{R}^n\to\mathbb{R}^n$ is C^1 with uniformly Lipschitz continuous partial derivatives and vanishes at the origin 0, which is a hyperbolic stationary point. Suppose that

(A1) The characteristic equation of $Df(0)$ has eigenvalues (λ_i) $i=1,\dots,n$ with

$$0 < \lambda_1 < -\mu \tag{22}$$

where $\mu = \max\{\mathrm{Re}\,\lambda_2, \dots, \mathrm{Re}\,\lambda_n\}$;

(A2) For all λ_i, λ_j we have $\mathrm{Re}\,\lambda_j \neq \lambda_1 + \mathrm{Re}\,\lambda_i$;

(A3) There are two homoclinic orbits, Γ_0 and Γ_1, bi-asymptotic to 0 and bounded away from any other stationary point.

Systems satisfying (A1)-(A3) will be called critical, and our aim is to characterise the behaviour of systems which are sufficiently smooth perturbations of critical systems. (A1) is similar to the conditions applied in Theorem 2, where only one periodic orbit can be generated, and (A2) is a resonance condition needed for the linearisation theorems used in the proof. The periodic orbits which may arise can be described in a simple way by exploiting an analogy with rotations.

Let S denote the circle, a rotation $R_a: S\to S$ can be written as

$$R_a(x) = x + a \pmod 1$$

for $a\in[0,1)$. Thus for any point $x\in[0,1]$ we can associate a sequence $I_a(x)$ of "0"s and "1"s such that $I_a(x)=(s_n)$, $n\in\mathbb{N}$ where $s_n=1$ if $R_a^n(x)\geq 1-a$ and $s_n=0$ if $R_a^n(x)< 1-a$. Sequences which can be constructed in this way are called rotation compatible. There are many ways of characterising rotation compatible sequences (Gambaudo, Lanford and Tresser, 1984, Series, 1985), but for our purposes the equivalence given below, which uses a renormalisation type approach, is most useful.

Lemma

A sequence $X \in \{0,1\}^{\mathbb{N}}$ is rotation compatible for a rotation R_a if and only if

(i) by cutting X after each 0 we obtain a sequence of blocks of length n and $n+1$, with $1-a = 1/(n+b)$ for some $b \in [0,1)$, except perhaps the first block which may be smaller;

(ii) the new sequence Y obtained by replacing blocks of length $n+1$ by "1" and n by "0" in X (omitting the first block) is rotation compatible for the rotation R_b.

□

The number $a \in [0,1)$ is called the rotation number of the rotation R_a, and we shall talk of sequences which are rotation compatible with rotation number a if this needs to be specified. The rotation number gives the proportion of 1s in the sequence. Note that all these ideas can be extended to more general maps of the circle. Finally, we say that two rational numbers, p/q and r/s are Farey neighbours if $|ps - qr| = 1$.

We are now in a position to state a theorem due to Gambaudo, Glendinning and Tresser (1985a) which is the analogue of Theorem 2 for two homoclinic loops:

Theorem 6

For every sufficiently small, smooth perturbation of a critical system there exists a neighbourhood of the homoclinic orbits in phase space such that

(i) there are zero, one or two closed invariant curves (not counting the origin); if any of these is a periodic orbit then it is attracting;

(ii) a rotation compatible code (which determines a unique closed curve) can be associated with each closed invariant curve;

(iii) if there are two closed invariant curves then their associated rotation numbers are Farey neighbours.

Problem: Generalise Theorem 6 to flows in $\mathbb{R}^n$ with $m\ (>1)$ unstable directions.

Theorem 6 limits the complexity of behaviour near critical systems, but given more information about the flow near 0 we can find selection rules for which rotation compatible sequences exist. To illustrate how these selection rules work we shall look at one example; the full list of possibilities is described in Gambaudo, Glendinning, Rand and Tresser (1986).

4.2 A particular example.

Suppose $\dot{x}=F(x)$ is a critical system and $\dot{x}=f(x;r_0,r_1)$ a continuous two parameter family such that $f(x;0,0)=F(x)$. As we observed in the previous sub-section, the first intersections of the two branches of $W^u(O)$ with the cylinder C in a neighbourhood of O are on the hyperplane $x_1=0$ (where x_1 is the eigenvector associated with the positive eigenvalue, λ, in (22)), thus forming the homoclinic orbits Γ_0 and Γ_1. For (r_0,r_1) sufficiently small these manifolds, $W^{u,0}(O)$ and $W^{u,1}(O)$ say, will strike C with $x = \mu_0(r_0,r_1)$ and $x= -\mu_1(r_0,r_1)$ respectively. Assuming that these relations are locally invertible, these parameters provide a geometrically appealing way of describing the two-parameter bifurcation diagram, and we use them from here onwards. Note that when $\mu_i=0$, there is a homoclinic orbit that follows a tubular neighbourhood of Γ_i once, $i=0,1$. Some features of the bifurcation diagram are the same in all the cases (Coullet, Gambaudo and Tresser, 1984).

Proposition 1

For (μ_0,μ_1) sufficiently small and for a small neighbourhood V of $\Gamma_0 \cup \Gamma_1$ the parameter space can be decomposed into six regions (see Fig. 4):

<u>Region 1</u>: there exist two stable periodic orbits in V with codes 0 and 1;
<u>Region 2 (resp. 3)</u>: there is a single stable periodic orbit in V with code 1 (resp. 0);
<u>Region 4</u>: there is a single stable periodic orbit in V with code 01;
<u>Region 5</u>: is defined to lowest order by $-c_1\mu_1^\delta < \mu_0 < c_2\mu_1^\delta$, $\mu_1>0$;
<u>Region 6</u>: is defined to lowest order by $-c_3\mu_0^\delta < \mu_1 < c_4\mu_0^\delta$, $\mu_0>0$;
with $c_i>0$, $i=1,2,3,4$, $\delta = -\lambda_3/\lambda_1>1$ (see the beginning of section 3.1 for definitions). Nothing can be said about the dynamics in regions 5 and 6 without more details of the system under consideration.

□

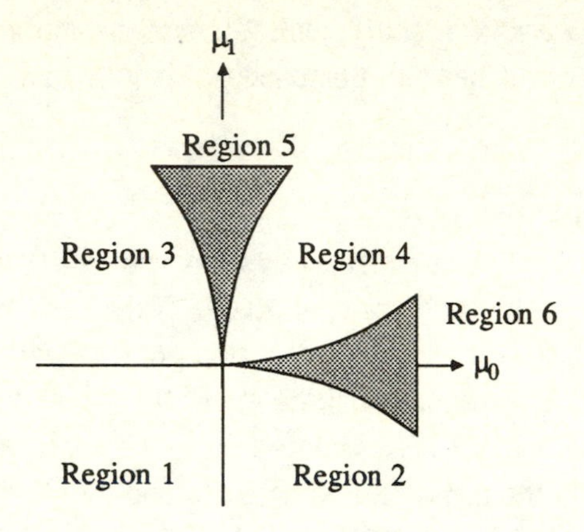

Figure 4: Parameter space near a codimension two gluing bifurcation (see Proposition 1).

The different possible selection rules for periodic orbits in regions 5 and 6 are described in Gambaudo, Glendinning and Tresser (1984,1985b) and Gambaudo, Glendinning, Rand and Tresser (1986). Here we shall consider the periodic orbits in a neighbourhood of critical systems in $\mathbb{R}^3$ which can be written in the form (13) with $0<\lambda_1<\lambda_3<\lambda_2$. There are two different configurations possible for the pair of homoclinic orbits depending on whether they approach O along the same branch, or different branches of the z-axis; these are called the <u>butterfly configuration</u> and the <u>figure eight</u> respectively. Here we shall concentrate on the same configuration that we described in the previous section on symmetric systems: the butterfly. In Gambaudo, Glendinning, Rand and Tresser (1986) we show that there is a stable foliation generically for these cases, and hence that the behaviour in a small neighbourhood of a critical flow can be analysed via one-dimensional maps as we did, with less justification, in section 3.

By a slight modification of the derivation in section 3.1 the one-dimensional return map can be written as

$$x' = \begin{cases} -\mu_1 + ax^{\delta} + \text{h.o.t. for } x > 0 \\ \mu_0 - b|x|^{\delta} + \text{h.o.t. for } x < 0 \end{cases} \tag{23}$$

where a and b are positive constants. From these equations it is easy to see that in the part of region 5 with $\mu_0<0$ and region 6 with $\mu_1<0$ the behaviour of systems is as in region 3 and 2 respectively (see Fig. 4). However, when μ_0 and μ_1 are both positive the situation is considerably more complicated:

Proposition 2

For a sufficiently small neighbourhood of $(\mu_0,\mu_1)=(0,0)$ in parameter space and $\Gamma_0 \cup \Gamma_1$ in phase space with Γ_0 and Γ_1 in the butterfly configuration

(i) there is at most one periodic orbit;

(ii) the rotation number of periodic orbits varies continuously and monotonically with one parameter when the other is held fixed.

Note that statement (ii) implies that there are parameter values in a neighbourhood of $(0,0)$ with attracting sets which have rotation compatible codes with irrational rotation numbers; i.e. there are aperiodic orbits that are stable and not chaotic. From the geometry of the flow it is clear that these orbits lie on a torus with a hole: this is precisely the property of Cherry flows (see Palis and de Melo, 1984).

Proof of Proposition 2

Given the remarks made before the statement of the proposition, we will only consider the one-dimensional model (23) for $\mu_i > 0$. By the change of coordinates $z = (\mu_1 + x)/(\mu_0 + \mu_1)$ we obtain a map $F(z;\mu_0,\mu_1)$ such that all the recurrent dynamics lies in $z \in [0,1]$ and with the properties that

- $F(z;\mu_0,\mu_1)$ is piecewise increasing with a single discontinuity at $a = \mu_1/(\mu_0 + \mu_1)$: $F(a^+;\mu_0,\mu_1) = 0$ and $F(a^-;\mu_0,\mu_1)=1$;
- $F(z;\mu_0,\mu_1)$ is injective;
- labelling orbits of $F(z;\mu_0,\mu_1)$ with 0 if $x \in [0,a)$ and 1 if $x \in [a,1)$ yields the same code as for the corresponding orbit of the flow.

Thus $F(z;\mu_0,\mu_1)$ can be viewed as a discontinuous map of the circle to itself (the discontinuity being at 0, not a) and we can associate a lift $\mathcal{F}(\cdot;\mu_0,\mu_1)$ with F and so define a rotation number in the usual way. Thus we have, from Theorem 1 of Gambaudo and Tresser (1985) that for each $x \in [0,1)$ there is a unique rotation number, $\rho(\mu_0,\mu_1)$, which is independent of x. This follows from the fact that F is monotonic and injective. Furthermore, for every such $\rho(\mu_0,\mu_1)$, there is a rotation compatible orbit of $F(z,\mu_0,\mu_1)$ with that rotation number.

Gambaudo and Tresser (1985) also show that if the lift of a map depending on a single parameter is monotonic in the parameter, then the rotation number increases continuously with the parameter. This remark completes the proof.

□

Outside the region of validity of this local analysis, the appearance of chaotic behaviour is possible. Tresser (1983) extends the analysis of one-dimensional maps like $F(z;\mu_0,\mu_1)$ above to show how chaotic motion arises.

The codimension two unfolding of critical systems involving a saddle-focus is also very pretty (see Gambaudo, Glendinning and Tresser, 1984).

4.3 Lorenz-like equations.

The transition to complicated dynamics in the return map of the symmetric Lorenz equations is abrupt (Theorem 5), but the codimension two unfolding of this situation is more gentle. When two parameters can be varied to control the two branches of the unstable manifold of the origin independently, the strange invariant set described in section 3.2 may be created by infinitely many sequences of homoclinic bifurcations instead of all at once.

Consider a two parameter family of o.d.e.s in $\mathbb{R}^3$ such that for $(\mu_0,\mu_1)=(0,0)$ there is a pair of homoclinic orbits, Γ_0 and Γ_1, biasymptotic to the origin 0 in the butterfly configuration. Suppose that the eigenvalues of the characteristic equation of the linearised flow at 0 are $(\lambda_1,-\lambda_2,-\lambda_3)$ with $0<\lambda_3<\lambda_1<\lambda_2$, so we have a similar configuration to that described in section 3.2 but without the symmetry. As in section 4.2 the full return map can be reduced to the study of a two-parameter family of one dimensional maps (23) with $\delta = \lambda_3/\lambda_1<1$. This reduction is only rigorously valid if there is a stable foliation (Robinson, 1981) which, although persistent, is less likely than in the case $\delta>1$ described above. Despite these caveats we shall work with this simple model rather than the full two-dimensional return map. (Afraimovich and Shil'nikov (1983) use a similar formalism to the general description of the gluing bifurcation to get around this problem.) In a neighbourhood of $(\mu_0,\mu_1)=(0,0)$ and $\Gamma_0 \cup \Gamma_1$ there is no attracting behaviour: almost all trajectories eventually leave a tubular neighbourhood of $\Gamma_0 \cup \Gamma_1$. The parameter space can be divided into eight regions as shown in Fig. 5. Using the standard symbolic description of

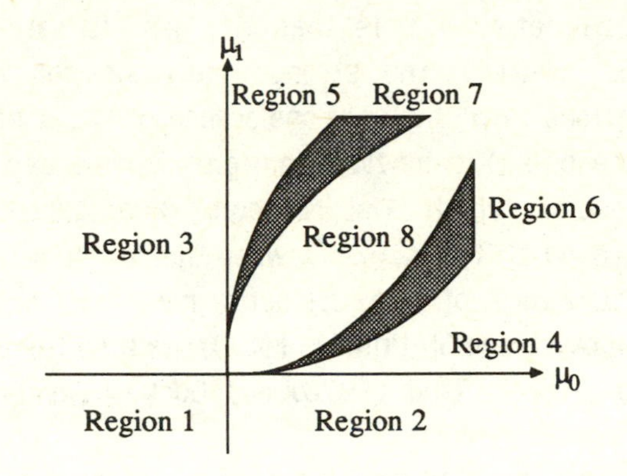

Figure 5: Parameter space for Lorenz-like equations (see Proposition 3).

0s and 1s, and denoting the codes of the two branches of the unstable manifold of 0 by k_+ and k_- (referred to as the kneading sequences of (23)) we have

Proposition 3

For (μ_0,μ_1) sufficiently small and for a small neighbourhood V of $\Gamma_0 \cup \Gamma_1$ the parameter space can be decomposed into eight regions (see Fig. 5):

Region 1: there are no periodic orbits in V, $k_+ = 111\ldots$ and $k_- = 000\ldots$;
Region 2: there is a single non-stable periodic orbit with code 0 in V, $k_+ = 111\ldots$ and $k_- = 0111\ldots$;
Region 3: as region 2 with 0s and 1s and + and - exchanged; Region 4: there are two non-stable periodic orbits, with codes 0 and 1 in V, $k_+ = 10111\ldots$ and $k_- = 0111\ldots$;
Region 5: as region 4 with 0s and 1s and + and - exchanged;
Region 6: there is an infinite sequence of homoclinic bifurcations with k_+ varying monotonically, $k_- = 0111\ldots$ and the topological entropy increases from zero to $\ln(2)$ passing through the region;
Region 7: as region 6 with 0s and 1s and + and - exchanged;

<u>Region 8:</u> the invariant set is homeomorphic to a full shift on two symbols, $k_+ = 1000...$ and $k_- = 0111....$

□

Afraimovich and Shil'nikov (1983) and Glendinning (1986) give more details of these bifurcations. Note that in a generic path into region 8 which avoids $(\mu_0,\mu_1)=(0,0)$, the strange invariant set is created by homoclinic bifurcations involving only one branch of the unstable manifold of the origin: k_+ (resp. k_-) is constant in region 7 (resp. region 6). Part of the strange invariant set which exists in region 8 can become an attractor outside a neighbourhood of $(\mu_0,\mu_1)=(0,0)$, when the curves bounding region 8 intersect. The structure of the attractor here has been beautifully described by Afraimovich and Shil'nikov (1983), see also the work on Lorenz maps of Williams (1979), Rand (1978) and Guckenheimer and Williams (1979).

Some other possible codimension two homoclinic and heteroclinic bifurcations which have been discussed in the literature (see e.g. Broer and Vegter, 1984, Glendinning and Sparrow, 1986, and Gaspard, Kapral and Nicolis, 1984), and many others which may be interesting await analysis.

Problem: Can anything interesting be said for the codimension two unfolding of a pair of homoclinic orbits bi-asymptotic to a saddle-focus with $\delta<1$?

5. OTHER GLOBAL MECHANISMS

The general idea used in this chapter can been applied to many other geometric situations. All that is needed is some "privileged" trajectory or set of trajectories near which the dynamics is well-understood, and some reinjection principle by which trajectories leaving the privileged set are taken back to it. As an example of this more general approach to global bifurcations we shall consider saddle node reinjections. Once again, the analysis of this mechanism is due to Shil'nikov (1969).

Consider systems which, after a linear change of coordinates, can be written in the form

$$\begin{aligned}\dot{x} &= \lambda x + P(x,y,z;\mu)\\ \dot{y} &= -\gamma y + Q(x,y,z;\mu) \qquad (25)\\ \dot{z} &= \mu + az^2 + R(x,y,z;\mu)\end{aligned}$$

where $\mu \in \mathbb{R}$, P, Q and R are sufficiently smooth and vanish with their partial derivatives at the origin O and the second partial derivative of R with respect to z also vanishes at the origin. This is the canonical form for a saddle-node bifurcation: there are no stationary points in a small neighbourhood of O for $\mu > 0$ and two for $\mu < 0$. Shil'nikov (1969) proves

Theorem 7

Given a one-parameter family of systems satisfying the conditions above and such that for $\mu=0$ there are p homoclinic orbits, Γ_i, $i=1,2,\ldots,p$, which tend to O tangent to the z axis as $t \to \pm\infty$, then the dynamics of the flow in the union of sufficiently small neighbourhoods of the homoclinic orbits is homeomorphic to the full shift on p symbols for $\mu \geq 0$, sufficiently small.

Shil'nikov's statement of the theorem is in fact more general: in (25) x and y can be taken as elements of $\mathbb{R}^n$ and $\mathbb{R}^m$ respectively, with λ (resp. $-\gamma$) an $n \times n$ (resp. $m \times m$) matrix with eigenvalues with positive (resp. negative) real parts.

Problem: Find a simple example of such a one-parameter family.

The proof of theorem 7 relies on the same techniques as used in section 3.1.

6. CONCLUSION

Theorem 7 demonstrates that the hyperbolicity of the stationary point is unnecessary for the general method used here to be useful. It seems likely that this geometric approach can be used in a variety of problems which do not involve stationary points of flows. For example, this technique has been applied to homoclinic orbits of diffeomorphisms (and hence trajectories biasymptotic to periodic orbits of flows) with interesting results (e.g. Newhouse, 1979) and may well prove useful in understanding some results in boundary layer theory, such as the Falkner-Skan equations (Hastings and Troy, 1985). Applications of these results to partial differential equations are also interesting. Moore et al. (1983) have shown numerically that bifurcations similar to those described in Theorem 3 occur in partial differential equations, and it would certainly be interesting to see how far these low dimensional results extend to infinite dimensional systems.

In a general article such as this it is impossible to give a complete historical account of the development of ideas in this subject, nor is it possible to treat all the results with the thoroughness they deserve. However, I hope that this chapter gives some indication of the way geometric ideas can be used in global bifurcation theory, and that the reader can take these ideas further by consulting the references. I am grateful to the SERC for support and to King's College, Cambridge for its hospitality. Finally I would like to thank J.M. Gambaudo, C. Sparrow, C. Tresser and N.O. Weiss for many stimulating conversations, during which I have learnt a great deal.

References

Afraimovich V.S. and L.P. Shil'nikov (1983) "Strange attractors and quasiattractors" in *Nonlinear dynamics and Turbulence* eds. G.I.Barenblatt, G.Iooss and D.D.Joseph, Pitman

Arnéodo A., P.Coullet and C.Tresser (1981a) "A possible new mechanism for the onset of turbulence" *Phys. Lett.* **81A** 197-201

Arnéodo A., P.Coullet and C.Tresser (1981b) "Possible new strange attractors with spiral structure" *Comm. Math. Phys.* **79** 573-579

Belitskii G.R. (1978) "Equivalence and normal forms of germs of smooth mappings" *Russ. Math. Surv.* **33** 107-177

Broer H.W. and G.Vegter (1984) "Subordinate Sil'nikov bifurcations near some singularities of vector fields having low codimension" *Ergod. Thy. and Dynam. Sys.* **4** 509-525

Collet P., P.Coullet and C.Tresser (1985) "Scenarios under constraint" *J. de Phys. Lettres* **46** 143-147

Collet P. and J.P.Eckmann (1980) *Iterated maps of the interval as dynamical systems* Birkhauser, Boston

Collet P., J.P.Eckmann and O.Lanford (1980) "Universal properties of maps on an interval" *Comm. Math. Phys.* **76** 211-254

Coullet P., Gambaudo J.M. and C.Tresser (1984) "Une nouvelle bifurcation de codimension 2: le collage de cycles" *C. R. Acad. Sci. (Paris) Serie I* **299** 253-256

Coullet P., C.Tresser and A.Arnéodo (1979) "Transition to stochasticity for a class of forced oscillators" *Phys. Lett* **72A** 268-270

Feigenbaum M.J. (1978) "Quantitative universality for a class of nonlinear transformations" *J. Statist. Phys.* **19** 1-25

Fowler A.C. and C.Sparrow (1984) "Bifocal homoclinic orbits in four dimensions" *Preprint*, University of Oxford

Gambaudo J.M., P.Glendinning and C.Tresser (1984) "Collage de cycles et suites de Farey" *C. R. Acad. Sci. (Paris) Serie I* **299** 711-714

Gambaudo J.M., P.Glendinning and C.Tresser (1985a) "The gluing bifurcation I: symbolic dynamics of closed curves" submitted to *Comm. Math. Phys.*

Gambaudo J.M., P.Glendinning and C.Tresser (1985b) "Stable cycles with complicated structure" *J. de Phys. Lettres* **46** L653-L658

Gambaudo J.M., P.Glendinning, D.A.Rand and C.Tresser (1986) "The gluing bifurcation II: stable foliations and periodic orbits" *Preprint*

Gambaudo J.M., O.Lanford and C.Tresser (1984) "Dynamique symbolique des rotations" *C. R. Acad. Sci. (Paris) Serie I* **299** 823-826

Gambaudo J.M. and C.Tresser (1985) "Dynamique régulière ou chaotique. Applications du cercle ou de l'intervalle ayant une discontinuité" *C. R. Acad. Sci. (Paris) Serie I* **300** 311-313

Gambaudo J.M. and C.Tresser (1987) "On the dynamics of quasi-contractions" *Preprint*, Université de Nice

Gaspard P. (1984a) "Generation of a countable set of homoclinic flows through bifurcation" *Phys. Lett.* **97A** 1-4

Gaspard P. (1984b) "Generation of a countable set of homoclinic flows through bifurcation in multidimensional systems' *Bull. Class. Sci. Acad. Roy. Belg. Serie 5* **LXX** 61-83

Gaspard P., R.Kapral and G.Nicolis (1984) "Bifurcation phenomena near homoclinic systems: a two parameter analysis" *J. Statist. Phys.* **35** 697-727

Glendinning P. (1984) "Bifurcations near homoclinic orbits with symmetry" *Phys. Lett.* **103A** 163-166

Glendinning P. (1986) "Asymmetric perturbations of Lorenz-like equations" to appear, *Dynamics and Stability of Systems*

Glendinning P. (1987) "Subsidiary bifurcations near bifocal homoclinic orbits", submitted to *Math. Proc. Cambridge Philos. Soc.*

Glendinning P. and C.Sparrow (1984) "Local and global behaviour near homoclinic orbits" *J. Statist. Phys.* **35** 645-696

Glendinning P. and C.Sparrow (1986) "T points: a codimension two heteroclinic bifurcation" *J. Stat. Phys.* **43** 479-488

Guckenheimer J., and P.Holmes (1983) *Nonlinear oscillations, dynamical systems and bifurcations of vector fields* Appl. Math.Sci. 42, Springer

Guckenheimer J. and R.F.Williams (1979) "Structural stability of Lorenz attractors" *Inst. Hautes Etudes Sci. Publ. Math.* **50** 307-320

Hastings S.P. and W.Troy (1985) "Oscillatory solutions of the Falkner-Skan equation" *Proc. Roy. Soc. London* **397A** 415-418

Holmes P. (1980) "A strange family of three-dimensional vector fields near a degenerate singularity" *J. Differential Equations* **37** 382-403

Holmes P. and D.Whitley (1984) "Bifurcations of one- and two-dimensional maps" *Phil. Trans. Roy. Soc. London* **311A** 43-102

Kaplan J.L. and J.A.Yorke (1979) "Preturbulence: a regime observed in a fluid flow model of Lorenz" *Comm. Math. Phys.* **67** 93-108

Lorenz E.N. (1963) "Deterministic nonperiodic flows" *J. Atmospheric. Sci.* **20** 130-141

Lyubimov Y. and M.A.Zaks (1983) "Two mechanisms of the transition to chaos in finite-dimensional models of convection" *Physica* **9D** 52-64

Moore D.R., J.Toomre, E.Knobloch and N.O.Weiss (1983) "Period-doubling and chaos in partial differential equations for thermosolutal convection" *Nature* **303** 663-667

Newhouse S. (1979) "The abundance of wild hyperbolic sets" *Inst. Hautes Etudes Sci. Publ. Math.* **50** 101-151

Palis J. and W. de Melo (1982) *Geometric theory of dynamical systems* Springer

Procaccia I., S.Thomae and C.Tresser (1986) "First return maps as unified renormalisation scheme for dynamical systems" *Preprint*, Weizmann Institute, Israel

Rand D. (1978) "The topological classification of Lorenz attractors" *Math. Proc. Camb. Phil. Soc.* **83** 451-460

Robinson C. (1981) "Differentiability of the stable foliation for the model Lorenz equations" in *Dynamical systems and turbulence: Warwick 1980* eds. D.A.Rand and L.S.Young, LNM 898, Springer

Rodriguez J.A. (1986) "Bifurcations to homoclinic connections of the focus-saddle type" *Arch. Rational Mech. Anal.* **93** 81-90

Series C. (1985) "Geometry of Markov numbers" *Math. Intelligencer* **7** 20-29

Shil'nikov L.P. (1965) "A case of the existence of a denumerable set of periodic motions" *Soviet Math. Dokl.* **6** 163-166

Shil'nikov L.P. (1967) "Existence of a countable set of periodic motions in a four-dimensional space in an extended neighbourhood of a saddle-focus" *Soviet Math. Dokl.* **8** 54-58

Shil'nikov L.P. (1968) "On the generation of a periodic motion from trajectories doubly asymptotic to an equilibrium state of saddle type" *Math. USSR Sb.* **6** 427-438

Shil'nikov L.P. (1969) "On a new type of bifurcation of multidimensional dynamical systems" *Soviet Math. Dokl.* **10** 1368-1371

Shil'nikov L.P. (1970) "A contribution to the problem of the structure of an extended neighbourhood of a rough equilibrium of saddle-focus type" *Math. USSR Sb.* **10** 91-102

Sparrow C. (1982) *The Lorenz equations: bifurcations, chaos and strange attractors* Appl.Math.Sci. 41, Springer

Tresser C. (1981) "Modéles simples de transitions vers la turbulence" Thèse d'Etat, Université de Nice

Tresser C. (1983) "Nouveaux types de transitions vers une entropie topologique positive" *C. R. Acad. Sci. (Paris) Serie I* **296** 729-732

Tresser C. (1984a) "About some theorems by L.P.Sil'nikov" *Ann. Inst. H. Poincaré* **40** 441-461

Tresser C. (1984b) "Homoclinic orbits for flows in $\mathbb{R}^3$" *J. Physique Lett.* **45** L837 - L841

Williams R.F. (1979) "The structure of Lorenz attractors" *Inst. Hautes Etudes Sci. Publ. Math.* **50** 321-347

KNOTS AND ORBIT GENEALOGIES IN NONLINEAR OSCILLATORS

Philip Holmes*

Departments of Theoretical and Applied Mechanics and
Mathematics and Center for Applied Mathematics
Cornell University, Ithaca, NY 14853 U.S.A.

1 INTRODUCTION: NONLINEAR OSCILLATORS

In this paper we describe recent work on the classification of knotted periodic orbits in periodically forced nonlinear oscillators, specifically Duffing's equation and the pendulum (Josephson junction) equation. After reviewing the Birman & Williams template construction, elementary knot theory and kneading theory for one dimensional maps, we give existence and uniqueness results for families of torus knots which arise in these two oscillators. We indicate how these results enable one to study bifurcation sequences occurring during the creation of complicated invariant sets such as Smale's horseshoe, and how they imply infinitely many distinct "routes to chaos" even in families of two dimensional maps such as that of Hénon. We also include a number of general results on knot and link types, including iterated knots, cablings and prime knots.

Our aim is two-fold; to introduce new and useful techniques to the field of "applied" dynamical systems, and to suggest new applications and sources to the "pure" topologist or knot theorist. In keeping with its introductory nature, the paper is informal and many technical details are

---

*Work partially supported by NSF under CME 84-02069, AFOSR under 84-0051 and ARO under DAAG 29-85-C-0018. [Mathematical Sciences Institute]

omitted; we hope to convey the spirit of our methods and provide a sampling of the results they have yielded so far. In this respect the present paper can be viewed as a guided tour of a series of recent papers which contain full proofs of the results sketched here as well as more complete technical and background information (Birman & Williams [1983a,b], Williams [1983], Franks & Williams [1985a,b], Holmes & Williams [1985], Holmes [1986a,b]).

For simplicity and definiteness, we restrict ourselves to the case of flows arising from periodically forced second order ordinary differential equations, although the original results of Birman & Williams [1983a] were obtained for the autonomous third order Lorenz [1963] system. Forced oscillators model a wide range of mechanical and electrical processes (cf. Hayashi [1964], Guckenheimer & Holmes [1983]). They take the form

$$\ddot{x} + g(x, \dot{x}, t) = 0 \tag{1}$$

where $(\dot{\ })$ denotes $\frac{d}{dt}(\ )$ and g is smooth and T-periodic in t. Equation (1) is easily rewritten as a first order autonomous system:

$$\begin{aligned} \dot{x} &= y \\ \dot{y} &= -g(x, y, \theta) \\ \dot{\theta} &= 1 \end{aligned} \tag{2}$$

with phase space $(x,y;\theta) \in \mathbb{R}^2 \times S^1$. More generally, the models studied in this paper apply to systems of the form $\{\dot{x} = f_1(x,y,t),\ \dot{y} = f_2(x,y,t)\}$, where the f_i are T-periodic in t. We denote the flow of (2) by $\varphi_t: \mathbb{R}^2 \times S^1 \to \mathbb{R}^2 \times S^1$ and assume that the usual existence and uniqueness results hold and moreover that solutions are defined globally in time, specifically that they enter and remain in a solid torus $D^2 \times S^1$ as $t \to \infty$, although this is not essential.

The two main examples considered here are the Duffing equation:

$$\ddot{x} + \delta\dot{x} - x + x^3 = \gamma\cos\omega t, \tag{3}$$

and the perturbed pendulum or Josephson junction equation:

$$\ddot{\theta} + \delta\dot{\theta} + \sin\theta = \nu - \beta\cos\omega t. \tag{4}$$

In the former case the displacement $x \in \mathbb{R}$ and thus the phase space is $\mathbb{R}^2 \times S^1$ (or $D^2 \times S^1$, since the positive cubic stiffness and linear damping terms imply that all solutions are uniformly bounded and approach an attracting set as $t \to \infty$). However, in the latter case θ is an angular variable ($\theta \in S^1 = \mathbb{R} \bmod 2\pi/\omega$) and thus the phase space of the flow is $(\theta, \dot{\theta}; t) \in S^1 \times \mathbb{R} \times S^1$ or, since $|\dot{\theta}|$ also remains bounded, $(\theta, t, \dot{\theta}) \in T^2 \times I$, where $I = [a,b] \subset \mathbb{R}$ is a closed interval. The differing topologies of these manifolds will profoundly affect the knot types which occur.

Associated with the flow φ_t, we define a Poincaré or time-T map P for (1.2) by selecting a cross section $\Sigma = \{(x, y, \theta)|_{\theta=0}\}$ and defining

$$(x_1, y_1) = P(x_0, y_0) \equiv \Pi \cdot \varphi_T(x_0, y_0, 0) \tag{5}$$

where Π denotes projection onto the first two factors. By virtue of its definition, P is an orientation preserving diffeomorphism. Moreover, if (1-2) has a linear damping factor δ, as do our examples, then P_δ has constant determinant $\det(DP_\delta) = e^{-\delta T} \equiv \varepsilon$. As δ ranges from 0 (a Hamiltonian system) to ∞, so P_δ ranges from an area preserving diffeomorphism to a singular, non-invertible map. We will return to this aspect in section 7.

In this paper we will appeal to geometrical properties of the Poincaré maps of (3-4) without providing proofs that these specific equations possess the properties in question. That they do possess such properties for certain (open) sets of $(\delta, \gamma, \omega, \nu, \beta)$-parameter values can be proved using global perturbation techniques such as that of Melnikov [1963]; see Guckenheimer & Holmes [1983], Greenspan & Holmes [1983, 1984], or Hockett & Holmes [1986a,b] for specific results relevant to (3-4).

Aside from their intrinsic interest as new families of "template knots", a major reason for studying knot types of orbits in parametrized families of differential equations is that knot type is a topological invariant, and thus serves to identify a given branch of orbits as parameters are varied. Used in conjunction with period, rotation number and symbolic dynamical invariants such as kneading sequences, a knowledge of knot types enables us to obtain certain ordering results on bifurcation sequences. However, while this is the main "application" described in the paper, we feel that there are many more potential uses of the technique introduced here.

It has been known that knots arise in differential equations for many years, perhaps the simplest example being the pair of (uncoupled) linear oscillators

$$\dot{x}_1 = px_2, \ \dot{x}_2 = -px_1; \ \dot{y}_1 = qy_2, \ \dot{y}_2 = -qy_1, \qquad (6)$$

where (p,q) are co-prime integers. Take the phase space $(x_1, x_2; y_1, y_2) = \mathbb{R}^4$ and consider $S^3 \subset \mathbb{R}^4$ given by $p(x_1^2 + x_2^2) + q(y_1^2 + y_2^2) = 1$. It is clear that S^3 is an invariant manifold for the flow of (6) (it is a Hamiltonian energy surface), moreover, it is filled with the one parameter family of two-tori $\{p(x_1^2 + x_2^2) = c, \ q(y_1^2 + y_1^2) = 1-c; \ c \in (0,1)\}$. Each such torus is filled with periodic orbits, every one of which is a (p,q) torus knot in S^3. However, this example is very degenerate and in some sense trivial--although there are infinitely many knots they are all isotopic--and it was not until Williams [1977,1979] recognized that the inverse limit process (Williams [1979]) preserves periodic orbits and subsequently developed the idea of knot holders or templates (Birman and Williams [1983a,b]) that a serious study of knots in three dimensional flows could begin. (But see Franks [1981]).

The paper is organized as follows. In section 2 we review the Birman & Williams template construction and describe the templates for the Duffing and Josephson equations. The construction requires that the flows possess hyperbolic limit sets. A canonical example of a nontrivial hyperbolic set is Smale's horseshoe [1963,1967], which arises naturally in the Duffing equation. Thus the Duffing template is perhaps more properly called the horseshoe template (Holmes & Williams [1985], Holmes [1986a]). Section 3 addresses the question of parametrized families in which the limit sets may bifurcate, thus becoming non-hyperbolic, and even disappear completely. It turns out that, even if this happens, the knot types are invariants as long as the orbits in question exist. Section 4 briefly reviews some results from knot theory, including simple invariants and the notions of companionship, iterated knots and cablings.

The templates, being semiflows on branched two-manifolds, naturally induce non-invertible one dimensional return maps: in a sense these are one dimensional limits of the Poincaré maps defined in (5). In section 5 we recall relevant symbolic and kneading theoretic techniques for the study of such maps. This concludes the review of methods and tools.

In sections 6-11 we give samples of existence and uniqueness theorems, without proofs, for specific families of knots which arise in our two examples: primarily torus and iterated torus knots. We also indicate how the uniqueness results can be used to study bifurcation sequences occurring during the creation of horseshoes and we state results of a more general nature due to Franks and Williams. We comment on the main ideas of the proofs and conclude in Section 12 with some general remarks on the relevance of this work to dynamical systems theory.

Related work on knots and braids in 3-flows which we do not consider here has been done by Boyland [1985] and Matsuoka [1983, 1984, 1986a,b].

For background on nonlinear oscillations, iterated mappings and dynamical systems, see Andronov et al. [1966], Arnold [1983], Guckenheimer & Holmes [1983] and Devaney [1985], or the other papers in this volume. Rolfson [1977] provides an introduction to knot theory and Birman [1975] to braid theory.

2 TEMPLATES

Let φ_t: $M^3 \to M^3$ be a flow on a 3-manifold having a hyperbolic invariant or chain recurrent set Ω with neighborhood $N \subset M^3$ (Bowen [1978], Newhouse [1980], Guckenheimer & Holmes [1983]). Let $\sim$ denote the equivalence relation $z_1 \sim z_2$ if $\|\varphi_t(z_1) - \varphi_t(z_2)\| \to 0$ as $t \to +\infty$ and $\varphi_t(z_i) \in N$, $\forall t > 0$. Equivalently, $z_1 \sim z_2$ if both z_1 and z_2 lie in the same connected component $W^s(x) \cap N$ of some local stable manifold of a point $x \in N$. Under $\sim$, the flow φ_t on M becomes a semiflow $\bar{\varphi}_t$ on a branched two-manifold $\mathcal{K} \in M^3$. The pair $(\mathcal{K}, \bar{\varphi}_t)$ is the <u>template</u>. Effectively, $\sim$ collapses along (strong) stable manifolds and identifies orbits with the same future asymptotic behavior, regardless of their past. The key result for our purposes is:

Proposition 2.1. (Williams [1977,1979], Birman & Williams [1983a,b]). The collapsing map $\sim$ is one to one on the union of periodic orbits in Ω.

This implies that knot and link types of periodic orbits in Ω survive unchanged in the template representation. In fact, almost all information about (Ω, φ_t) can be reconstructed from $(\mathcal{K}, \varphi_t)$; hence the name "template" (Franks & Williams [1985a]) is now preferred to the earlier "knot-holder" of Birman & Williams [1983a,b].

To give an intuitive feeling for template construction, we sketch the derivations for the two nonlinear oscillators of equations (3-4). We start by assuming that the Poincaré maps satisfy certain geometric conditions, here presented merely pictorially, and then suspend these maps in the appropriate way to produce the flows $\varphi_t: D^2 \times S^1$ and $\varphi_t: T^2 \times I$. Finally we collapse using $\sim$ to obtain the templates.

2.1 The Duffing or Horseshoe Template

In Greenspan & Holmes [1983] (cf. Guckenheimer & Holmes [1983]) it was proved that for $\gamma > 4\delta \cosh(\pi\omega/2)/3\sqrt{2}\pi\omega > 0$ and γ, δ sufficiently small, some iterate P^N of the Poincaré map of (3) possesses a hyperbolic nonwandering set Ω_D conjugate to a shift on two symbols: a Smale horseshoe. (See Smale [1963,1967], Moser [1973], Guckenheimer & Holmes [1983]). Numerical studies due to Ueda [1981] indicate that parameter values can be found for which the first iterate of P possesses horseshoes, and that, perhaps after an appropriate change of variables, a "rectangular" region $S \subset \Sigma$ can be found such that $P(S) \cap S$ is two "vertical" strips and $S \cap P^{-1}(S) = H_x \cup H_y$ is two "horizontal" strips. The monotonic twisting of the flow due to the first component $\dot{x} = y$ of (3) implies that the flow carries S to $P(S)$ as sketched in Figure 1a (cf. Holmes & Whitley [1983] for more details). The set Ω_D is defined as $\Omega_D = \bigcap_{n\in\mathbb{Z}} P^n(S)$ and $DP|_{\Omega_D}$ is uniformly contracting in the "horizontal" direction and uniformly expanding in the "vertical" direction (see arrows). Thus, $\sim$ collapses out the horizontal dimension (Figure 1b) and yields the branched manifold $\mathcal{K}_D$ of Figure 1c after identification of the cross sections at $t = 0$ and $t = 2\pi/\omega$. The choice of embedding of $\mathcal{K}_D \subset \mathbb{R}^3$ is somewhat arbitrary, and that of Figure 1d is sometimes more convenient, since it makes the orbits into braids (section 4, below).

The semiflow $\bar{\varphi}_t$ naturally induces a one dimensional return map f on the branch line I, as shown in Figure 1e. The map f stretches the intervals I_x and I_y so that they both cover I, the orientation of $f(I_y)$ being reversed. By analogy with the well known quadratic family of maps $x \to \lambda x(1-x)$ or $x \to \mu - x^2$, it is natural to think of f as a member of a one-parameter family f_μ of maps which vary with μ as indicated in Figure 1f. The bifurcations of this family help us organize knot information and we will return to this question in section 5.

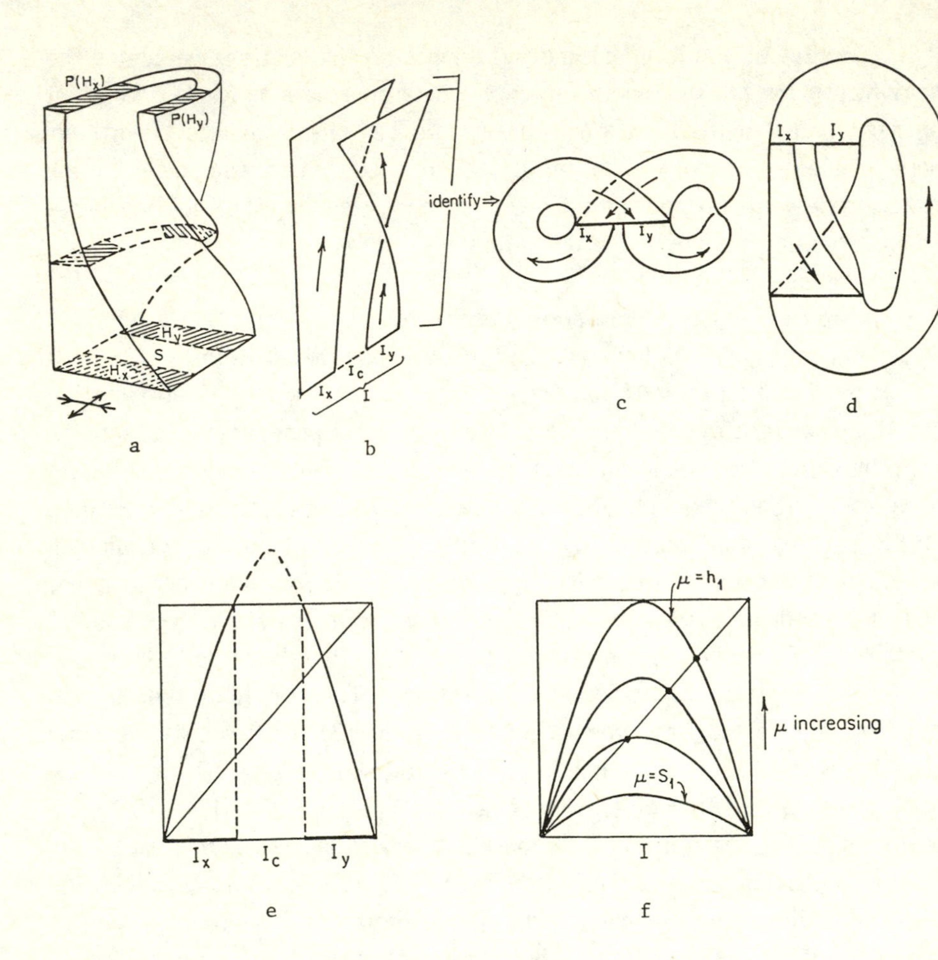

Figure 1. The Duffing template.

2.2 The Josephson Template

Hockett & Holmes [1986a,b] and Hockett [1986] show that in certain parameter ranges, equation (4) has a nonwandering set Ω_J conjugate to a shift on 3 symbols, the behaviour of the Poincaré map being that of Figure 2a. Strictly speaking, some iterate P^N has the behavior indicated for $\pi\beta > |(\pi\nu - 4\delta)\cosh(\pi\omega/2)|$ and β, ν, δ sufficiently small. Taking $\omega \to 0$ reduces N and numerical evidence suggests that we can choose $\omega > 0$ small enough such that $N = 1$, but this remains unproven.

There are four fixed points s_0, s_1, s_2, and a_0, and, recalling that θ is a 2π-periodic variable and identifying the vertical boundaries of the (annular) strip, we have the associated flow of Figure 2b. Collapsing by $\sim$ and identifying time 0 and $2\pi/\omega$ sections yields the template $\mathcal{K}_J$ of Figure 2d. The three 1-periodic orbits corresponding to the fixed points s_i

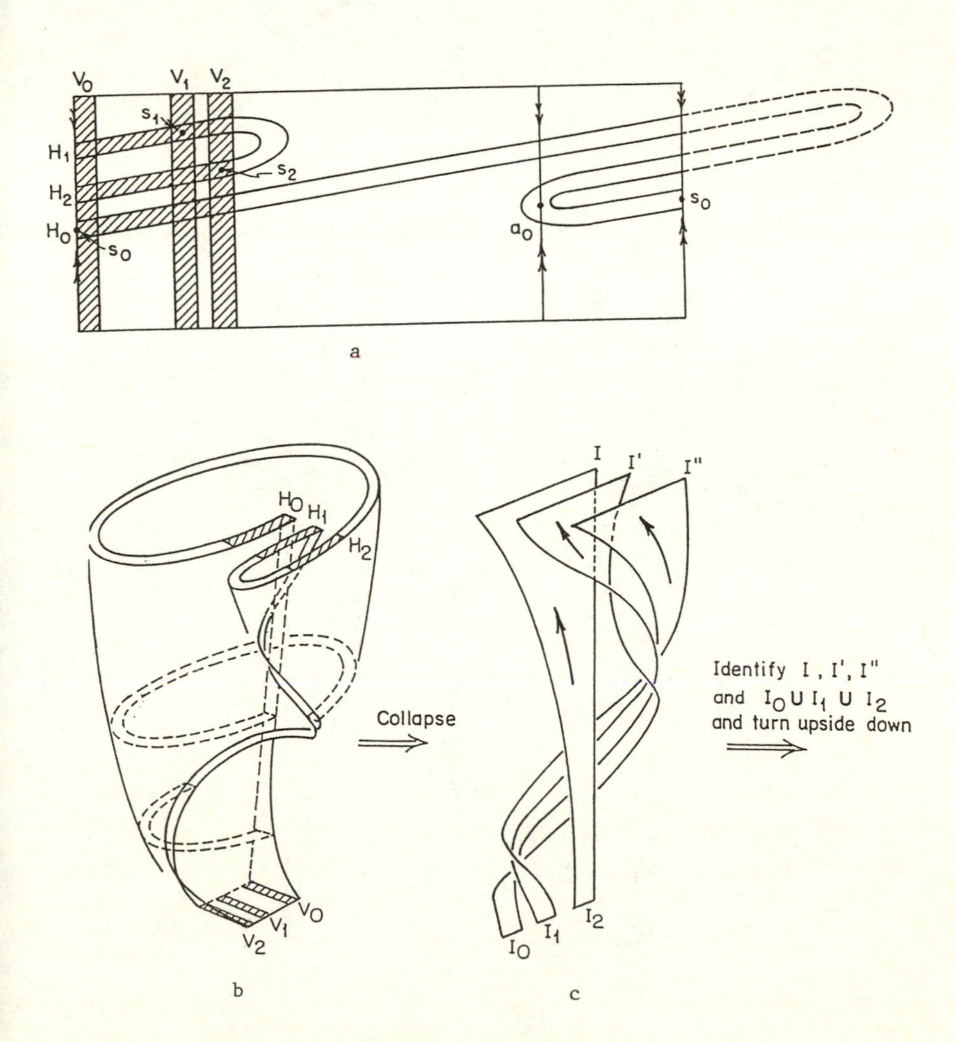

Figure 2. The Josephson template.

in Ω_J lie on $\mathcal{K}_J$ as indicated in Figure 2e, forming a non-trivial 3 component link. In this case the one dimensional return map takes the form of Figure 2f, having 3 branches, and one can also embed this map in a one parameter family of circle maps, as described by Hockett [1986], cf. Hockett & Holmes [1986b].

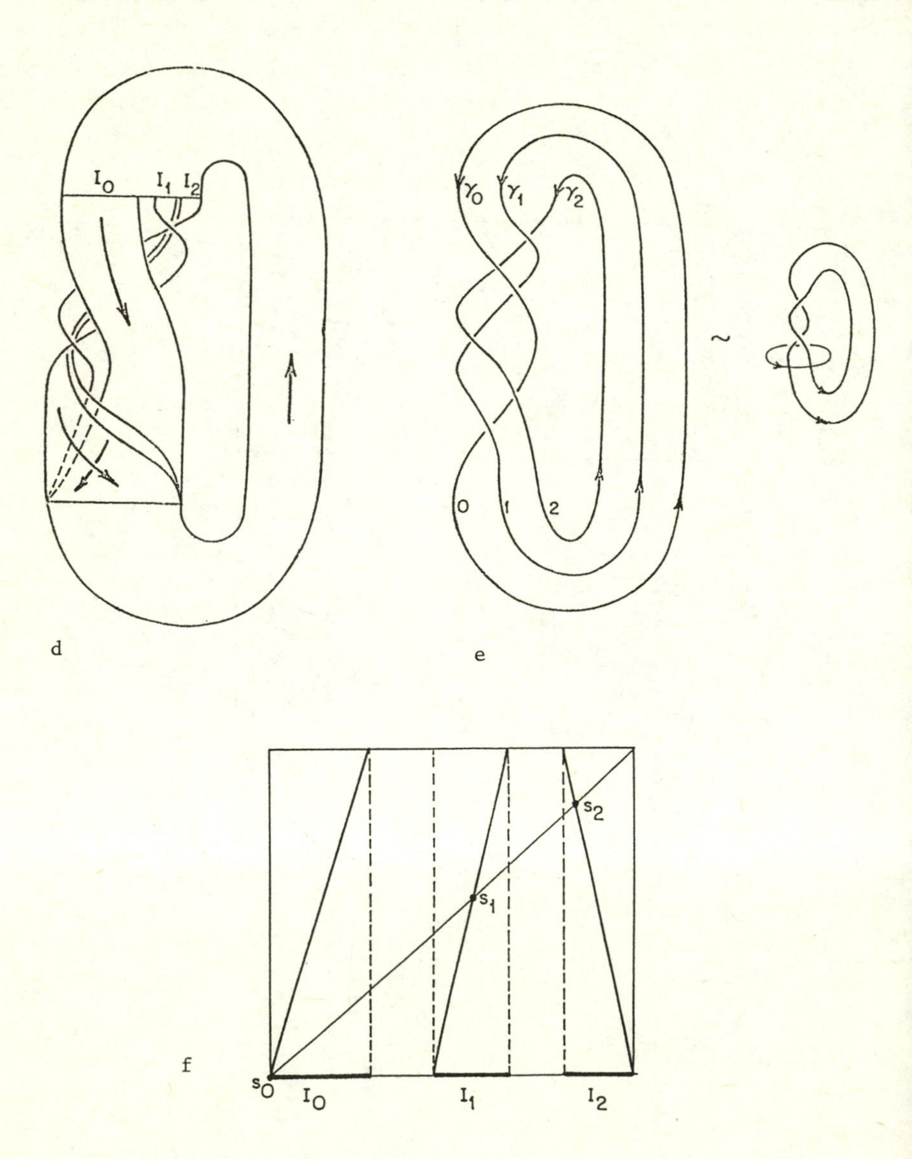

Figure 2, continued.

2.3 The Lorenz template

For completeness, we include the original Lorenz template $\mathcal{K}_L$ derived by Birman & Williams [1983a] along with its braid representation (Figures 3a,b). In the latter, a neighborhood of the saddle point p has been removed to reveal the similarity of the Lorenz template to the Duffing or horseshoe template: it lacks the half twist on the I_y band. Thus, $\mathcal{K}_L$ appears to be the simplest of all three templates, but it turns out that the additional twisting and consequent crossings of orbits induced on the Duffing and Josephson templates actually makes certain knot computations easier. For more information and references on the Lorenz system see Guckenheimer & Williams [1979], Williams [1979] and Sparrow [1982].

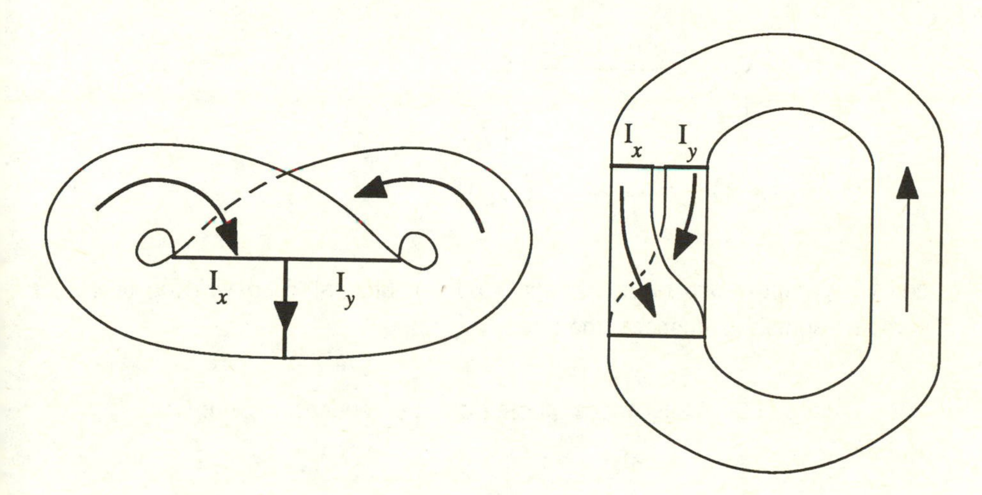

Figure 3. The Lorenz template.

2.4 Naming the knots and links

Proposition 2.1 implies that the template $(\mathcal{K}, \bar{\varphi}_t)$ and the associated return map f completely determine the knot and link types of all periodic orbits in the hyperbolic set Ω. It is now necessary to uniquely name these knots and links. This is most conveniently done by the conventional <u>symbolic dynamics</u> of the hyperbolic set Ω (cf. Guckenheimer & Holmes [1983], Chapter 5). We illustrate for the horseshoe. To each point $z \in \Omega = \bigcap_{n \in \mathbb{Z}} P^n(S)$ we associate a bi-infinite sequence of symbols $\underline{\psi}(z) = \{\psi_j(z)\}_{j \in \mathbb{Z}} \in \{x, y\}^{\mathbb{Z}}$ by the rule

$$\psi_j(z) = \left\{ \begin{array}{l} x \text{ if } P^j(z) \in H_x \\ y \text{ if } P^j(z) \in H_y \end{array} \right\} \tag{7}$$

It is then possible to show that $z \to \underline{\psi}(z)$ is a homeomorphism (after endowing $\{x, y\}^{\mathbb{Z}}$ with a suitable metric) and thus to each sequence $\underline{\psi}$ there corresponds one and only one point of Ω. Moreover, by (7) the action of P on Ω induces the shift on $\{x, y\}^{\mathbb{Z}}$:

$$\psi_j(P(z)) = \psi_{j+1}(z) \equiv (\sigma(\underline{\psi}(z)))_{j+1} \tag{8}$$

Thus we have the commuting diagram

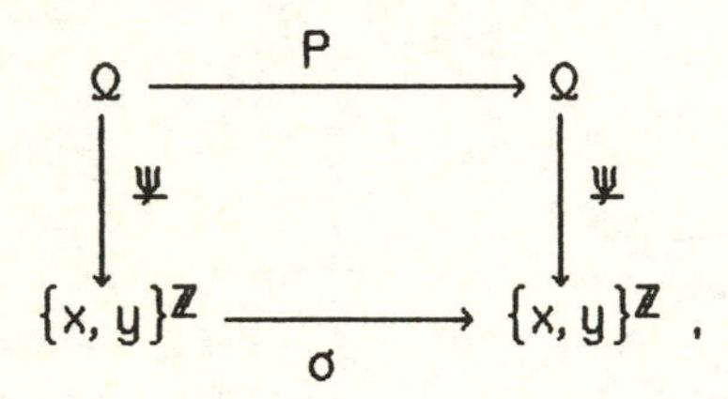

and the periodic orbits of Ω are easily enumerated by writing down all periodic symbol sequences, the first few being:

period	sequences (periodically extended)	number
1	$x,\ y$	2
2	xy	1
3	$x^2y;\ xy^2$	2
4	$x^3y,\ x^2y^2;\ xy^3$	3
5	$x^4y,\ x^3y^2;\ x^2yxy;\ x^2y^3,\ xy^2xy,\ xy^4$	6
k	...	$\approx 2^k/k$

In this symbolic context, the collapsing process $\sim$ corresponds to removal of the negative going portion of the sequence: we ignore histories. (Since each distinct periodic orbit has its unique history (identical to its future), no information is lost by $\sim$.) Thus, via symbolic dynamical coding and Proposition 2.1, we have

Proposition 2.2. To each finite acyclic word w with letters $\{x, y\}$ (resp. $\{0, 1, 2\}$) there corresponds a periodic orbit of ψ_t and hence a unique knot K_w on $(\mathcal{K}_D, \bar{\varphi}_t)$ (resp. $(\mathcal{K}_D, \bar{\varphi}_t)$). Let $\bar{w}$ denote the periodic extension of w. If $\bar{w}' = \sigma^{\ell}(\bar{w})$ for some ℓ then w' and w correspond to the same knot . A collection of N shift-inequivalent, finite, acyclic words corresponds to an N-component link $L = \{K_{w_1}, \dots, K_{w_n}\}$.

We refer to the knots and links as Duffing or (more frequently) horseshoe knots, Josephson knots and Lorenz knots respectively, as they live on $\mathcal{K}_D$, $\mathcal{K}_J$, or $\mathcal{K}_L$.

Given a knot $K_w \subset \mathcal{K}$ it is a simple matter to read the word w. First we note that the time direction of the semiflow $\bar{\varphi}_t$ provides a natural orientation for K_w. One then simply follows K_w, starting at, say, the leftmost point where K_w crosses the branch line and reading out the sequence of $\{x, y\}$ or $\{0, 1, 2\}$ as one crosses $\{I_x, I_y\}$ or $\{I_0, I_1, I_2\}$ until one returns to the starting point.

To go from w to K_w is more difficult, since it requires determination of the permutation of branch crossing points induced by the return map f. This information is provided by the appropriate ordering of sequences which must reflect the natural ordering of points $z \in I$. For the Lorenz system there is no problem, both branches of the return map preserve orientation and one simply uses lexicographical ordering $x < y$. Thus the "iterates" x^2yxy, $xyxyx$, $yxyx^2$, xyx^2y, yx^2yx of the word x^2yxy are ordered

$$x^2yxy < xyx^2y < xyxyx < yx^2yx < yxyx^2,$$

so that, numbering the points corresponding to these sequences as 0, 1, 2, 3 or 4, f induces the permutation $(0, 1, 2, 3, 4) \to (2, 3, 4, 0, 1)$ or $0 \to 2 \to 4 \to 1 \to 3$. Figure 4a shows the associated Lorenz knot, which turns out to be a trefoil or (2,3) torus knot.

For the Duffing and Josephson templates, the map f reverses orientation on one interval and thus one must count the parity of y (resp. 2) entries in w and use this in the ordering algorithm. We deal with this in Section 5. Meanwhile, Figure 4b shows examples of non-trivial Duffing knots, examples of Josephson knots will follow later.

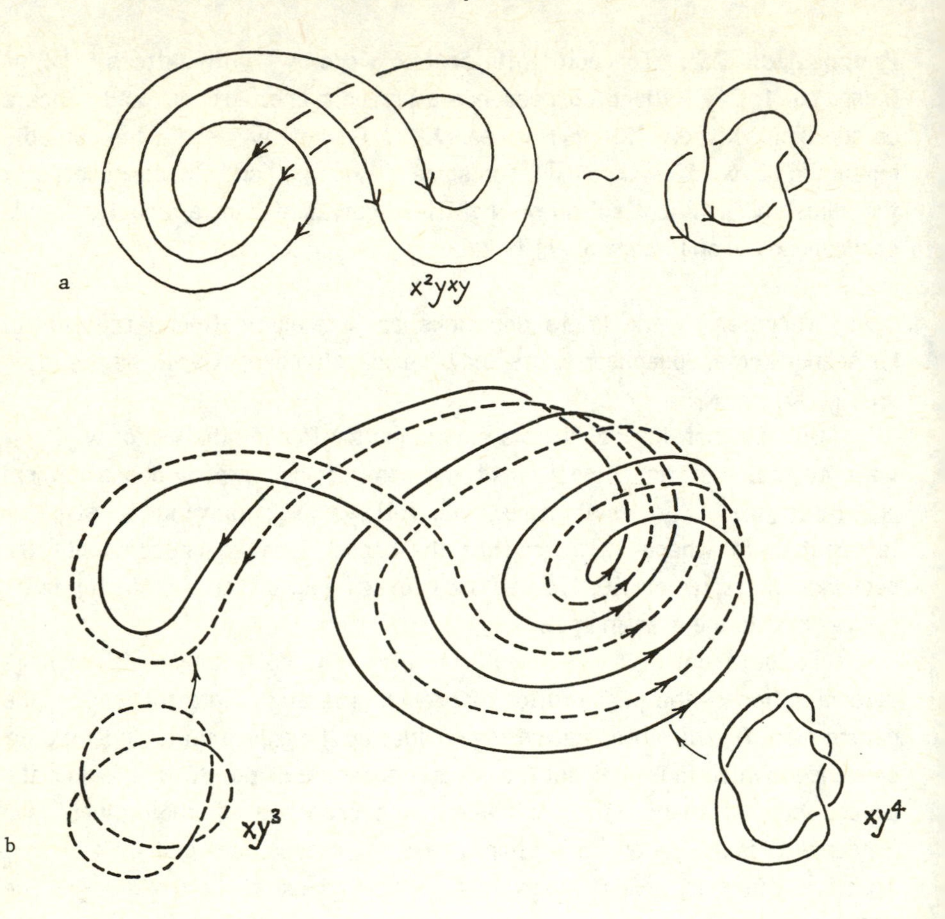

Figure 4. Some template knots and links. (a) Lorenz; (b) Duffing.

3 INVARIANCE OF KNOT TYPE

Having constructed the templates and given some preliminary examples of knots and links, we now come to a major property which makes the study of these objects worthwhile. Suppose that, for some $\mu = \mu_1$, L is a (possibly single component) link of periodic orbits of a parametrized flow φ_t^{μ} on a three manifold M^3. For example L may be a template link corresponding to a hyperbolic set Ω_{μ_1}. Assume that the vector field never vanishes on L (the periods of all orbits are finite, albeit they may be arbitrarily large). Now vary μ. The component(s) of L will deform and may collapse to points (Hopf bifurcations) or coalesce (saddle-node or pitchfork bifurcations) but no component of L can intersect itself or any other component of L, for if it did we would have a point $z \in M^3$ at which two distinct solutions were based, thus violating the local existence and uniqueness theorem for differential equations. The only way in which such an intersection can occur is for the tangent vectors to L to coincide at every intersection point and this happens in two ways: (1) when two (or more) isotopic knots of the same period coalesce in a saddle-node, pitchfork or multiply degenerate bifurcation or (2) when a knot K_1, whose period is an integer multiple of the period of a second knot K_2 collapses onto K_2 in a period multiplying bifurcation. The simplest example is period doubling. See section 8, below.

We summarise

Proposition 3.1. Knot and link types are topological invariants of (unions of) periodic orbits in parametrized three dimensional flows, provided the orbits have bounded periods.

In the periodically forced oscillator examples, the periods of all periodic orbits are necessarily integer multiples of the basic forcing period $2\pi/\omega$ and, since these flows possess no fixed points in view of the third component $\dot{\theta} = 1$ of (2), no Hopf or homoclinic bifurcations can occur. Thus the "bounded period" caveat of the Proposition is unnecessary in this case. In fact the period $2\pi q/\omega$ of a given orbit is a second invariant, and the integer q is simply the length of the word w corresponding to the knot in question.

This result suggests the following strategy. We construct the template $(\mathcal{K}, \bar{\varphi}_t)$ of the system in question for parameter values for which it possesses a "well understood", if complicated, hyperbolic set Ω_{μ_1}. Having thus established (implicitly) all the knot and link information, we vary parameters in such a way that family of sets $\{\Omega_\mu\}$, of which Ω_{μ_1} is the "final" member, undergoes a sequence of bifurcations to a "simpler" (possibly empty) set Ω_{μ_0}. The knots and links provide invariants which enable us to follow certain sets or orbits back to their birth (or death) in these bifurcations. In this way, as we shall see, we can establish orbit genealogies and even determine bifurcation sequences in some situations.

4 KNOTS AND BRAIDS

In this section we review relevant ideas from knot and braid theory. For more information see Rolfsen [1977] and Birman [1975].

We recall that a <u>link</u> L is a collection of pairwise disjoint, oriented, simple closed curves embedded in (oriented) S^3. Working in $\mathbb{R}^3$, where we have pictured our templates, presents no problems: one can simply add $\{\infty\}$. The link type of L is its equivalence class under the relation $L \approx L'$ if there is an orientation preserving homeomorphism $h: (S^3, L) \to (S^3, L')$. A <u>knot</u> is a link consisting of a single component. The trivial knot type is the knot type of the unit circle $S^1 \subset S^2 \subset S^3$. The <u>genus</u> of L is the smallest integer g such that L is the boundary of an embedded orientable surface M $\subset S^3$ where M has genus g. The surface M is a (minimal) Seifert surface for L. Genus is an isotopy invariant for L.

A (closed) <u>braid on p strands</u> is a presentation of an (oriented) knot or link so that its projection onto the plane passes in the same direction about the origin p times. We remark that orbits of period q for the templates of Figures 1d, 2d and 3b occur naturally as braids on q strands. However, we will be interested in attempting to minimise the number of strands in the braid presentation. The <u>braid number</u> of a braid B is the smallest integer k such that B is isotopic to a braid on k strands; moreover, k is also an isotopy invariant.

We illustrate with an example of a 4-braid: Figure 5a,b. There are six types of crossings, σ_i and σ_i^{-1}, (i = 1, 2, 3). If no σ_i^{-1} are involved one says the braid is <u>positive</u>. The presentations of the templates in Figures 1d, 2d and 3b make any template link a positive braid and henceforth we shall only be concerned with positive braids. A full twist (Figure 5b), denoted Δ^2, is quite special; it can be passed over (commutes with) each σ_i (and σ_i^{-1}). These σ_i (i = 1, . . . , n-1) generate the braid group on n strings, B_n;

see Birman [1975] for details. All our braids are connected top to bottom in the obvious "trivial" way, see Figure 5c. The defining relations of B_n are

$$\sigma_i \sigma_{i+1} \sigma_i = \sigma_{i+1} \sigma_i \sigma_{i+1} \quad (9a)$$

and

$$\sigma_i \sigma_j = \sigma_j \sigma_i \text{ if } |i-j| \geq 2. \quad (9b)$$

The statements are geometrically obvious: see Figure 5d.

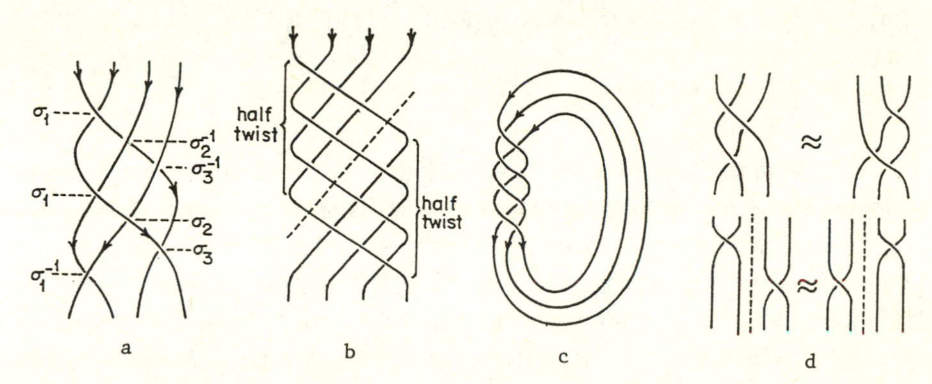

Figure 5. Braids. (a) crossings; (b) twists; (c) a closed positive 3-braid; (d) defining relations.

Our proofs will rely on the following

Theorem 4.1. (Franks and Williams [1985], cf. Morton [1985]) If B is a positive braid on p strands which contains a full twist then p is the braid number of B.

We will also use

Proposition 4.2. (Formula for positive braids; see Birman & Williams [1983a]). Positive braids which are knots have genus g given by

$$2g = \#cr - \#st + 1, \quad (10)$$

where $\#cr$ is the number of crossings and $\#st$ is the number of strands.

An (m,n) torus knot is a special braid which lies on a torus T^2, passing m times longitudinally and n times meridionally around the torus; (m,n) are assumed relatively prime and we generally take $n > m$. Theorem 4.1 implies that such a knot can be "minimally" presented as a braid on m strands.

Example. In the example of Figure 5c, we have $2g = 8-3+1 = 6$. This is in fact a (3,4) torus knot (one can visualize a torus passing along the top of the strands at the right, and through the middle of the strands at the left.) For a (m,n) torus knot we have the identity (Rolfsen [1977]):

$$2g = (m-1)(n-1). \tag{11}$$

The genus, braid number and period of the knot (as an orbit in the dynamical system) will provide our major invariants for orbit identification. (The period is simply the word length). However, none of these are complete invariants. In fact in Holmes & Williams [1985] and Holmes [1986b] we give examples of non-isotopic Duffing and Josephson knots for which all three invariants agree. Computation of the Alexander polynomials reveals that these knots are in fact non-isotopic. Such polynomial invariants, especially ones recently derived from von Neumann algebras (Jones [1985,1986]) are much sharper than the numerical invariants given above, but even they are not complete, and they are much harder to compute (cf. Birman & Williams [1983a, §9]).

Finally we need the notions of cablings and iterated knots (cf. Rolfsen [1971], Eisenbud & Neumann [1985], and Birman & Williams [1983a]). First, recall that the linking number $\ell(K_1, K_2)$ of two knots K_1 and K_2 is the net number of positive crossings of K_2 by K_1. It is easy to see that $\ell(K_1, K_2) = \ell(K_2, K_1)$ and that ℓ is a topological invariant. Let K_0 denote a knot in $\mathbb{R}^3$ with solid torus neighborhood V_0 and let $\ell_0 \subset \partial V_0$ denote a preferred longitude: i.e. ℓ_0 bounds an orientable 2-manifold in the complement $\mathbb{R}^3 \setminus V_0$. Let K_1 be a knot which lies on V_0 and winds a times longitudinally and b times meridionally around K_0, with (a,b) relatively prime, so that K_1 links K_0 b times (and crosses ℓ_0 b times). Then K_1 is an a-cabling or a type (a,b)-cabling of K_0.

If K_0 is the unknot, so that ℓ_0 bounds a disc, then an (a,b)-cabling is an (a,b) torus knot. Let K_1 be an (a_1,b_1) torus knot, and K_i be an

(a_i, b_i)-cabling of K_{i-1} for each $1 < i \le r$. Then K_r is an iterated torus knot of type $\{(a_i,b_i)\}_{1\le i\le r} = \{(a_1,b_1), \ldots, (a_r,b_r)\}$. Note that each K_i links K_{i-1} b_i times. We call K_0 the core of K_r. More general notions of "iterated" knots, such as companionship and satellites, are also available, see Rolfson [1977] §4D for details.

5 KNEADING AND ONE DIMENSIONAL BIFURCATIONS

We now sketch those aspects of kneading theory necessary for our purposes, omitting all details. See Milnor & Thurston [1977], Collet & Eckmann [1980], Guckenheimer [1977,1979,1980], Derrida et al. [1978], Jonker [1979] and Jonker & Rand [1981] for more details: Guckenheimer & Holmes [1983], §6.3, also has some information. Our requirements are two-fold: (1) to derive from the symbol sequences $\psi(z) \in \{x,y\}^{\mathbb{Z}}$ (resp. $\{0,1,2\}^{\mathbb{Z}}$) invariant coordinates which permit us to correctly order points on I when the return map f reverses orientation, and (2) to obtain a global description of the bifurcation sequences that the one-parameter families f_μ undergo as the "limiting" or "complete" hyperbolic return map f is created. The former is easy, the latter is more difficult and is only available in complete form for certain families of "quadratic like" maps such as those of Figure 1f. No completely satisfactory theory for circle maps exists, although Boyland [1983] and Hockett [1986] (cf. Hockett & Holmes [1986b,c]) have partial information.

5.1 Ordering

We start with the two-symbol sequences and the return map of figure 1e. Let $\underline{a} = a_0, a_1, a_2 \ldots = \underline{\psi}(z)$ denote the semi-infinite sequence corresponding to a point z in the nonwandering set $\Omega = \bigcap_{n\ge0} f^{-n}(I)$ of f. (Recall that $\sim$ has eliminated negatively indexed entries a_{-k}.) For example, $\underline{a}$ might be the periodic sequence $\overline{w} = w_0w_1\ldots w_kw_0w_1\ldots w_k \ldots$ corresponding to a template knot K_w. Since $\underline{a}$ codes the future behavior of z under iterates of f, we call $\underline{a}$ the itinerary of z. From $\underline{a}$ we derive the invariant coordinate $\underline{\theta}(z) = \theta_0\theta_1\theta_2\ldots$ by writing $\theta_j = a_j$ if the y-parity in the first j entries of $\underline{a}$ is even and $\theta_j = \hat{a}_j$ if the parity is odd. Here $\hat{a}_j$ denotes x if $a_j = y$ and y if $a_j = x$. Thus if $\underline{a} = xyyxyxx\ldots$ then $\underline{\theta} = xyxxyyy\ldots$. The following is then an easy consequence:

Proposition 5.1. $z_1 < z_2 \iff \underline{\theta}(z_1) \blacktriangleright \underline{\theta}(z_2)$, where $\blacktriangleright$ denotes lexicographical ordering $(y \blacktriangleright x)$.

A similar result holds for the 3-symbol sequences of the Josephson problem and the associated map. Here $\underline{\theta}(z)$ is derived from $\underline{\psi}(z)$ by counting the 2-parity: specifically, for given itineraries $\underline{s}$, $\underline{s}'$ we set $\underline{s} \blacktriangleleft \underline{s}'$ if $s_j = s_j'$ for $0 \leq j < k$ and $s_k < s_k'$ when the number of 2's in $\{s_0, \ldots, s_{k-1}\}$ is even, or $s_k > s_k'$ when the number of 2's in $\{s_0, \ldots, s_{k-1}\}$ is odd. Here $<$ denotes the usual number ordering $0 < 1 < 2$. Thus 10220... $\blacktriangleleft$ 10221... but 1021.. $\blacktriangleright$ 1022... .

Example. The three orbits of period one: 0, 1, and 2 are ordered $0 \blacktriangleleft 1 \blacktriangleleft 2$ and appear on the Josephson template as shown in Figure 2e. Note that they are nontrivially linked.

We have already seen that words w, w' which are shift equivalent correspond to points on the <u>same</u> (periodic) orbit. To avoid ambiguity, by the <u>word</u> of a given periodic orbit, we shall generally mean the minimum word with respect to the order $\blacktriangleleft$ among all shifts of that word. For example, the period 4 Josephson orbit corresponding to 0102, 1020, 0201 and 2010 will be called "0102" or (as a knot) K_{0102}.

The reader may now like to verify that the examples of Figure 4b are drawn correctly. The following table will help with the period 5 knot K_{xy^4}:

itinerary and shifts	invariant coordinate	order
xy^4	xyxyx	0
y^4x	yxyxx	2
y^3xy	yxyyx	3
y^2xy^2	yxxyx	1
yxy^3	yyxyx	4

5.2 Bifurcations of "quadratic-like" maps

Here we merely provide a brief descriptive summary: for details see the references cited above or the review contained in Holmes & Whitley [1984]. Kneading theory is the appropriate tool for studying bifurcations of families f_μ of continuous maps with critical points, which take an interval $[a,b] \in \mathbb{R}$ into itself. The theory is easier if f_μ has negative Schwarzian derivative and easiest if, in addition, there is only one critical point, c. In that case (with some other, technical, restrictions), f_μ possesses at most one attracting periodic orbit for each μ and the orbit $\{f_\mu{}^n(c)\}_{n\geq 0}$, and hence the itinerary and invariant coordinate $\underline{\theta}(c, f_\mu)$, of the critical point almost completely characterizes the dynamics of f_μ. (To define these sequences, we must either take limits (Holmes & Whitley [1984]) or introduce a third symbol, c, to take care of orbits containing c.) We call $\underline{\theta}(c, f_\mu)$ the kneading invariant and denote it $\underline{\nu}(\mu)$.

The prototypical family is the quadratic functions

$$z \rightarrow \mu - z^2 \tag{12}$$

the behavior of which is illustrated in Figure 6a (cf. Figure 1f). The five panels correspond to $\mu < -\frac{1}{4}$, $\mu = -\frac{1}{4}$, $\mu \in (-\frac{1}{4}, 2)$, $\mu = 2$ and $\mu > 2$. For $\mu < -\frac{1}{4}$ the nonwandering set Ω_μ is empty: all orbits escape to $-\infty$, while for $\mu > 2$, Ω_μ is a hyperbolic 2-shift: a "one-dimensional" horseshoe. $\Omega_{\mu>2}$ is created in a sequence of bifurcations which we now describe.

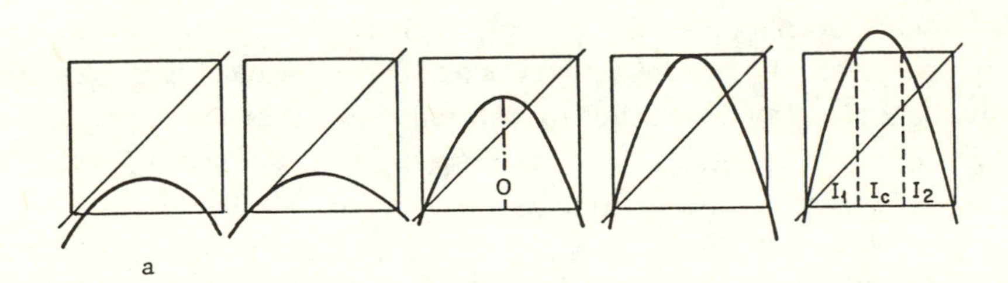

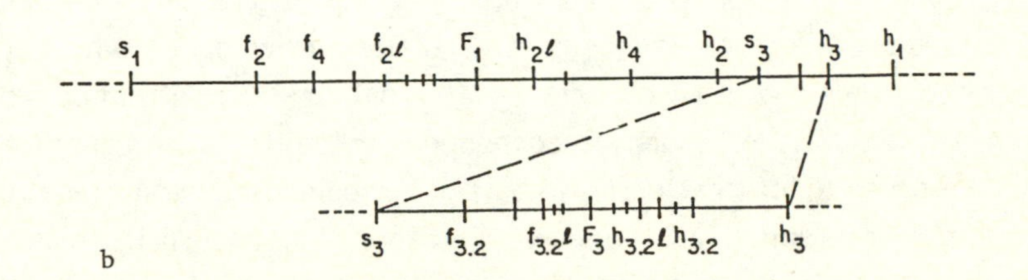

Figure 6. Bifurcation sequence for a "quadratic-like" map.

At $\mu = -\frac{1}{4}$ a saddle-node bifurcation occurs and two fixed points, a source $z_1 = -(1+(1+4\mu)^{\frac{1}{2}})/2$ and a sink $z_2 = -(1-(1+4\mu)^{\frac{1}{2}})/2$, the first members of Ω_μ, are created. Subsequently, at $\mu = 5/4 = f_1$, z_2 undergoes a period doubling (or flip) bifurcation which is the first in a cascade f_{2^ℓ} of such bifurcations accumulating at the "Feigenbaum point" $\mu = F_1$ ($\approx$1.40...). As μ continues to increase we traverse a countable sequence of "windows"; at the opening of each window a saddle-node bifurcation to an orbit of period k occurs. Let the jth such open at $\mu = s_k^j$. Each window is closed at a parameter value $\mu = h_k^j$, for which the kth iterate of f maps a collection of k disjoint intervals onto themselves. The fourth panel of Figure 6a shows f_{h_1}, where the whole sequence ends (when there is only a single window of period k, as in the cases k = 1, 3, and 4 the index j is omitted). Each $[s_k^j, h_k^j]$ period k window in parameter space repeats the structure of the $[s_1, h_1] = [-\frac{1}{4}, 2]$ period 1 window in miniature and this nesting process continues ad infinitum: (Gumowski and Mira [1980a,b] refer to "boîtes emboîtées". See Figure 6b for an impression of the structure.

The description above was known, at least in part, to Metropolis et al. [1973], May [1976] and others, but the full description was obtained only after development of the kneading theory (also see Douady & Hubbard [1982]). The global structure of the sequence is obtained by studying the monotonic (lexicographical) decrease of $\underline{\nu}(\mu)$ as μ increases. Intuitively, as μ increases, $f_\mu(c)$ moves rightward and, since no point in Ω_μ can lie to the right of $f_\mu(c)$, the orbit of c determines which orbits belong to Ω_μ for particular μ values. As μ increases (and $\underline{\nu}(\mu)$ decreases), more and more orbits appear at bifurcations, building the sequence described above.

In Holmes [1986b] we extract information from the kneading theory relevant to the nesting behavior of Figure 6b. This permits us to classify the kneading sequences associated with periodic orbits as <u>prime</u> or <u>factorizable</u> (with respect to a certain $*$-decomposition). Prime sequences correspond to itineraries or <u>words</u> in $\{x,y\}$ and hence orbits which appear in pairs in saddle node bifurcations s_k^j (but note that k need not be a prime number). Factorizable sequences correspond to orbits which appear <u>either</u> in period doubling bifurcations $f_{k\cdot 2^\ell}$ <u>or</u> in saddle-nodes which are nested within a more "primitive" window. Factorizable sequences must have non-prime period. The important part as far as the knots corresponding to factorizable words are concerned, is that nesting also occurs in the phase space of the interval I and consequently also on the template $\mathcal{K}_D$. In Figure

7 we illustrate a subtemplate $\mathcal{L}(\nu) \subset \mathcal{K}_D$ on which all orbits which factor as products of a certain kneading sequence $\nu_0 = xy^3x$ lie. The two prime orbits (the saddle-node pair) which govern the construction of $\mathcal{L}(\nu_0)$ have words $w_0 = x^2y^3$ and $w_0' = x^2yxy$ and correspond to the knots K_0, K_0' of the figure. Each window $[s_k^j, h_k^j]$ in the bifurcation set corresponds to such a subtemplate and the main point is that each such $\mathcal{L}(\nu_k^j)$ is a "poorly embedded" copy of $\mathcal{K}_D$ itself and thus all the knots on $\mathcal{K}_D$ reappear on each $\mathcal{L}(\nu)$ in iterated form. The nesting and self-similarity of the one dimensional dynamics of f_μ is reproduced in a precisely analogous way. We shall return to these ideas in section 8.

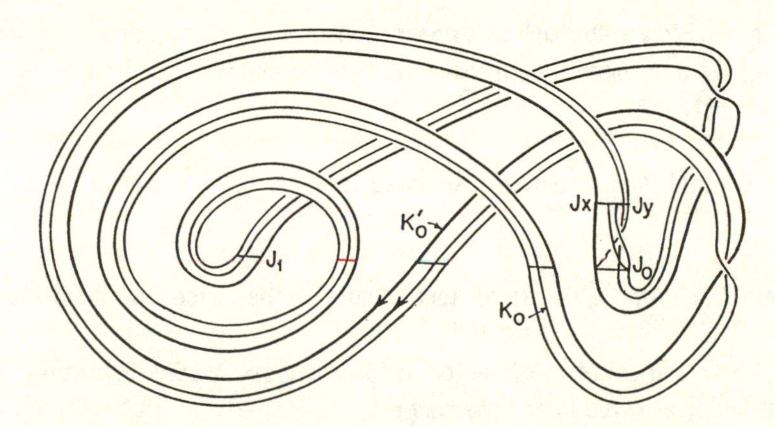

Figure 7. The subtemplate $\mathcal{L}(\nu_0)$ associated with a nested bifurcation set. Here $\nu_0 = xyyyx$, $w_0 = x^2y^3$, and $w_0' = x^2yxy$.

We have now introduced the basic tools for our study: templates, knot theoretic methods and the symbolic dynamics and kneading theory of one dimensional maps. The remainder of the paper provides samples of results obtained with these methods. Perhaps the most basic question is, given a template $(\mathcal{K}, \bar{\varphi}_t)$, which knot and link types occur? This appears to be very difficult (Birman and Williams [1983a]) and one can only hope for partial results. The most successful work thus far concerns special families of knots, in particular torus knots, which are well understood. We shall review some results of this nature and then move onto more general results.

6 EXISTENCE AND UNIQUENESS OF RESONANT TORUS KNOTS

We start with examples of the most complete and satisfactory classification results obtained to date for template knots. We have already met examples of torus knots in sections 2 and 4. Here we will need a special notion adapted to template knots:

Definition. If K is a (p,q) torus knot of (dynamical) period q then we call it a resonant torus knot. If K is a type $\{(p_i, q_i), 1 \leq i \leq \ell\}$ iterated torus knot of period $q = \prod_{1 \leq i \leq \ell} q_i$ we call it a resonant iterated torus knot.

Example. In Figure 4b, xy^3 is a non-resonant (2,3) torus knot of period 4, while xy^4 is a resonant (2,5) torus knot (of period 5). Iterated knots will appear in section 8.

The first major uniqueness results were by Holmes & Williams [1985]:

Theorem 6.1. Among the (p,q) torus knots on the horseshoe template $\mathcal{K}_D$ there are

(i) two and only two of period q for each $q > 2p$, and infinitely many if the period is allowed to be arbitrary;

(ii) none of period q if $q < 2p$;

(iii) none at all if $q \leq 3p/2$.

The situation for the Lorenz template is not quite as clear:

Theorem 6.2. The number $N_{p,q}$ of (p,q) torus knots of period p+q in the Lorenz template $\mathcal{K}_L$ satisfies $2[q/p] \leq N_{p,q} \leq q - p + 1$ where $[\cdot]$ denotes integer part. There are infinitely many if the period is allowed to be arbitrary.

Prior to these results, Birman Williams [1983a] (Theorem 6.1) had already shown that all torus knots occur on the Lorenz template. The analogous result for the Josephson problem is

Theorem 6.3. (Holmes [1986b]). Among the Josephson knots there are two and only two (p,q)-resonant torus knots for each $0 < p/q \leq 1$. They are the only order preserving orbits of period q and rotation number p/q.

A notable consequence of these results is:

Corollary 6.4. Each of the templates $\mathcal{K}_D$, $\mathcal{K}_J$ and $\mathcal{K}_L$ support infinitely many distinct knots.

In fact, this conclusion turns out to be more generally true of template knots: see section 9.

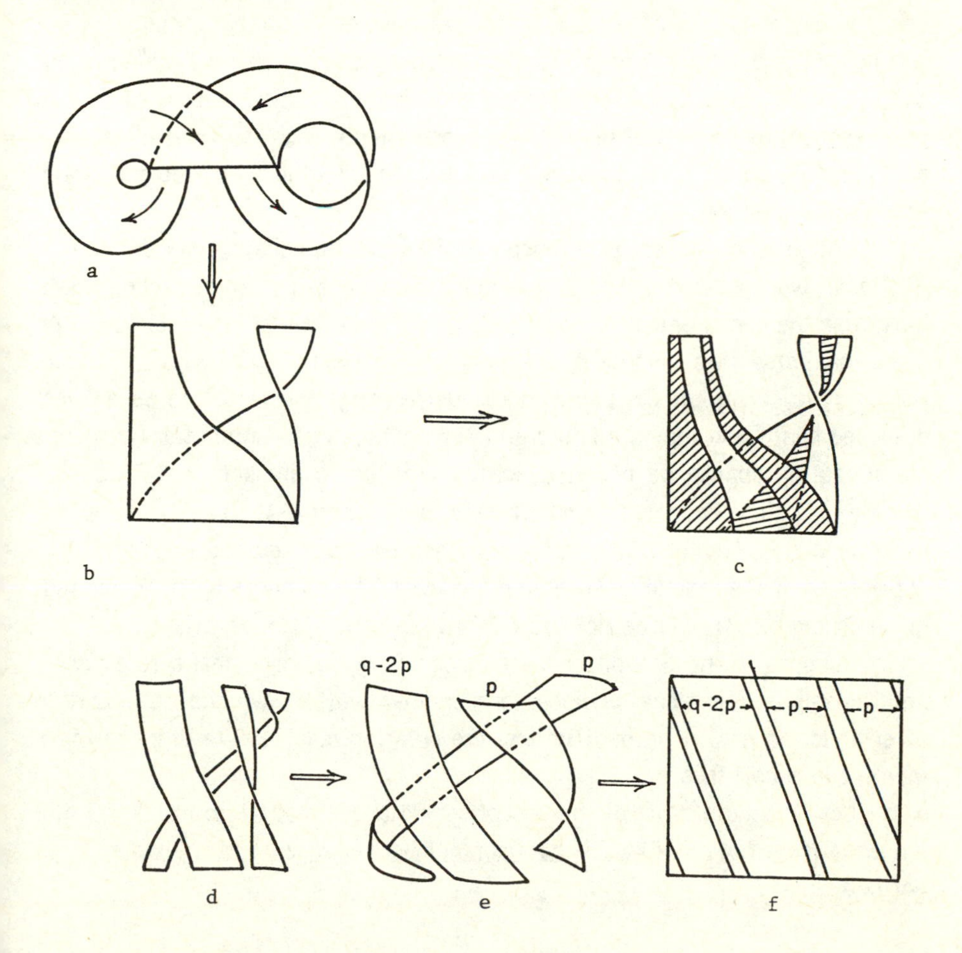

Figure 8. Existence of torus knots on $\mathcal{K}_D$.

In all cases the existence proofs for torus knots are relatively simple, once one sees how the template fits onto the torus which carries the knot in question. Since the templates are twisted and branched, they cannot fit directly onto tori and surgery must be performed to remove "bad" pieces. It then remains to show that the desired orbits lie on the remaining sections, and hence on the torus. Figure 8 shows the construction of a (p,q) resonant torus knot ($p \leq q/2$) for the horseshoe template $\mathcal{K}_D$. We first cut along the branch line and display the template like a braid in that the flow is always downward (b). Next, the outer part of the front strip and the middle part of the back one are shaded in and discarded (c). Finally, we show how the remainder lies on a tube (e), which is a braid representation of a torus (f). For a (p,q) torus knot of period q the resulting braid has q strands. The leftmost strand of the first group of p strands can be "lifted over" onto the $q-2p$ strand group to form a second, isotopic knot. To produce a non-resonant torus knot of period $q + \ell$ one simply adds ℓ "trivial" strands at the extreme left. The existence results for the other templates proceed in a similar fashion.

Uniqueness proofs are much harder and the basic idea is due to Williams, who observed that in the horseshoe case the (p,q) torus knots maximise the genus among all p-braids of period q. Since no other knots or links maximise this invariant, the pair of resonant torus knots is unique. However, to carry out the proof, one must arrange the template as a "well disposed braid", so that the knots and links which are candidates for genus maximization appear as positive braids with braid number p. This is a complex surgical procedure and appeals to Theorem 4.1: The twist on the right hand side (I_y) of $\mathcal{K}_D$ happily provides just the required full twist for application of the theorem. (It also prevents all torus knots from occurring; cf. Theorem 6.1 (iii)). See Holmes & Williams [1985] for details.

Surgery in the Josephson case is simpler, and here there is a pretty relationship between braid number and another well known characterization of behavior in nonlinear oscillators, the rotation number. We digress for a moment to recall this.

Let $\bar{P}$ denote a lift of the Josephson Poincaré map P to $\bar{A} = \mathbb{R} \times I$, the universal cover of $S^1 \times I$. The rotation number ρ of a point z under P is defined as

$$\rho(z, P) = \lim_{n\to\infty} \frac{\pi_1 \circ \bar{P}(z) - \pi_1(z)}{n},$$

when the limit exists. Here π_1 denotes projection onto $\mathbb{R}$, the first factor of $\bar{A}$. Loosely, ρ is the average speed with which the orbit $\{P^n(z)\}$ moves around the annulus. We define the symbolic rotation number $r(\underline{s})$ for $\underline{s} \in \{0,1,2\}^{\mathbb{Z}}$ as

$$r(\underline{s}) = \lim_{n\to\infty} \sum_{i=0}^{n-1} \mathrm{rot}(s_i), \tag{14}$$

where $\mathrm{rot}(s_i) = 0$ if $s_i = 0$ and $\mathrm{rot}(s_i) = 1$ otherwise. Since the action of the map P carries z around the annulus once if $x \in V_1 \cup V_2$ (Figure 2a) it is not hard to see that the following holds:

Proposition 6.5. (Hockett & Holmes [1986a]): Either $\rho(z, P) = r(\underline{\varphi}(z))$ (both limits exist), or neither rotation number is defined.

Thus, for a q-periodic orbit having word w (of length q), the rotation number is p/q where p = # of 1's and 2's in w.

The key observation for our purposes is:

Proposition 6.6. Any Josephson orbit of period q and rotation number $0 \le p/q \le 1$ on $(\mathcal{K}_A, \bar{\varphi}_t)$ can be arranged as a positive braid with braid number p. (p and q need not be relatively prime).

Figure 9 shows the (pictorial) proof. The period q survives on the new template as the number of intersections with the branch line.

Here it turns out that, on the p-braid template of Figure 9c, the (p,q) resonant torus knots minimise genus among all p-braids of period q: they are the braids whose strands twist and cross as little as possible.

Other torus knots occur in the Josephson problem, among them ones which are essentially horseshoe torus knots "with an extra twist", knots whose words contain only 1's and 2's and which therefore lie on a subtemplate which is that of figure 1d with a full twist added (cf. Figure 2d, with the I_0 band removed.) See Holmes [1986b].

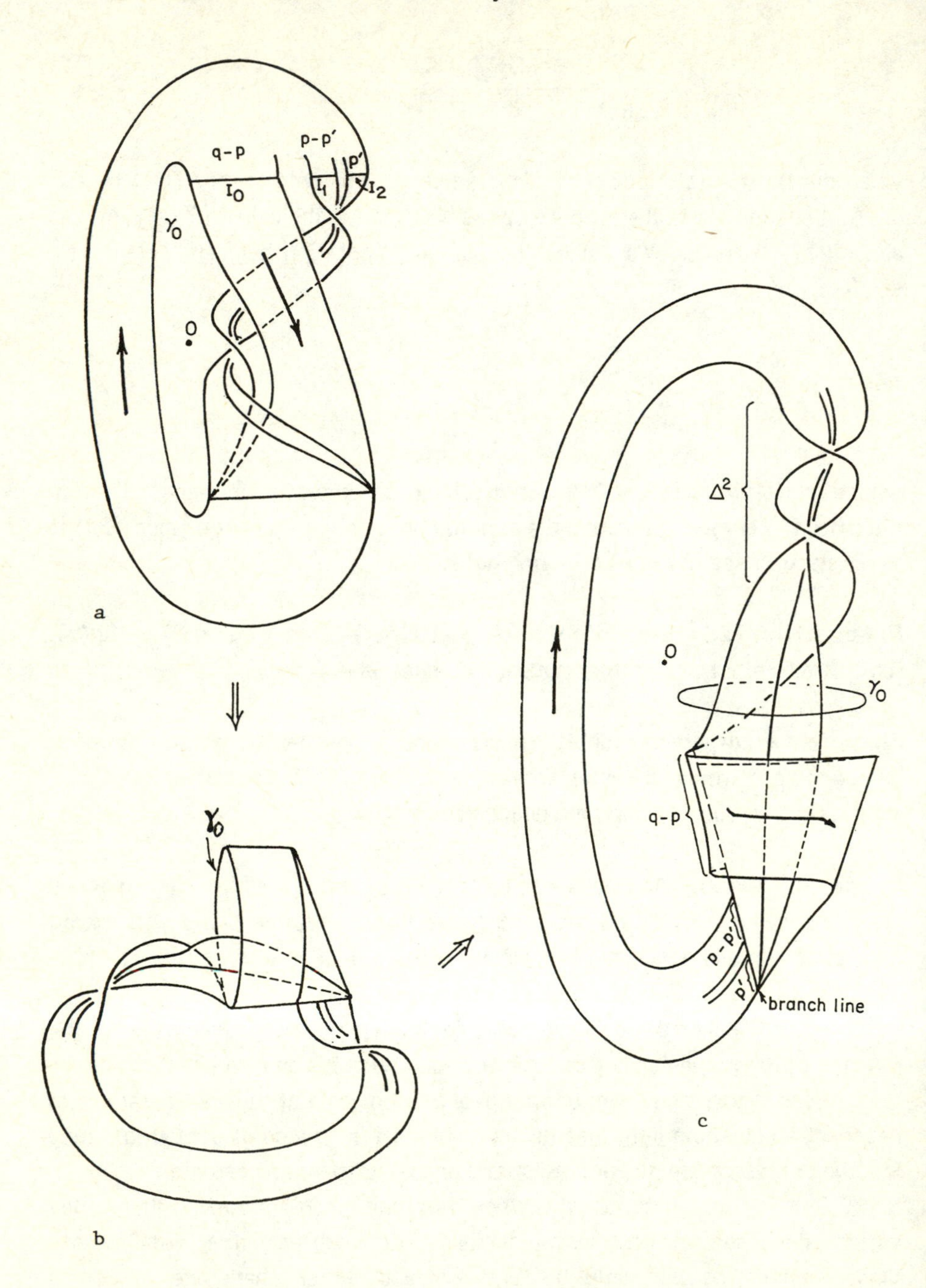

Figure 9. Josephson orbits of rotation number p/q are p-braids.

7 BIFURCATION SEQUENCES FOR TORUS KNOTS

The uniqueness results for horseshoe torus knots of Theorem 6.1 can be used in a surprising way, as we now indicate. We have already seen that the family of one dimensional maps $\{f_\mu\}$ undergoes a countable set of bifurcations to periodic orbits during the creation of the hyperbolic two-shift. Now consider a two parameter family of maps of the form

$$(u, v) \to F_{\mu,\varepsilon}(u,v) = \big(v, -\varepsilon u + f_\mu(v)\big)\ ; \quad (u,v) \in \mathbb{R}^2, \tag{15}$$

having (constant) determinant $\det DF_{\mu,\varepsilon} = \varepsilon$. Letting ε vary from 0 to 1 we mimic the reduction of linear damping in the Poincaré map of a nonlinear oscillator (cf. the discussion below equation (5)). For $\varepsilon = 0$, all orbits are immediately attracted to the curve $v = f_\mu(u)$ and the map's behavior collapses to that of the one dimensional map $v \to f_\mu(v)$. For simplicity take the specific case $f_\mu(v) = \mu - v^2$. In this case, for μ sufficiently large and all $\varepsilon \in [0, 1]$, $F_{\mu,\varepsilon}$ possesses a hyperbolic horseshoe (Devaney & Nitecki [1979]), while for $\mu < -(1+\varepsilon)^2/4$, Ω_μ is empty and all orbits escape to ∞. We ask: In what order are specific families of periodic orbits created as μ increases for fixed ε?

For $\varepsilon = 0$, we have the complete answer of section 5 (Figure 6b). For $\varepsilon = 1$ the map is area preserving and there is a partial answer. In particular, for $\mu \in (-1, 3)$, the quadratic family

$$(u, v) \to (v, -u + \mu - v^2) \tag{16}$$

has an elliptic fixed point $\bar{u} = \bar{v} = -1 + (1+\mu)^{\frac{1}{2}}$ whose eigenvalues traverse the unit circle from +1 to -1 as μ goes from -1 to 3. In doing so they pass monotonically through all roots of unity of the form $e^{\pm 2\pi i p/q}$ with $0 < p/q < \frac{1}{2}$. Normal form theory (e.g. Chenciner [1983]) reveals that at each such passage a pair of q-periodic orbits bifurcates from $(\bar{u}, \bar{v})$, each having rotation number p/q with respect to $(\bar{u}, \bar{v})$. In the "natural" suspension of (16), it is not hard to see that they are (p,q) resonant torus knots.

Let $\mu(\varepsilon, p/q)$ denote the parameter value for which a pair of (p,q) torus knots first appears for $F_{\mu,\varepsilon}$, as μ increases for fixed ε. Confining ourselves to a neighborhood of $(\bar{u}, \bar{v})$, the behavior of the eigenvalues of $DF_{\mu,\varepsilon}$ outlined above implies that if $p_1/q_1 < p_2/q_2$ then $\mu(1, p_1/q_1) < \mu(1,$

p_2/q_2): low rotation numbers appear first. In contrast, examination of the kneading invariants associated with the words of resonant torus knots shows that in the one dimensional limit $\varepsilon = 0$ we have $\mu(0, p_2/q_2) < \mu(0, p_1/q_1)$: high rotation numbers appear first. (With care, the words of the knots can be read off the existence picture of Figure 8).

For μ large the non-wandering set Ω_μ is hyperbolic and we have the uniqueness result of Theorem 6.1. It remains to show that for $\varepsilon = 0$ and $\varepsilon = 1$ the branches of resonant torus knots can be continued uniquely back to $\mu(\varepsilon, p/q)$ as μ decreases. For $\varepsilon = 0$ the kneading theory and invariant coordinates take care of this, while for $\varepsilon = 1$ we have to appeal to symmetry properties of the area preserving map (16) and the proof is quite messy. See Holmes & Williams [1985], §7 for details. The final result is

Theorem 7.1. Let (p_i, q_i), $i = 1, 2$ be two pairs of relatively prime integers with $p_i < q_i/2$, $q_i \geq 5$ and $p_1/q_1 < p_2/q_2$. Let $\mu(\varepsilon, p_i/q_i)$ denote the μ value for which the unique pair of (p_i, q_i) torus horseshoe knots of period q appear in a (saddle-node) bifurcation for the natural suspension of the Hénon map (15). Then $\mu(1, p_1/q_1) < \mu(1, p_2/q_2)$ and $\mu(0, p_2/q_2) < \mu(0, p_1/q_1)$.

Corollary 7.2. Infinitely many saddle-node bifurcation curves cross each other on the (μ, ε) parameter plane between $\varepsilon = 0$ and $\varepsilon = 1$. In particular, each resonant torus bifurcation sequence of period q for the area preserving suspension $(\varepsilon = 1)$ is precisely reversed for the one dimensional suspension $(\varepsilon = 0)$.

For related results see Holmes & Whitley [1984]. Gumowski & Mira [1980a,b] and their colleagues have similar "bifurcation reversal" results for the Hénon map which they obtain by different techniques. See Holmes & Williams [1985] for references. Matsuoka [1985b] also has a rather general ordering theorem for braids which can be seen as an analogue of Sarkovskii's [1964] order theorem (cf. Stefan [1977]) for continuous one dimensional maps.

8 ITERATED TORUS KNOTS, CABLINGS AND ORBIT GENEALOGIES

The results sketched in the proceeding section show that knot theory can be used to determine bifurcation sequences; we now indicate how it can help in the more general question of orbit genealogies.

In Section 5.2 we indicated how certain families of orbits of the one dimensional map f_μ are "factorizable" in the sense that they are created as the parameter μ passes through a periodic window $[s_k^j, h_k^j]$. For example, the 6 period 5-orbits having words

$$w_5^1 = xy^2xy, \qquad \tilde{w}_5^1 = xy^4;$$
$$w_5^2 = x^2y^3, \qquad \tilde{w}_5^2 = x^2yxy;$$
$$w_5^3 = x^4y, \qquad \tilde{w}_5^3 = x^3y^2;$$

are all prime (or merely 1-factorizable), and are created in the pairs indicated at the bifurcation points s_5^j, $j = 1, 2, 3$. The <u>parents</u> of these orbits are the fixed points y and x. Among the 2,182 orbits of period 15, there is precisely one pair "divisible" by each of these period 5 pairs of parents. These 6 period 15 orbits are thus grandchildren of y and x; the other 2,176 orbits of period 15 are prime, or first generation children of y and x. In this way each pair of parents gives birth to a countable family of children, grandchildren, etc. There is even a precise notion of sex: orbits whose words have odd y-parity are female, those with even y-parity are male. Male-female pairs, such as the w_5^j, $\tilde{w}_5^j$ pair listed above, all of whom are parents to be, are created in saddle-node bifurcations. Females are also created singly in period doubling bifurcations.

This somewhat incestuous picture carries over in full to horseshoe knots and links via the subtemplate construction illustrated in Figure 7. Non-trivially factorizable orbits are called iterated horseshoe knots. The technical statement is

Proposition 8.1. (Holmes [1986b]): Corresponding to any set $\underline{\nu} = \{\nu_{k_i}{}^i\}_{0 \le i \le \ell}$ of k_i periodic kneading invariants, there exists a pair of knots K_0 and $\tilde{K}_0$ (a mother and a father) and a subtemplate $\mathcal{L}(\underline{\nu})$ which is an embedded copy of $\mathcal{K}_D$ and which carries a set of iterated horseshoe knots determined by $\underline{\nu}$ (the children, grandchildren, etc.).

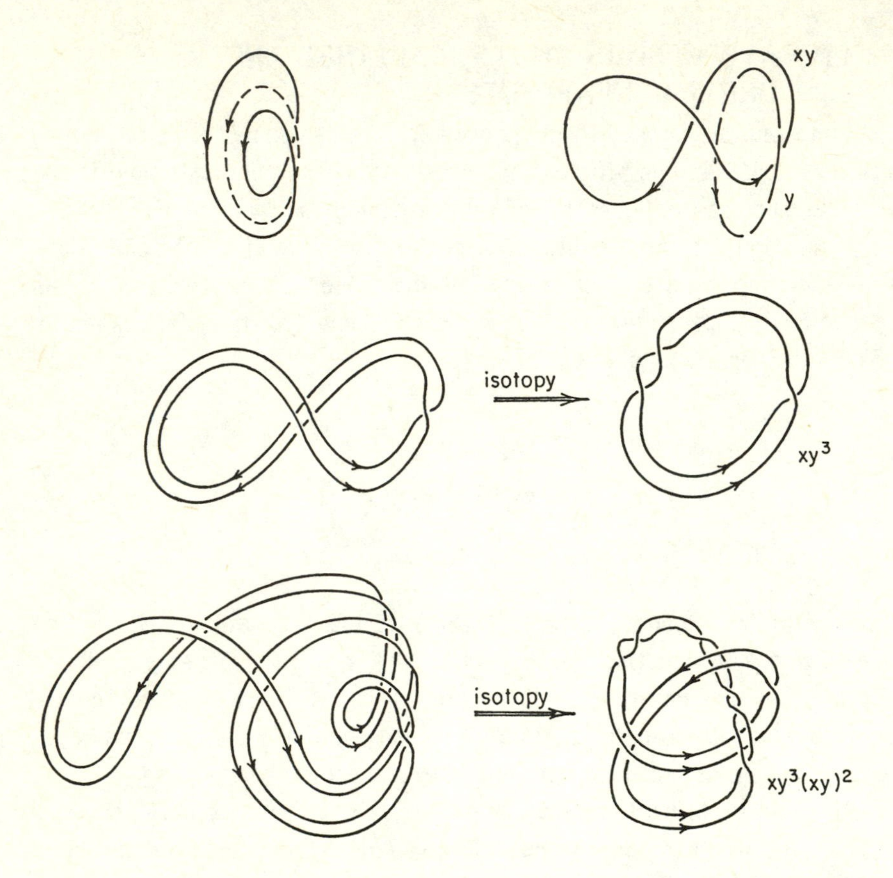

Figure 10. Period doubling.

The simplest such family, containing only children, arises from the basic period doubling sequence bifurcating from y and contains the members {x, y; xy; xy^3; xy^3xyxy; ...); (cf. Feigenbaum [1978]). Figure 10 shows the iterated knot thus created. At each stage we have a 2-cabling of the previous knot (cf. Crawford & Omohundro [1984]).

A second insight into such iterated knots is provided by the generic, self-similar orbit structure near an elliptic fixed point of a two dimensional area preserving diffeomorphism (Arnold & Avez [1966], Zehnder [1973]): This is shown in Figure 11. Visualization of the suspension of this picture leads us to expect families of iterated period multiplying knots which generalize the period doubling families of Figure 10. In fact our wildest dreams are fulfilled: for the horseshoe template we have

Theorem 8.2. (Holmes [1986]): Fix $\ell < \infty$ and an arbitrary sequence $\zeta_\ell = \{(p_j, q_j)\}_{0 \le j \le \infty}$ of relatively prime, positive integer pairs each satisfying $0 < p_j/q_j < \frac{1}{2}$. Let $\underline{\nu}_{q_j}{}^j$ denote the kneading invariant corresponding to the even (p_j, q_j) resonant torus knot. Then corresponding to the kneading invariants $N_\ell = \underline{\nu}_{q_0}{}^0 * \ldots * \underline{\nu}_{q\ell}{}^\ell$ and $N_\ell * \overline{xy}$ there exist iterated horseshoe knots $K_\ell, \tilde{K}_\ell$ which are resonant iterated torus knots of type $\{(q_0, p_0), (q_1, p_0 q_0 q_1 - p_1), \ldots, (q_\ell, b_\ell)\}$ with the b_ℓ computable from the formula $b_j = q_j q_{j-1} b_{j-1} + (-1)^j p_j$. Moreover, K_ℓ and $\tilde{K}_\ell$ are the only iterated torus knots of period $q = \prod_{0 \le j \le \infty} q_j$ with these type numbers.

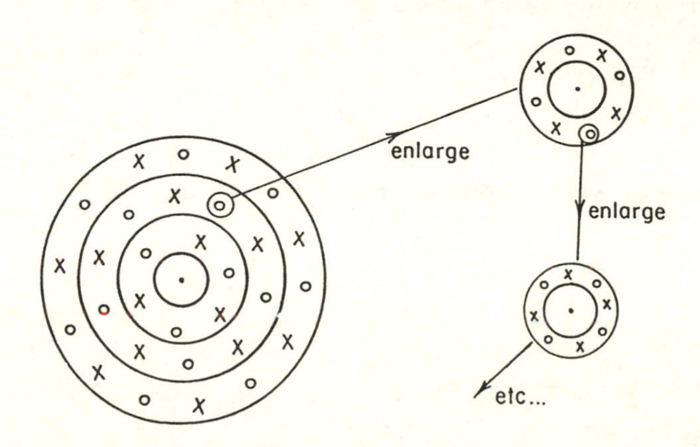

Figure 11. **Generic structure near an elliptic fixed point.**

This theorem is completely "global": it is not restricted to some neighborhood of an elliptic orbit, as is the KAM theorem; more significantly, it shows that the "cycles on cycles" picture extends to dissipative maps.

The iterated horseshoe knots of Theorem 8.2, being built out of resonant torus knots, turn out to be iterated torus knots. There are also more general results which show that <u>any</u> female horseshoe knot can be q-cabled for every $q \ge 2$. These results, applied iteratively with selection of appropriate (female) cables at each stage, yield the sequences of torus knots of Theorem 8.2. As in Section 6, the existence proof involves spotting the torus hidden in a twisted subtemplate such as that of Figure 7, and then computing linking numbers and genera. The torus knot sequences alternately maximise and minimise genus in a surprising and elegant way.

In particular cases one can give closed form expressions for type numbers and linking numbers. See Holmes [1986a] for details.

The number of crossings, and hence the genus, of such knots rises exponentially fast, but numerical work has revealed some of these iterated knots in specific nonlinear equations: cf. Uezu & Aizawa [1982], Beiersdorfer et al. [1983].

The fact that linking and type numbers of such iterated knots are topological invariants implies that the genealogical structure of the one dimensional family $\{f_\mu\}$ extends to the three dimensional horseshoe flow. Informally, we might say that these n^{th} generation children are so badly twisted that only their mothers can recognize them!

9 ENTROPY AND KNOTS

So far we have concentrated on special families of knots which live on specific templates. In this and the next section we turn to more general results, due to John Franks and Bob Williams. The first is:

Theorem 9.1. (Franks & Williams [1985a]): If φ_t is a C^r flow on $\mathbb{R}^3$ or S^3 and either (i) $r \geq 1$ and φ_t has a hyperbolic periodic orbit with a transverse homoclinic point, or (ii) $r \geq 2$ and φ_t has a compact invariant set with positive topological entropy, then φ_t has infinitely many distinct knot types among its periodic orbits.

We sketch the ideas of the proof. A theorem of Katok [1980] shows that (ii) implies (i) and so we may assume the presence of a transverse homoclinic orbit. The Smale-Birkhoff homoclinic theorem (Guckenheimer & Holmes [1983], §5.3) implies that the flow has a hyperbolic nonwandering set with a cross section on which the induced Poincaré map is conjugate to a subshift of finite type. Franks & Williams show that this implies that there is a semi-flow $\bar{\varphi}_t$ and a (possibly poorly embedded) template $\mathcal{K}$ of Lorenz or Duffing type. For a "standard" template the torus knot results of Sections 6 and 8 would immediately imply infinitely many distinct knot types (cf. Corollary 6.4). The problem here is that the "natural" embedding of $\mathcal{K}$ may not even be a braid. However, a trick of Alexander implies that any (orientable or non orientable) template can be arranged as a braid, possibly with some half twists. The braid still need not be positive, but here a recent result of Bennequin [1982] comes to the rescue :

Theorem 9.2. If a knot K is arranged as a braid then its genus satisfies $2g > |c| - n$, where c is the algebraic crossing number, and n is the number of strands in the braid.

Thus we can bound the genera of our knots from below in terms of crossing number, c. Franks and Williams show that c can be made arbitrarily large by taking periodic orbits which follow suitable routes through the template (roughly speaking - the poor embedding makes for extra crossings, and so the genera of knots on the "standard" templates are minimal). Thus they produce knots of arbitrarily large genus, implying the existence of infinitely many knot types. In fact their result can be obtained without Theorem 9.2 by appealing to the bounds on polynomial exponents in Franks-Williams [1985b].

For a result in the direction converse to Theorem 9.1, see Boyland [1985].

10 PRIME KNOTS

We have already used the term "prime" in connection with factorization of kneading sequences. To a knot theorist the word has a different meaning:

Definition. A knot K is prime if it is not the connected sum of any two knots. $K \subset \mathbb{R}^3$ is the connected sum of K_1, K_2 if there is a smooth 2-sphere $S^2 \subset \mathbb{R}^3$ such that S^2 intersects K transversely to exactly 2 points: $K \cap S^2 = \{a, b\}$ and if α is an arc in S^2 joining a to b then $\alpha \cup \gamma_i = K_i$, where $\gamma_1 = K \cap$ (interior of S^2) and $\gamma_2 = K \cap$ (exterior of S^2).

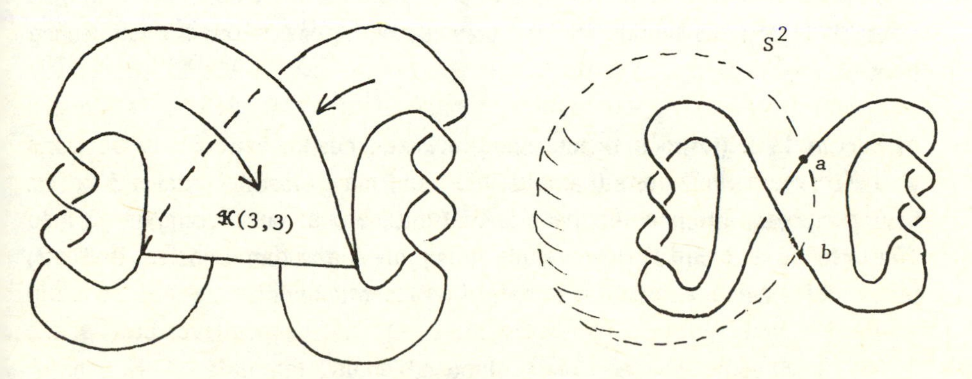

Figure 12. A non-prime knot.

Figure 12 shows an example, due to Williams [1983], of a non-prime knot which is the connected sum of two trefoils or (3, 2) torus knots. It lives on a "generalized horseshoe" template which has 3 positive (=left handed) half-twists each band. Let $\mathcal{K}(n, m)$ denote such a template having n (positive or negative) half twists on the left and m on the right. Thus $\mathcal{K}_D = \mathcal{K}(0, 1)$ and $\mathcal{K}_L = \mathcal{K}(0, 0)$. Let $\mathcal{K}'(n,m)$ denote the mirror image of $\mathcal{K}(n,m)$. Williams [1983] proves the following:

Theorem 10.1. For any $m \geq 0$, the knots on $\mathcal{K}(0, m)$ and $\mathcal{K}'(0, m)$ are prime.

11 CLASSIFICATION OF TEMPLATE KNOTS

The ultimate aim of the work outlined so far is the classification of all knots occurring on specific templates. An obvious question is: do all knots occur on a particular template $\mathcal{K}$? Theorem 6.1 already shows that certain knots might be excluded, and, since the three templates $\mathcal{K}_D$, $\mathcal{K}_J$, and $\mathcal{K}_L$ considered here are all positive braids, knots with "essential" negative crossings like the square or reef knot cannot occur. The prime knot theorem 10.1 of Williams is also relevant in this respect. Leaving aside the more general question of the existence of some "grandfather" template on which all knots might live, we turn to specific classification results of $\mathcal{K}_L$ and $\mathcal{K}_D$. As we have indicated, the problem is very difficult, but limited results are available: among other things they show that torus knots and iterated torus knots are not the only families which occur.

Since all single component 2-braids are torus knots of type (2,2n+1), one moves directly to knots which are 3-braids. Bedient [1985a] has classified Lorenz 3-braids and the corresponding result for the horseshoe template is:

Theorem 11.1. (Holmes, unpublished). Any horseshoe knot K_w with word $w = x^k y^{a_1} x^{\ell} y^{a_2} x^m y^{a_3}$, $k \geq \ell \geq m$; $a_i = 0,1$ or 2, is a closed, positive 3-braid, having representation $\sigma_1^{2b}(\sigma_1\sigma_2)^2 \Delta^{2n}$, $b \geq 0$, $n \geq 1$ or $(\sigma_1\sigma_2)\Delta^{2n}$, $n \geq 1$. The latter is a $(3, 3n+1)$ torus knot. When $b = 0$ the former is a $(3, 3n+2)$ torus knot; when $n = 1$ it is a $(2b+5, 3, -2)$ pretzel knot.

The proof uses the "well disposed braid" template of Holmes & Williams [1985] and involves a case by case study of the 3-braids which occur. In the process one obtains the braid word exponents b, n in terms of the exponents k, ℓ, m, a_i of the knot word w. For example, the pair of knots with words $x^{k+1}y^2x^{k+1}y^2x^ky^2$ and $x^{k+1}y^2x^{k+1}y^2x^{k+1}y$; $k \geq 0$ are $(3, 3k+8)$ resonant torus knots ($b = 0$, $n = 2$). Additional x's can be added to the first syllable without changing the knot type: thus x^ny^6 is a $(3,7)$ torus knot for all $n \geq 1$, but is only resonant for $n = 1$. The lowest period non-torus knots to occur are the pair with words xy^2xy^3, xy^2xyxy, which have braid representation $\sigma_1{}^2(\sigma_1\sigma_2)^2\Delta^2$ and are $(7, 3, -2)$ pretzel knots. For pictures see Birman & Williams [1983b] or Holmes & Williams [1985, §8]. Details will be published later. The 3-braids $\sigma_1{}^{2b}(\sigma_1\sigma_2)^2\Delta^{2n}$ have crossing number (= exponent sum) $6n + 2b + 4$ and thus, from equation (10), genus $g = 3n + b + 1$. That this is not a complete invariant is revealed by computation of Jones polynomials for these braids; b and n occur independently in the polynomials and thus both b and n are knot invariants. Knots with $3n_1 + b_1 = 3n_2 + b_2$, $n_1 \neq n_2$ are non-isotopic knots of the same genus.

Although similar methods will work for $(p \geq 4)$-braids, the calculations become extremely long and no clear patterns have emerged so far. Nonetheless, several interesting classification issues arise, one being the questions of non-uniqueness. As we have indicated, one way in which non uniqueness of knot types occurs is that extra "trivial" loops can be added as one passes around a non-twisted band of $\mathcal{K}$. This suggests that "well-twisted" templates, such as the twisted horseshoe subtemplate of the Josephson template (section 6) will have better uniqueness properties than templates lacking a full twist. A second way in which pairs of isotopic knots arise is when one loop can be lifted from left to right or vice versa, corresponding to an orbit which has crossed the critical point during parameter variation for the one dimensional family $\{f_\mu\}$. Such pairs arise in saddle-node bifurcations. Although no bifurcation pairs exist in the Lorenz problem, there one has $x \leftrightarrow y$ symmetry: the words $\sum x^{a_i}y^{b_i}$ and $\sum x^{a_i}y^{b_i}$ correspond to isotopic knots.

12 CONCLUSIONS AND OPEN PROBLEMS

In this paper we have tried to give a taste of recent work on knotted orbits in dynamical systems, with samples of results and an indication of the methods used in their derivation. The references we have cited contain considerably more: for example one can prove that there are infinitely many template knots which are neither torus knots nor iterated torus knots (for example, the families of pretzels of Theorem 11.1) and one can obtain bounds on the genera of knots in terms of their periods and braid or trip numbers. Birman & Williams [1983a] listed several conjectures, questions and problems arising from their study of Lorenz knots and links. At least three of these have now been settled: positive topological entropy implies (infinitely many) knotted orbits (Section 9); Lorenz trip number = braid number and Lorenz knots are prime (Section 10). Moreover, the Duffing and Josephson template studies of Holmes & Williams [1985] and Holmes [1986a,b] and in particular the subtemplates discussed in Sections 5 and 8, address their question of "variants" of the Lorenz template which occur naturally in differential equations. However, several substantial open problems remain.

A major requirement in the theory is the development of algebraic invariants, zeta functions or link polynomials, which compactly encode all the knots and links of a given template. There should at least be a systematic way of computing polynomial invariants. Williams [1977] introduced a type of zeta function for the Lorenz template and Birman & Williams [1983a] continued this work, but at present computation must still proceed case by case, although Bedient [1985b] has implemented the computation procedure for Alexander polynomials in general two letter Lorenz templates $\mathcal{K}(n, m)$ as a Macintosh Pascal program. (He also has a program for braids.) The recent development of new polynomial invariants (Jones [1985, 1986]) has radically changed the situation for knot theorists. At present, few general results for specific knot and link families are available, and the polynomials must be computed iteratively using "Conway moves". Obvious questions are: do the special template structures (positive braids, "regular" twisting) promote simpler iterative formulae for such computations? Do iterated horseshoe and similar knots and links satisfy multiplicative formulae of the type derived by Seifert [1950] for Alexander polynomials of companions and satellites? Until answers to such general questions are available, piecemeal computations of specific cases are probably worthwhile: one should assemble a zoo of template knots and links.

A second area requiring work is the study of multi-component links of periodic orbits. The families of iterated horseshoe knots of section 8, linked about their parents, provide examples, but a more general study might yield useful ordering information on bifurcation sequences, since linking number provide additional topological invariants to those of individual knots. Linking will also be useful in determining bifurcation sequences. While we presently see this as a major application of knot theoretic methods, we confidently expect that many more uses will emerge, and, in any case, the subject is a delightful cat's cradle.

REFERENCES

V.I. Arnold [1982]. *Geometrical Methods in the Theory of Ordinary Differential Equations.* Springer Verlag, Berlin, Heidelberg, New York (Russian original, Moscow, 1977).

V.I. Arnold & A. Avez [1968]. *Ergodic Problems of Classical Mechanics.* New York, W.A. Benjamin Inc.

R.E. Bedient [1985a]. Classifying 3-trip Lorenz knots. *Topology Appl.* **20**, 89-96.

R.E. Bedient [1985b]. MacBraid, A Macintosh Pascal Program.

P. Beierdorfer, J.-M. Wersinger & Y. Trève [1983]. Topology of the invariant manifolds of period-doubling attractors for some forced nonlinear oscillators. *Phys. Lett.* **96A**, 269-272.

D. Bennequin [1982]. "Entrelacements et équations de Pfaff". Thèse de Doctorat d'Etat, Université de Paris VII.

J.S. Birman [1975]. *Braids, Links and Mapping Class Groups.* Princeton University Press, Princeton, N.J.

J. Birman & R.F. Williams [1983a]. Knotted periodic orbits in dynamical systems I: Lorenz's equations. *Topology* **22**, 47-82.

J. Birman & R.F. Williams [1983b]. Knotted periodic orbits in dynamical systems II: knot holders for fibered knots. *Contemporary Mathematics* **20**, 1-60.

R. Bowen [1978]. *On Axiom A Diffeomorphisms.* In CBMS Regional Conference Series in Mathematics 35. AMS Publications, Providence, RI.

P. Boyland [1983]. Bifurcations of circle maps, Arnold tongues, bistability and rotation intervals. Preprint.

P. Boyland [1985]. Braid types and a topological method of proving positive entropy. Preprint.

A. Chenciner [1983]. Bifurcations de difféomorphismes de $\mathbb{R}^2$ au voinsinage d'un point fixe élliptique. In *Les Houches Summer School Proceedings*, ed. R. Hellemen, G. Iooss, North Holland.

P. Collet & J.P. Eckmann [1980]. *Iterated Maps on the Interval as Dynamical Systems.* Birkhauser, Boston.

J.D. Crawford & S. Omohundro [1984]. On the global structure of period doubling flows. *Physica* **13D**, 161-180.

B. Derrida, A Gervois & Y. Pomeau [1978]. Iteration of endomorphisms on the real axis and representation of numbers. *Ann. Inst. H. Poincaré* **A29**, 305-356.

R. Devaney [1985] *An Introduction to Chaotic Dynamical Systems.* Benjamin Cummings, Menlo Park, CA.

R. Devaney & Z. Nitecki [1979]. Shift automorphisms in the Hénon mapping. *Comm. Math. Phys.* **67**, 137-148.

A. Douady & J.H.. Hubbard [1982]. Itération des polynomes quadratrques complexes. *C.R. Acad. Sci. Paris* **294** Série I, 123-126.

D. Eisenbud and W. Neumann [1977] preprint. Fibering iterated torus links.

M.J. Feigenbaum [1978]. Quantitative universality for a class of nonlinear transformations. *J. Statist. Phys.* **19**, 25-52.

J. Franks [1981]. Knots, links and symbolic dynamics. *Ann. of Math* **113**, 529-552.

J. Franks & R.F. Williams [1985a]. Entropy and Knots. *Trans. Amer. Math. Soc.* **291**, 241-253.

J. Franks & R.F. Williams [1985b]. Positive braids via the Jones polynomial. *Trans. Amer. Math. Soc.* (to appear).

B.D. Greenspan & P.J. Holmes [1983]. Homoclinic orbits, subharmonics, and global bifurcations in forced oscillations. Chapter 10, pp. 172-214 in *Nonlinear Dynamics and Turbulence* ed. G. Barenblatt, G. Iooss and D.D. Joseph, Pitman, London.

B.D. Greenspan & P.J. Holmes [1984]. Repeated resonance and homoclinic bifurcation in a periodically forced family of oscillators. *SIAM J. Math. Anal.* **15**, 69-97.

J. Guckenheimer [1977]. On the bifurcation of maps of the interval. *Invent. Math* **39**, 165-178.

J. Guckenheimer [1979]. Sensitive dependence on initial conditions for one dimensional maps. *Comm. Math. Phys.* **70**, 133-160.

J. Guckenheimer [1980]. Bifurcations of Dynamical Systems. In *Dynamical Systems*, ed. J.K. Moser, Birkhauser, Boston.

J. Guckenheimer & P.J. Holmes [1983]. *Nonlinear Oscillations, Dynamical Systems and Bifurcations of Vector Fields.* Springer Verlag, New York.

J. Guckenheimer & R.F. Williams [1979]. Structural Stability of Lorenz Attractors. *Inst. Hautes Etudes Sci. Publ. Math.* **50**, 59-72.

I. Gumowski & C. Mira [1980a]. *Dynamique Chaotique,* Editions Cepadue, Toulouse, France.

I. Gumowski & C. Mira [1980b] *Recurrences and Discrete Dynamical Systems.* Springer Lecture Notes in Mathematics Vol. 809, Springer Verlag, Berlin, Heidelberg, New York.

C. Hayashi [1964]. *Nonlinear Oscillations in Physical Systems.* McGraw Hill, New York.

M. Hénon [1976]. A two dimensional mapping with a strange attractor. *Comm. Math. Phys.* **50**, 69-77.

K. Hockett [1986]. Bifurcations of Rational and Irrational Cantor Sets and Periodic Orbits in Maps of the Circle. Ph.D. Thesis, Cornell University.

K. Hockett and P.J. Holmes [1986a]. Josephson's junction, annulus maps, Birkhoff attractors, horseshoes and rotation sets. *Ergod. Th. and Dyn. Sys.* **6**, 205-239.

K. Hockett and P.J. Holmes [1986b]. Bifurcation to rotating Cantor sets in maps of the circle. Preprint, Cornell University.

P.J. Holmes [1979]. A nonlinear oscillator with a strange attractor. *Philos. Trans. Roy. Soc. London* **A292**, 419-448.

P.J. Holmes [1986a]. Knotted periodic orbits in suspensions of Smale's horseshoe: period multiplying and cabled knots. *Physica* **21D**, 7-41.

P.J. Holmes [1986b]. Knotted periodic orbits in suspensions of annulus maps. *Proc. Roy. Soc. London.* (to appear).

P.J. Holmes and D.C. Whitley [1983]. On the Attracting Set for Duffing's Equation II: A Geometrical Model for Moderate Force and Damping. In *Proc. Order in Chaos, Los Alamos National Laboratory,* May 1982 - also *Physica* **7D**, 111-123.

P.J. Holmes & D.C. Whitley [1984]. Bifurcations of one and two dimensional maps. *Philos. Trans. Roy. Soc. London* **A311**, 43-102.

P.J. Holmes & R.F. Williams [1985]. Knotted periodic orbits in suspensions of Smale's horseshoe: torus knots and bifurcation sequences. *Arch. Rational Mech. Anal.* **90**, 115-194.

V.F.R. Jones [1985]. A Polynomial Invariant for knots via von Neumann algebras. *Bull. Amer. Math. Soc.* **12**, 103-111.

V.F.R. Jones [1986]. A new knot polynomial and von Neumann algebras. *Notices Amer. Math. Soc.* **33**, 2, 214-225.

L. Jonker [1979]. Periodic orbits and kneading invariants. *Proc. London Math. Soc.* **39**, 428-450.

L. Jonker & D.A. Rand [1981]. Bifurcations in one dimension I: the nonwandering set and II: A versal model for bifurcations. *Invent. Math.* **62**, 347-365 and 63, 1-15.

E.N. Lorenz [1963]. Deterministic non-periodic flow. *J. Atmospheric Sci.* **20**, 130-141.

T. Matsuoka [1983]. The number and linking of periodic solutions of periodic systems. *Invent. Math* **70**, 319-390.

T. Matsuoka [1984]. Waveform in the study of ordinary differential Equations. *Japan J. Appl. Math* **1**, 417-434.

T. Matsuoka [1985a]. The number and linking of periodic solutions of non-dissipative systems. Preprint.

T. Matsuoka [1985b]. Braids of periodic points and a 2-dimensional analogue of Sarkovskii's ordering. Preprint.

V.K. Melnikov [1963]. On the stability of the center for time-periodic perturbation. *Trans. Moscow Math. Soc.* **12**, 1-57.

J. Milnor & R. Thurston [1977]. On iterated Maps of the Interval I and II. Unpublished notes, Princeton University, Princeton, New Jersey.

J. Moser [1973]. *Stable and Random Motions in Dynamical Systems.* Princeton University Press, Princeton, New Jersey.

S.E. Newhouse [1980]. Lectures on Dynamical Systems. In *Dynamical Systems*, ed. J.K. Moser, Birkhauser, Boston.

D. Rolfsen [1977]. *Knots and Links.* Publish or Perish, Berkeley, California.

A.N. Sarkovskii [1964]. Coexistence of cycles of a continuous map of a line into itself. *Ukr. Math. Z* **16**, 61-71.

H. Seifert [1950]. On the homology invariants of knots. *Quart. J. Math. Oxford* (2) **1**, 23-32.

S. Smale [1963]. Diffeomorphisms with many periodic points. In *Differential and Combinatorial Topology*, ed. S.S. Cairns, pp. 63-80, Princeton University Press, Princeton, New Jersey.

S. Smale [1967]. Differentiable dynamical systems. *Bull. Amer. Math. Soc.* **73**, 747-817.

C.T. Sparrow [1982]. *The Lorenz Equations: Bifurcations, Chaos and Strange Attractors.* Springer Verlag: Berlin, Heidelberg, New York.

P. Stefan [1977]. A theorem of Sarkovskii on the existence of periodic orbits of continuous endomorphisms of the real line. *Comm. Math. Phys.* **54**, 237-248.

Y. Ueda [1981]. Personal communication to P. Holmes

T. Uezu & Y. Aizawa [1982]. Topological character of a periodic solution in three dimensional ordinary differential equation system. *Progr. Theoret. Phys.* **68**, 1907-1916.

R.F. Williams [1974]. Expanding attractors. *Inst. Hautes Etudes Sci. Publ. Math.* **43**, 169-203.

R.F. Williams [1977]. The structure of Lorenz attractors, pp. 94-116. In *Turbulence Seminar, Berkeley 1976/77,* A. Chorin, J.E. Marsden, S. Smale (eds), Springer Lecture Notes in Math. Vol. 615.

R.F. Williams [1979]. The structure of Lorenz attractors. *Inst. Hautes Etudes Sci. Publ. Math.* **50**, 73-99.

R.F. Williams [1983]. Lorenz knots are prime. *Ergod. Th. & Dyn. Sys.* **4**, 147-163.

LIMIT CYCLES OF POLYNOMIAL SYSTEMS–SOME RECENT DEVELOPMENTS

N.G. Lloyd

Department of Mathematics
The University College of Wales
Aberystwyth
Dyfed, Wales

1 INTRODUCTION

This paper is concerned with systems of the form

$$\dot{x} = P(x, y), \quad \dot{y} = Q(x, y) \tag{1}$$

in which P and Q are polynomials. An account is given of some recent work concerning the number of limit cycles of such systems and their relative configurations, twin questions often referred to as Hilbert's sixteenth problem. (Lest there should be any confusion, a limit cycle is, of course, an <u>isolated</u> closed orbit.)

Two-dimensional flows of the form (1) continue to be investigated intensively, partly because of the many applications in which they arise, but also because of their intrinsic interest. Since they do not display the complicated dynamical features which can arise only in higher dimensions, it is reasonable to expect to be able to describe their orbit structure relatively completely. In part, this paper is concerned with the degree to which this expectation is fulfilled.

To describe Hilbert's problem precisely, let S_n be the collection of systems of the form (1) with P and Q of degree at most n, and let $\pi(P, Q)$ be the number of limit cycles of (1). Denoting the system (1) by (P, Q), define the so-called Hilbert numbers

$$H_n = \sup \{ \pi(P, Q) ; \ (P, Q) \in S_n \}. \tag{2}$$

The question is to estimate H_n in terms of n and to obtain information about the possible relative configurations of limit cycles. This is the second part of the sixteenth of the list of problems enunciated by Hilbert in his address to the International Congress of Mathematicians in Paris in 1900, and described in print in [34]. It has proved to be a remarkably intractable question.

Part of the fascination of the problem is the simplicity of its statement, a simplicity that belies the difficulty of making appreciable progress. Another striking aspect is that the hypothesis is algebraic, while the conclusion is topological. The problem as I have stated it is open ended, requiring, as it does, information about the possible configurations of limit cycles. It is therefore more realistic to think in terms of a whole area of investigation rather than a single problem to which a concise solution may be attainable.

This paper is not designed to be an encyclopaedic survey of the subject. My aim is the modest one of seeking to communicate some of the charm of the problem by describing some aspects of the theory which have attracted my interest and on which my colleagues, my students and myself have worked during the last few years. I shall highlight some particular questions which, though as yet unresolved, appear to merit further investigation, and I shall suggest some lines of thought which appear promising. Generally speaking, I shall omit technical details, but rather refer the interested reader to the specialist literature. I emphasise that this is very much a personal view of the subject. Without doubt I shall fail to refer to many valuable contributions: for this I apologise in advance.

Apart from the classical work of Dulac [27], relatively little work of direct relevance to Hilbert's sixteenth problem appeared until the fifties. Then Bautin [5] proved that $H_2 \geq 3$, a genuinely significant result to which I shall return. Two papers by Petrovskii & Landis ([63], [64]) then appeared; in [63] it was claimed that $H_2 = 3$, and in [64] an upper bound for H_n was given when $n > 2$. These authors considered (1) with both independent and dependent variables complex, and used some sophisticated geometric ideas. However, it was soon realised that the proofs were incomplete, and the authors withdrew their results [65]. Nevertheless, it seems to have been widely believed that H_2 was indeed equal to 3, and it was not until 1979 that the first examples appeared of quadratic systems with at least four limit cycles; these were given by Shi [77], and Chen & Wang [12]. Interest in

Hilbert's problem has quickened rapidly since the appearance of these examples, and has been further stimulated by the notable work of several other Chinese mathematicians, a large proportion of which has been concerned with quadratic systems.

The literature which impinges in some way or other on Hilbert's sixteenth problem is very extensive. At this stage I refer to two survey articles by Coppel on quadratic systems: [20], written in 1966, which is widely cited, and another which is about to appear [22]. Three more general short surveys of particular aspects are also worthy of note: those of Chicone & Tian [17], Tian [85] and Ye [87]. Of great usefulness, especially for some of the technical results which have originally appeared in Russian or Chinese, is the recent translation of the second edition of Ye's book 'Theory of Limit Cycles' [89].

The remainder of the paper is organized into six sections:

Section 2: General remarks and Dulac's theorem
Section 3: Small-amplitude limit cycles
Section 4: Quadratic systems
Section 5: Cubic systems
Section 6: Liénard equations
Section 7: Systems with homogeneous nonlinearities.

At the end of each section I shall briefly note some open questions.

2 GENERAL REMARKS AND DULAC'S THEOREM

Flows in the plane cannot be completely understood without taking their behaviour at infinity into account. Consequently, in the classification of any class of systems (1), one has to work over $\mathbb{R}^2$ with infinity adjoined in some way. Two compactifications of $\mathbb{R}^2$ are used: the sphere and the projective plane. The flow defined by a polynomial system can be extended continuously to a flow on either, a fact which was known to Poincaré and exploited by him.

To extend the flow to the sphere is relatively straightforward, and I shall not describe the details ([23] is a suitable reference). The north pole of the sphere is an isolated critical point of the extended flow, but it is degenerate. Consequently, it is often preferable to work in the projective plane, and this is the setting for much of the work on Hilbert's problem. The technical details of the transformation are given in Gonzales [33], for example, and also in a number of papers to be cited later (for instance [4] and [19]). Most authors use the disc as a model for the projective plane; the line at infinity is the boundary of the disc with diametrically opposed points identified This is often described as the Poincaré compactification. Supposing that the origin is a critical point, system (1) in polar coordinates takes the form

$$\left.\begin{aligned} \dot{r} &= r f_1(\theta) + r^2 f_2(\theta) + \dots + r^n f_n(\theta) \\ \dot{\theta} &= g_1(\theta) + r g_2(\theta) + \dots + r^{n-1} g_n(\theta) \ , \end{aligned}\right\} \qquad (3)$$

where, for $k=1,\dots,n$, $f_k(\theta)$ and $g_k(\theta)$ are homogeneous polynomials of degree $(k+1)$ in $\cos\theta$ and $\sin\theta$. The critical points on the line at infinity are given by the zeros of g_n. It is clear that if P and Q do not have a common factor, then (1) has finitely many critical points in total (that is, in the finite plane and at infinity).

Having defined the Hilbert numbers H_n, the obvious first question is whether they are finite. Remarkably, this is unknown, even for $n = 2$. What, then, about individual systems? - can polynomial systems have infinitely many limit cycles? Dulac considered this question in [27], a long article which appeared in 1923 and had considerable influence on the development of the whole subject. He came to the conclusion that analytic systems have finitely many limit cycles; unfortunately, however, it became clear that his proof was incomplete (see [39]). It has recently been shown by Bamón (see [4]) that quadratic systems cannot have infinitely many limit cycles, but Dulac's 'theorem' remains unproved for $n > 2$.

The whole question of finiteness is more subtle than one might expect. It is easy to see that limit cycles cannot accumulate on a closed orbit. For suppose that Γ is a closed orbit; take a line ℓ transverse to Γ and a coordinate x on ℓ such that $x = 0$ corresponds to Γ. For small x, the return map $h(x)$ is defined: $h(x)$ is the next crossing of the positive semi-orbit through x with ℓ. Closed orbits correspond to zeros of $H(x) = h(x) - x$. By standard results on the dependence of solutions on initial conditions, H is analytic at $x = 0$, and so its zeros are isolated or H is identically zero in an interval. Consequently, limit cycles cannot accumulate on Γ.

The remaining possibilities are that limit cycles accumulate at a critical point or on a separatrix cycle (which may include segments of the line at infinity). That there are indeed no other options is proved rigorously by Perko [62]; he also shows that if there is a band of closed orbits, then the outer boundary is a separatrix cycle and the inner boundary is either a critical point or a separatrix cycle. At first sight, it would appear that an argument using the analyticity of the return map would exclude the possibility that limit cycles can accumulate at a critical point. This fails, however, because the return map, though analytic in a deleted neighbourhood of the origin, may not be analytic at the origin itself. Using the polar form (3), the difficulty can occur when, for example, $g_k(\theta) \equiv 0$ for $k = 1, \ldots, j$ for some j, but not all the f_k with $k \leq j$ are identically zero. A critical point at which limit cycles accumulate is called a centre-focus. That polynomial systems can have such a critical point has not been excluded completely, though it is widely assumed that it is not possible. Perko [61] gives a variety of conditions on P and Q which ensure that (1) cannot have a centre-focus; in particular, he shows that limit cycles cannot accumulate at a critical point if $n = 4$.

It remains to consider the accumulation of limit cycles on a separatrix cycle. As well as noting the incompleteness of Dulac's proof in general, it is noted in [16] that Sotomayor pointed out that the argument in [27] is correct in the case of a homoclinic orbit at a non-degenerate critical point. More generally, Il'yashenko [38] proves that (1) has finitely many limit cycles if all the critical points of the system, including those at infinity, are non-degenerate. His arguments are extremely interesting; he works with the complexification of the return map and its continuation in the appropriate universal covering surface. Il'yashenko also deduces that in the space of systems (1) in which P and Q are of degree at most n, there is an algebraic submanifold such that systems in its complement have finitely many limit cycles. This was conjectured by Sotomayor & Paterlini [82], and proved by them for $n = 2$.

The results of [38] apply to analytic, and not just polynomial, systems, and in a more recent paper [39], Il'yashenko generalises some of these results to the case of smooth vector fields, and also presents some valuable work on the classification of critical points.

We turn now to quadratic systems. In a notable paper, Chicone & Shafer [16] proved that limit cycles cannot accumulate on a bounded separatrix cycle. Using this fact and Il'yashenko's theorem, Bamón [4], by a careful consideration of cases, has shown that Dulac's theorem is true for quadratic systems. I shall have more to say about this in Section 4.

The work of Petrovskii and Landis was mentioned in Section 1. Despite the difficulties which arose, the ideas which they introduced, from complex analytic geometry and the theory of holomorphic foliations, are deep and well worth careful consideration. In a series of papers, Il'yashenko has developed these techniques skilfully (see [36,37] for example, and also the book of Jouanolou [41]).

Nothing has yet been said about the periods of limit cycles. Reparameterisation of time, as occurs, for instance, when extending a flow to a compactification of the plane, obviously does not preserve the periods of limit cycles, though the orbit structure is, of course, invariant. Thus, the time taken to traverse orbits tends to attract little attention. However, there is one recent result involving the periods of limit cycles which is not only of intrinsic interest but suggests what may be an extremely useful approach to the whole question of Dulac's theorem. Françoise & Pugh [29], working from the start in the Poincaré compactification, prove that, given T, there is $b(n,T)$ such that a system (1) in S_n has at most $b(n,T)$ limit cycles of period less than T. They use techniques from algebraic geometry, and also prove that the arc length of any closed orbit of (1) is less than $2\pi(n+1)$, again in the Poincaré compactification.

Leaving aside the issue of finiteness of H_n, it is certainly reasonable to ask for lower bounds for H_n. It is easy to see that $H_n \geq [\frac{1}{2}(n-1)]$, where $[\cdot]$ denotes "integer part". For consider the system

$$\dot{x} = y + x(r^2-1)\dots(r^2-m)$$
$$\dot{y} = -x + y(r^2-1)\dots(r^2-m)$$

where $r^2 = x^2+y^2$ and $m = [\frac{1}{2}(n-1)]$. In polar coordinates,

$$\dot{r} = r(r^2-1)(r^2-2)\dots(r^2-m), \quad \dot{\theta} = -1.$$

Clearly, the circles $r = 1, \sqrt{2}, \dots, \sqrt{m}$ are limit cycles, and there are no others. It will be seen in Section 3 how this bound can be improved to be $O(n^2)$.

Open questions. Proving that a polynomial system cannot have infinitely many limit cycles is likely to require some sophisticated geometric ideas. A proof in particular cases should be possible; cubic systems are the obvious candidates, but the proof for $n=2$ uses so many of the special properties of quadratic systems that even this case is likely to be difficult. Another possibility is to consider systems in which the nonlinearities are homogeneous (such systems will be discussed in Section 7).

Once a proof that a polynomial system cannot have infinitely many limit cycles is forthcoming, attention will switch to the question of whether the Hilbert numbers H_n are finite.

Improved lower bounds for H_n should be sought. This will give information on the asymptotic growth of H_n (an interesting question recently posed to me by John Toland). It will be seen in Section 3 that H_n grows at least as fast as n^2. I conjecture that $H_n = O(n^3)$. My reasoning is simply that $O(n^2)$ critical points can be encircled by limit cycles, and that there are likely to be at most $O(n)$ limit cycles around each critical point.

It should be possible to prove that limit cycles cannot accumulate at a critical point, perhaps by considering the complexified return map (Perko and I have already done some work along these lines).

In general, some technique for tracking limit cycles is required. This might involve a classification of the kinds of bifurcations which limit cycles of polynomial systems can undergo, but this seems to be a realistic goal only for particular classes of equations - see the work of Rousseau [71,72], for example, on homoclinic loop bifurcations in quadratic systems. To make significant overall progress it is probably necessary to work with the appropriate complexified forms - we shall explain the reason for this in Section 7. Petrovskii and Landis, and more recently Qin [68] have complexified both the dependent and independent variables. The trouble then is that one has solution surfaces rather than curves, and my own inclination is to complexify the dependent variables only (see Section 7).

Added in proof. In a remarkable development Ecalle, Martinet, Moussu & Ramis have announced that they have proved Dulac's theorem that no polynomial system can have infinitely many limit cycles. A brief announcement is contained in [96] and [97] (see also [100]). The proof exploits the properties of certain formal series which, though not convergent, are quasi-analytic in a defined sense. Applying this theory to the return map, it is deduced that limit cycles cannot accumulate. I understand that the complete proof runs to well over a hundred pages: a full understanding of it will itself present a formidable challenge!

3 SMALL-AMPLITUDE LIMIT CYCLES

Much of the recent progress on Hilbert's sixteenth problem has involved the construction of systems with small-amplitude limit cycles. These are limit cycles which bifurcate out of a critical point under perturbation of the equations. The pioneering work of Bautin [5] was concerned with such limit cycles - he showed that for quadratic systems at most three small-amplitude limit cycles can bifurcate out of one critical point. Several other instances will be noted in later sections, including the examples of quadratic systems with four limit cycles given in, for example, [12], [67], [77], [44] and [8].

In this section, I describe how these small-amplitude limit cycles can be generated. The idea is to start with a system (1) for which the origin, say, is a critical point of focus type which is as near as is possible to a centre, and to make a sequence of perturbations of the coefficients of P and Q each of which reverses the stability of the origin, thereby causing a limit cycle to bifurcate.

Suppose, then, that the origin is a critical point of focus type. It is said to be a *fine focus* (or a *weak focus*) if it is a centre for the linearised system

$$\dot{\xi} = \begin{bmatrix} P_x & P_y \\ Q_x & Q_y \end{bmatrix} \xi \qquad (\xi \in \mathbb{R}^2) .$$

We choose coordinates so that our system is of the form

$$\dot{x} = \lambda x + y + p(x, y) , \quad \dot{y} = -x + \lambda y + q(x, y) , \tag{4}$$

where p and q are polynomials without linear terms. Thus the origin is a fine focus if $\lambda = 0$. In polar coordinates, we have

$$\dot{r} = \lambda r + O(r^2) , \quad \dot{\theta} = -1 + O(r) \quad \text{as } r \to 0.$$

Clearly $\dot{\theta} \neq 0$ in a neighbourhood of the origin, and we have a 'swirling flow', at least locally. The return map h, say, is therefore defined for sufficiently small x: for $x > 0$, $(h(x), 0)$ is the next crossing of the orbit through $(x, 0)$ and the positive x-axis. Limit cycles of (4) correspond to the zeros of the displacement map

$$H : x \mapsto h(x) - x .$$

Since the zeros of H occur in pairs - one positive and one negative - and $H(0) = 0$, the multiplicity μ of $x = 0$ as a zero of H is odd. If $\mu = 2k+1$, then at most k limit cycles can bifurcate from the origin under perturbation of the coefficients of p and q. However, it may not be possible to generate the full complement of k limit cycles without introducing perturbations which go outside the class of systems under discussion.

The technique is to start with a system for which $x = 0$ is a zero of H of maximal multiplicity in the sense that H is identically zero in a neighbourhood of $x = 0$ if this multiplicity is exceeded (in which case the origin would be a centre for the full nonlinear system). A sequence of perturbations is chosen each of which reverses the stability of the origin.

How, then, can these rather theoretical ideas be implemented? A full account of the technique described here is that given by us in [8]. It differs from that used by Bautin [5] for quadratic systems and by Sibirskii [78,80] for certain cubic systems in that it uses the form (4) of the equations rather than the corresponding polar form. Shi [77] proceeds along similar lines, and Sleeman [81] describes a somewhat different, though related, approach; the paper of Göbber & Willamowski [32] is also relevant. The ideas described by Françoise [28] are from a different viewpoint and are instructive.

Rather than work directly with the function H, we seek a Liapunov function V in a neighbourhood of the origin such that $\dot{V}$, the rate of change of V along orbits is of the form $\eta_2 r^2 + \eta_4 r^4 + \ldots$.

This procedure is classical and is described in, for example, Nemytskii & Stepanov [59]. The η_{2k} are the *focal values* and are polynomials in the coefficients arising in p and q. It is clear that the origin is stable or unstable according to whether the first non-zero focal value is negative or positive; the origin is a centre if all the η_{2k} are zero. It is easily calculated that $\eta_2 = \lambda$.

Since it is only the first non-zero focal value that is of significance, we reduce each η_{2k} modulo the ideal $\langle \eta_2, \eta_4, \ldots, \eta_{2k-2} \rangle$ generated by the 'previous' η_{2j}. This is done by substituting $\eta_2 = \ldots = \eta_{2k-2} = 0$ into the expression for η_{2k}. The polynomials which are obtained in this way are the *Liapunov quantities*: L(0), ..., L(M), say. The set of Liapunov quantities is called the *focal basis*.

Definition. Suppose that the origin is a fine focus of (4). Its *order* is k if $\eta_2 = \eta_4 = \ldots = \eta_{2k} = 0$ but $\eta_{2k+2} \neq 0$.

It can be shown that at most k limit cycles can bifurcate out of a fine focus of order k. The proof is given in [8] and proceeds by establishing a connection between the order of the fine focus as defined above and the multiplicity of $x = 0$ as a zero of the displacement function H.

Given a class of systems, the first task is to ascertain the maximum possible order of a fine focus. This is done by setting successive focal values to be zero until it appears that all focal values are necessarily zero. One then has to confirm that the critical point is indeed a centre; this is often far from an easy task, and a variety of conditions for a centre are needed, including generalisations of the classical divergence and symmetry criteria (see [1]). Suppose that the maximum order of a fine focus is known to be k. A system in which $\eta_2 = \ldots = \eta_{2k} = 0$ and $\eta_{2k+2} \neq 0$ is taken, and a perturbation sought such that after perturbation $\eta_2 = \ldots = \eta_{2k-2} = 0$ but $\eta_{2k}\eta_{2k+2} < 0$. Thus the stability of the origin has been reversed, and a limit cycle bifurcates. This procedure is continued until all the focal values are non-zero. Choosing these perturbations can be a complicated exercise, for it is essential that only one focal value becomes non-zero at each stage. It is not necessarily the case that such perturbations can be chosen - for example, it may be that for some j, $\eta_{2j} = 0$ whenever $\eta_{2j-2} = 0$, in which case fewer than the expected number of small amplitude limit cycles are produced.

To describe the calculation of the focal values, let

$$p(x, y) = \sum_{i=2}^{n} p_i(x, y) \quad \text{and} \quad q(x, y) = \sum_{i=2}^{n} q_i(x, y) \, ,$$

where p_i and q_i are homogeneous polynomials of degree i. We seek V of the form

$$\tfrac{1}{2}(x^2 + y^2) + V_3(x, y) + \ldots + V_i(x, y) + \ldots$$

with, say,

$$V_k = \sum_{i=0}^{k} V_{k-i,i}\, x^{k-i} y^i \ .$$

Let D_k denote the collection of terms of degree k in $\dot{V}$. Setting $D_k = 0$ gives a set of linear equations for the coefficients $V_{i,j}$. It turns out that these equations have a unique solution when k is odd, but not when k is even. For even values of k, the quantities η_k are introduced and it is found that $V_{i,j}$ can be so chosen that $D_{2k} = \eta_{2k}(x^2+y^2)^k$. Full details of the reasoning behind these remarks are given in [8].

It rapidly becomes clear that any attempt to perform the calculations by hand is bound to fail, and sooner rather than later. The usefulness of this approach therefore depends entirely on whether an efficient algorithm can be devised which can be implemented on a computer. We did this, initially using symbolic manipulation techniques developed locally [58]. More recently, we have used the REDUCE package, running on a VAX 11/750. I shall not describe the details of the algorithm save to say that it consists of two parts. In the first, the $V_{i,j}$ and the focal values are computed by a linear process. In the second, the focal values are reduced by substituting the relations $\eta_{2i} = 0$ $(i < k)$ into the expression for η_{2k}; this sometimes requires some subtlety and is partly interactive.

We have no qualms about the reliability of using a computer in this way. Since we work in integer arithmetic, there are, of course, no rounding errors. However, there is always the danger of systematic programming errors, and so the computations should always be carried through using two distinct programs. One thing is certain - calculation by hand would be immeasurably slower and much less reliable.

The scope of the technique which I have described, in terms both of the number of focal values computed and the degree of the systems considered, is restricted by the relatively rapid exhaustion of heap space in the computer - a problem which is encountered eventually in many applications of 'computer algebra.' Considerable ingenuity is required to extend the range of the computations significantly, and there are a number of interesting technical problems in this area.

In this description of the bifurcation of small-amplitude limit cycles, I have supposed that the origin is a non-degenerate critical point. It is possible to develop a similar theory for degenerate critical points. This is highly desirable in one particular context. If there are no critical points on the line at infinity (in which case n is odd), it is useful to be able to bifurcate limit cycles from infinity as well as from a finite critical point. Extending the flow generated by (1) to the sphere, the critical point at infinity is, however, degenerate. It is certainly possible to adapt the theory to apply to degenerate critical points (see [59]), but it is quite another matter to develop an efficient computer program. This is being investigated (see [7]), and we hope that the technical difficulties which arise will be resolved soon. Limit cycles 'at infinity' are discussed in a slightly different context by Chicone & Sotomayor in [18], where a class of polynomial systems which define a global flow is described.

Another natural extension is to bifurcate limit cycles out of several fine foci simultaneously. Interestingly, there seem to be quite severe restrictions on the orders of the fine foci which can co-exist. For example, a quadratic system can have two fine foci only if both are of order one, and the 'homogeneous' cubic systems considered in [8] (see Section 5) cannot have more than one if that is of order three (or more). It would be useful to investigate the possibilities of fine foci co-existing in more generality. Lower bounds for H_n can be obtained by considering systems with as many fine foci of order one as possible. System (1) has at most n^2 finite critical points and at most $(n+1)$ pairs on the line at infinity. By a simple index argument, at most $\frac{1}{2}n(n-1)$ critical points can be nodes or foci. A proportion of these can be fine foci and by bifurcating a single limit cycle out of each, systems can be constructed with $O(n^2)$ small amplitude limit cycles.

Though we have concentrated in this section on bifurcating limit cycles out of a critical point, bifurcation from separatrix cycles is likely to be of considerable significance in future developments. Computations similar to those described here have arisen in the construction of systems in which several limit cycles bifurcate out of homoclinic orbits. The theoretical basis of these calculations is described by Roussarie in [102]. In [99] Guckenheimer, Richard Rand & Schlomiuk have implemented the procedure in the case of certain quadratic systems, using the MACSYMA symbolic manipulation system. They show that at most two limit cycles can bifurcate out of a homoclinic orbit of a Hamiltonian quadratic system. In [71] Rousseau gives an example of a system in which two such limit cycles occur, and also investigates simultaneous bifurcation from a critical point and a separatrix cycle. In the context of such computations, the recent paper of Rand & Keith [69] is interesting, though not directly relevant. They examine the behaviour of orbits in a neighbourhood of degenerate critical points at which the linearisation is $\dot{u} = v$, $\dot{v} = 0$, and use MACSYMA to bring the system to Takens normal form [84]. I shall return to the question of homoclinic bifurcation when discussing quadratic systems in Section 4.

Open questions. The major problem is to estimate $\hat{H}_n$, the maximum possible number of small-amplitude limit cycles which can bifurcate out of a single critical point of systems of degree n. So far, $\hat{H}_n$ is known only for $n = 2$, though the corresponding upper bound has been found for a number of subclasses of systems of higher degree, as will be seen in later sections.

An investigation of the constraints on the orders of co-existing fine foci would be worthwhile. I know of no general work on this question. From the few particular cases which have been studied, it appears that there may exist some kind of index which would lead to upper bounds for the sum of the orders of the fine foci of a system. Information on the number and orders of co-existing fine foci will lead to improved lower bounds for H_n and the rate of growth of H_n with n. the details of the calculations that there can be $O(n^2)$ small-amplitude limit cycles are being worked out, with particular attention paid to arranging that the constant implied by the $O(n^2)$ is as large as possible. A refinement of this idea involving the bifurcation of more than one limit cycle from several fine foci may give better estimates, but the possible improvement may not justify the considerable increase in the complexity of the calculations.

There is ample scope for interesting work on the technical aspects of the computation of the focal values. The first priority is for a more efficient use of computer power, and this may require the development of a more specialised symbolic manipulation technique, dedicated to this application. Work is continuing on developing a program to deal with bifurcation out of a degenerate critical point (see Blows [7]). In another direction, the computation involved in the consideration of homoclinic bifurcations present formidable technical difficulties - for example,

Guckenheimer, Rand & Schlomiuk report in [99] that their calculations taxed the capabilities of a Symbolics 3670 computer.

Since the bifurcation of small-amplitude limit cycles is a local phenomenon, it is natural to ask what happens to these limit cycles as the perturbations which gave rise to them increase. In the well-known example of Bogdanov:

$$\dot{x} = y\,, \quad \dot{y} = -1 - \mu y + x^2 + xy\,,$$

the limit cycle which arises in a Hopf bifurcation when $\mu = -1$ is absorbed into a homoclinic loop when $\mu = -5/7$. On the other hand, consider the system

$$\dot{x} = y - \mu F(x)\,, \quad \dot{y} = -x\,, \tag{5}$$

where $F'(x) = (x^2-1)(x^2-k)$ and μ and k are positive parameters. The limit cycle which is born in a Hopf bifurcation when $k = 0$ is destroyed in a saddle-node bifurcation. There are two limit cycles for sufficiently small k and also for sufficiently large k.

Some understanding of the evolution of small-amplitude limit cycles can be obtained initially by means of careful numerical experiments. One of my students, J.M. Abdulrahman, has done this for (5) (see Section 6 for more information).

More generally, the whole question of whether the local results to be described in later sections hold globally is one that should be considered. Some powerful ideas from algebraic geometry may well be needed to establish the connection, and as explained in Section 2, it is probably necessary to work with complexified versions of the equations.

The whole question of bifurcation from a homoclinic orbit is likely to be a fruitful area of research, and one that is already attracting a good deal of attention. The first question is to estimate how many limit cycles can bifurcate out of a homoclinic orbit for various classes of systems (certain quadratic systems have been considered in [99]). Simultaneous bifurcation - from a homoclinic orbit and a critical point or even from two homoclinic orbits - is then a natural question to investigate (see [72], for example).

4 QUADRATIC SYSTEMS

Quadratic systems have been studied intensively over the years, and are much better understood than other polynomial systems. Much of this work has been done by Russian and Chinese mathematicians and some of it has been rather inaccessible. The two excellent survey articles by Coppel ([20], [22]) have been valuable in this regard, and the second half of the book of Ye *et al* [89] is now a valuable source of information. I shall obviously not attempt to survey the literature on quadratic systems, but simply refer to those contributions which seem to me to be particularly relevant in the context of this paper.

The basic facts are well known (see [20]): a limit cycle contains exactly one critical point in its interior, and that critical point must be a focus; at most two critical points can be encircled by limit cycles; limit cycles around the same critical point have the same orientation, while limit cycles around different critical points have opposite orientations; and the interior domain of a limit cycle is convex. In particular, there is 'swirling flow' inside the outermost limit cycle around a critical point.

Since the critical point inside a limit cycle of a quadratic system is necessarily non-degenerate, limit cycles cannot accumulate at such a point. In [16], Chicone & Shafer proved that limit cycles can accumulate only on an unbounded separatrix cycle, and by the theorem of Il'yashenko [38] which was mentioned in Section 2, the critical points on such a cycle cannot all be non-degenerate. By a careful consideration of the remaining possibilities, Bamón [4] then deduced that limit cycles cannot accumulate even on an unbounded separatrix cycle, whence the number of limit cycles is finite. In an earlier paper [3], Bamón described a family of quadratic systems which had a limit cycle inside a separatrix cycle, and in a forthcoming paper, Catherine Holmes [35] gives simple algebraic conditions for the existence or otherwise of a limit cycle inside a separatrix cycle consisting of an invariant line joining a pair of critical points at infinity. Coppel has now been able to give a much shorter proof of Bamón's theorem, at the same time proving substantially more, namely that in the cases not ruled out by the results of Il'yashenko, and Chicone & Shafer, not only is the number of limit cycles finite but there can be no more than one. He uses some extremely interesting results which have appeared in the Chinese and Russian literature - full references are given in [22]. Coppel gives accessible proofs of three such theorems in [94]. Specifically, let P_2 and Q_2 be the quadratic parts of P and Q, respectively. There is at most one limit cycle if P_2 and Q_2 are proportional, or if both P_2 and Q_2 are divisible by the non-constant part of the divergence $\partial P/\partial x + \partial Q/\partial y$.

The third result proved in [94] is especially useful: if a quadratic system has an invariant line, then it has at most one limit cycle. This is one of a number of results in which it is supposed that there are invariant curves of various kinds; in particular, there are no limit cycles if there is an invariant hyperbola (though there can be a centre), while if there is an invariant ellipse, the only possible limit cycle is the ellipse itself. Detailed references are again to be found in [22]. One of the staple tools used in proving such results is the classical divergence criterion, that if D is a simply-connected region and there is a function φ such that

$$\frac{\partial}{\partial x}(\varphi P) + \frac{\partial}{\partial y}(\varphi Q) \neq 0 ,$$

then there are no limit cycles of (1) entirely contained in D. The key to success is to choose φ appropriately. A far less obvious device is the rather remarkable way in which quadratic systems can be transformed to Liénard equations. The uniqueness theorems of Zhang [91,92] and Cherkas & Zhilevich [14] are then available; I shall have more to say about these in Section 6.

Several 'canonical forms' have been used for quadratic systems with limit cycles. Essentially, one has one degree of freedom in the coefficients in response to non-singular linear transformations of the co-ordinates and dilation of the time-scale. We refer to two particular forms, namely

$$\left.\begin{aligned} \dot{x} &= \lambda x - y + ax^2 + bxy + cy^2 \\ \dot{y} &= x + dx^2 + exy \end{aligned}\right\} \qquad (6)$$

and

$$\left.\begin{aligned} \dot{x} &= \lambda x + y + Ax^2 + (B+2D)xy + Cy^2 \\ \dot{y} &= -x + \lambda y + Dx^2 + (E-2A)xy - Dy^2. \end{aligned}\right\} \qquad (7)$$

The coefficients of xy in the two equations of (7) are so chosen because the divergence of the vector field is then identically zero when $\lambda = B = E = 0$, and hence the origin is a centre.

The methods described in Section 2 can be used to compute the focal values for (6) and (7). For (7), we have $L(0) = \lambda$, $L(1) = -B(A+C)$,

$$L(2) = DE(A+C)(E-5(A+C))$$

and

$$L(3) = -DE(A+C)^2[(A+2C)C+D^2].$$

The corresponding formulae for (6) are more complicated and are given in the final section of [8]; the formulae for the general forms of quadratic systems (without first transforming to a canonical form) are extremely complicated. These calculations, which have now been done by several authors, are used to construct examples of quadratic systems with at least four limit cycles. Shi's example [77], which is now very well known, is of the form (6) with $a = -10$, $b = 5 - 10^{-13}$, $c = d = 1$, $e = -25 + 9{\cdot}10^{-13} - 8{\cdot}10^{-52}$ and $\lambda = -10^{-200}$.

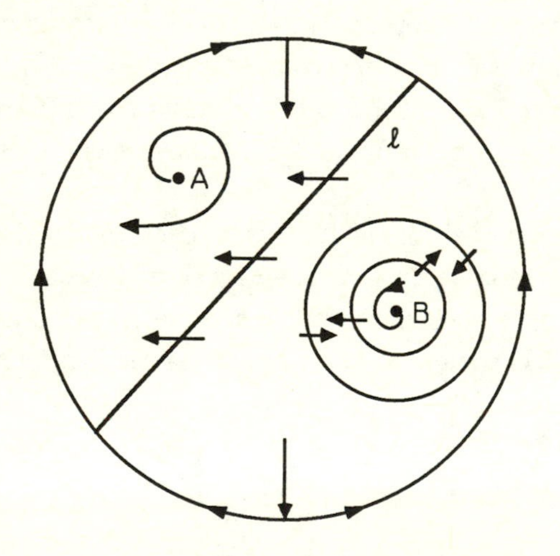

Figure 1. Shi's Example.

The salient characteristics of this example are shown in Figure 1. There are two finite critical points, one (A, say) an unstable focus and the other (B) a stable focus, and a single pair of critical points on the line at infinity. A transversal ℓ separates the plane into regions D_+ and D_-, say; D_+ contains A and is positively invariant, while D_- contains B and is is negatively invariant. There is obviously a limit cycle in D_+. Initially B is taken to be a fine focus of order three; suitable perturbations are then sought under which three small-amplitude limit cycles bifurcate out of B. (Recall that three is the maximum number that can be so generated - this was Bautin's fundamental contribution.)

Another way in which examples of systems with four limit cycles can be constructed is to start with foci A and B as above, but with B of order two. We then have the situation shown diagramatically in Figure 2(b) - two small-amplitude limit cycles surrounded by a limit cycle not of small-amplitude, in one half space, and a single limit cycle in the other half space.

Classes of systems with at least four limit cycles and whose phase portraits are as represented in Figure 2 have been given by Chen & Wang [12], Qin, Shi & Cai [67] and Li [44] (see also [8]).

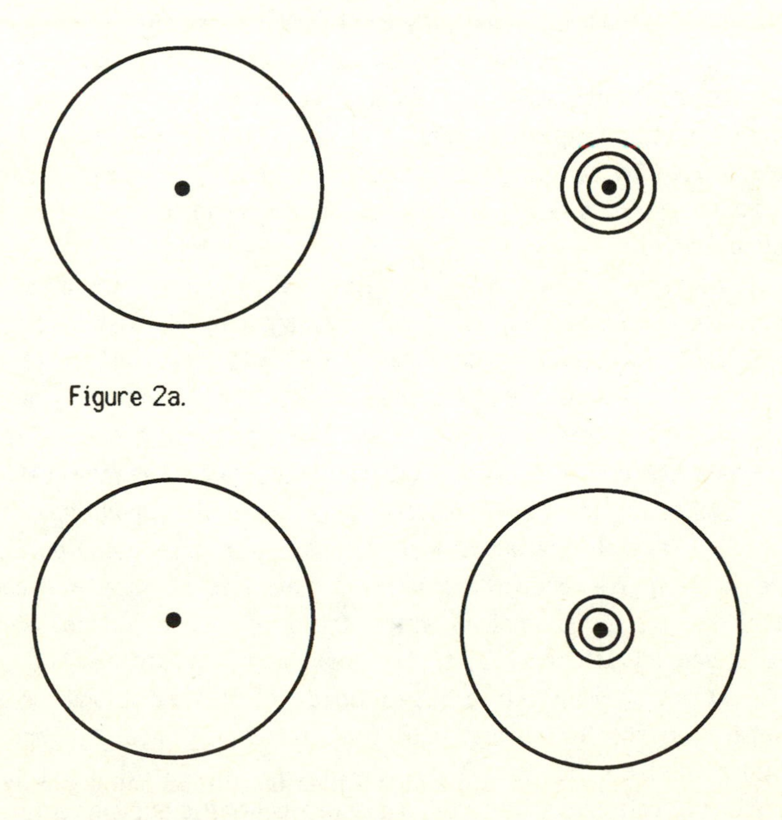

Figure 2a.

Figure 2b.

In all these examples, it is known only that there are *at least* four limit cycles. However, it has been shown by Li [45] that a third order fine focus cannot be encircled by a limit cycle. Claims have been made by Chin (= Qin) [68] that a proof has been found that quadratic systems can have no more than four limit cycles. These are so far unsubtantiated. In view of the remarks made by Tian in [85], the status of this work is unclear, and the suggestion that $H_2 = 4$ remains a conjecture, albeit one for which there is some evidence. Qin's ideas involve the complexification of x, y and t.

Turning to more global considerations, several classes of quadratic systems have been classified. Systems with two invariant lines can be written in the form

$$\dot{x} = x(ax + by + c), \quad \dot{y} = y(dx + ey + f) \tag{8}$$

and occur in many applications, the most familiar of which are in population dynamics. By a suitable use of the divergence criterion, it can be shown that there are no limit cycles. An exhaustive classification in the Poincaré compactification is given by Reyn in a forthcoming paper [70], and he also gives a nice account of recent applications. This work is part of a continuing programme to classify quadratic systems; [101] is another contribution to it.

Quadratic systems without finite critical points were classified by Gasull, Sheng & Llibre [30] – they call such systems 'chordal', and quadratic systems all of whose orbits are bounded were classified by Dickson & Perko [25]. The classification of homogeneous systems by Newton [60] can also be noted; such systems cannot, of course, have limit cycles. All these classifications are in terms of algebraic relations between the coefficients.

Quadratic systems with exactly one finite critical point have been the subject of several recent investigations. For example, Koditschek & Narendra [43] give conditions under which there is a unique limit cycle and describe other situations in which there are none. This work was generalised by Chicone [15] to systems of higher degree. By entirely different methods, which are explained in Section 7 and which are substantially shorter, Coppel [21], Coll, Gasul & Llibre [19] and I have obtained similar results. Those given by Coll, Gasull and Llibre are particularly detailed, and they also discuss quadratic systems all of whose orbits are bounded and which have two critical points; they exploit the formulae for the derivatives of the return map given in [53] and are able to show that limit cycles expand and contract monotonically as a parameter varies.

One of the topics in which there has been interesting work of late, and in which further progress is likely in the near future, is the bifurcation of limit cycles out of separatrix cycles. Some remarks have been made about these developments in Section 3, where the paper of Guckenheimer, Rand & Schlomiuk [99] has already been mentioned. They consider perturbations of Hamiltonian systems, which they show can be written, without loss of generality, in the form

$$\dot{x} = \frac{\partial H}{\partial y} + \varepsilon\beta xy \quad , \quad \dot{y} = -\frac{\partial H}{\partial x} + \varepsilon xy \quad ;$$

β is a constant and $H = xy + C(x,y)$, where C is a homogeneous cubic. It is shown that at most two limit cycles bifurcate out of the homoclinic loop which is attached to the origin. Now in this process, the loop itself is destroyed so the question arises whether a homoclinic orbit can contain two limit cycles in its interior domain. Paul de Jager, a graduate student at the University of Delft, reports an example which demonstrates that this is possible. His idea is to start with a quadratic system with a third order fine focus (A, say) and a col (B). A limit cycle Γ is bifurcated out of A and the relevant parameter increased so that Γ expands until it eventually becomes a homoclinic loop attached to B. Two limit cycles are then bifurcated out of A in the usual way; the key step is to show that the homoclinic loop is not destroyed in the process.

A different, but a very interesting and potentially useful approach is described by Drachman, van Gils & Zhang Zhifen [95]. They also consider perturbations of Hamiltonian systems and use the theory of Abelian integrals to investigate homoclinic bifurcations.

In [71] Rousseau gives an example of a quadratic system in which two limit cycles bifurcate out of a separatrix loop; the possibility of simultaneous bifurcation from a critical point and a separatrix loop is also considered, and it is shown that in total no more than four limit cycles can be constructed in this way. The investigation of simultaneous bifurcation is continued by Rousseau in [72]. Several phenomena are illustrated by example, and some possibilities are shown not to occur; for instance, a system is given in which there is bifurcation from two distinct homoclinic loops simultaneously. The author also relates that the known examples of four limit cycles in quadratic systems can all be constructed by two separate bifurcations out of critical points.

Open questions. The major question is, of course, whether $H_2 = 4$. The first task is to show that a critical point can be encircled by at most three limit cycles. This would be a global version of Bautin's theorem, and Li's result [45] that a third order fine focus cannot be encircled by a limit cycle is suggestive evidence. To convert a local result to a global one requires an effective continuation theory, and it is hard to see how to develop such a technique without complexifying the dependent variables in the system. It is shown in Section 7 how this can be done, but so far the strategy has not produced the results which were anticipated. It also appears that if one critical point is encircled by three limit cycles, then the other is encircled by at most one; why this might be so (if, indeed, it is true) requires investigation.

While the general question remains puzzling, it would be useful to be able to show that the examples described above (*à la* Shi) have no more than four limit cycles. It is worth trying to track the bifurcating limit cycles numerically, though great care must be exercised to ensure that any conclusions are reliable.

In all the known examples of systems with four limit cycles there are (at least) three limit cycles around one critical point and one around another. Can examples be found with a 2+2 configuration, or is this impossible? Ye [88] has shown that it is not possible to generate two limit cycles around each of two critical points by bifurcations of any kind. Thus it may be that such a configuration cannot occur.

The whole question of simultaneous bifurcation is likely to attract increasing attention, and may well lead to some new insights. As explained earlier, bifurcation from two critical points is not useful in this context, but bifurcations involving homoclinic loops and other kinds of separatrix cycles are certainly of considerable interest, and should be examined carefully. Of particular interest is the number of limit cycles which can occur inside a homoclinic loop and which can bifurcate from such an orbit.

The task of relating the number of limit cycles to algebraic conditions on the coefficients will continue. The significance of the existence of invariant curves of various kinds are being explored by Colin Christopher, one of my research students; for example, the geometric reason for the results mentioned above involving invariant conics is still not entirely clear. In an interesting recent development, Christopher has described a class of quadratic systems in which a limit cycle co-exists with an invariant parabola, a result which contrasts with those relating to an invariant ellipse or hyperbola.

5 CUBIC SYSTEMS

In comparison with quadratic systems, relatively little is known about general cubic systems. However, a number of particular classes have been investigated and it is some of these that I describe in this section, paying special attention to limit cycles which arise by bifurcation.

First, consider systems of the form

$$\dot{x} = \lambda x + y + p_3(x, y), \quad \dot{y} = -x + \lambda y + q_3(x, y), \tag{9}$$

where p_3 and q_3 are homogeneous cubic polynomials. In [8] we used the algorithm described in Section 3 to compute the Liapunov quantities (at the origin). Some lengthy machine calculations were involved; the focal values were very large polynomials, but the reduction phase of the algorithm led to considerable simplification. It was necessary to compute $L(k)$ for $k \leq 6$. The resulting formulae are too large to quote here: they are given in [8]. Sibirskii ([78], [80]) also considered (9) and used Bautin's method; see also a paper of Shi [76]. Using the criteria of Sibirskii & Lunkevich [79] for a centre, it was shown that the maximum number of small-amplitude limit cycles is five. We described a four-parameter family of systems having five small-amplitude limit cycles.

System (9) may have fine foci at points other than the origin; there would therefore appear to be some hope of constructing systems of this form with several sets of small-amplitude limit cycles. However, it turns out that this is not possible. In his doctoral dissertation, Kalenge [42] studied the co-existence of fine foci, and discovered that if the origin is of order three or more, then there are no other fine foci, while if the origin is of order two, there can be two other fine foci, but neither can be of order greater than one. Thus five is the maximum number of small-amplitude limit cycles of (9) even when all its critical points are considered simultaneously.

As for the Liapunov quantities for the general cubic

$$\left.\begin{aligned} \dot{x} &= \lambda x + y + p_2(x, y) + p_3(x, y) \\ \dot{y} &= -x + \lambda y + q_2(x, y) + q_3(x, y)\,, \end{aligned}\right\} \quad (10)$$

we calculated $L(k)$ for $k \leq 3$. They become so massive as to be of little value - $L(3)$ contains in excess of 600 terms. Thus it becomes necessary to restrict attention to particular classes, several of which were considered by Kalenge [42]. He found an example of systems of the following form with six small-amplitude limit cycles:

$$\begin{aligned} \dot{x} &= \lambda x + y + Cy^2 + Hxy^2 + Ky^3 \\ \dot{y} &= -x + \lambda y + D(x^2-y^2) + Lx^3 - Hx^2y + Nxy^2\,. \end{aligned}$$

Systems of odd degree differ from those of even degree in that there need not be critical points on the line at infinity. It is therefore possible to bifurcate limit cycles from infinity and a finite critical point simultaneously. Note, however, that bifurcation from infinity presents technical difficulties not present in the usual bifurcation from a non-degenerate focus. In [10] (see also [6]), we consider systems of the form

$$\left.\begin{aligned} \dot{x} &= \lambda x + y + p_2(x, y) - \delta y(x^2 + y^2)y + \mu x(x^2 + y^2) \\ \dot{y} &= -x + \lambda y + q_2(x, y) + \delta x(x^2 + y^2)y + \mu y(x^2 + y^2)\,. \end{aligned}\right\} \quad (11)$$

There are no critical points at infinity, and it is shown that $L(0)$, $L(1)$ and $L(2)$ at the origin are independent of δ and μ. We then describe a family of systems of this form which have three small-amplitude limit cycles around the origin, a large limit cycle (having bifurcated from infinity) and another limit cycle around a critical point other than the origin. The interest of the example lies in the fact that the large limit cycle has opposite orientation to those which bifurcate from the origin; indeed, all the 'nice' properties of limit cycles of quadratic systems described in the second paragraph of Section 4 are violated simultaneously. In a recent preprint, Rousseau [73] also considers systems of the form (11) and the bifurcation of limit cycles from infinity; he reports examples of cubic systems with seven limit cycles, three of which bifurcate from infinity.

Other examples of cubic systems with several limit cycles are given by Wang & Luo [86] and Li, Tian & Xu [48]. Especially interesting is the work of Li and co-authors ([46], [47], [48]) which is reported by Tian in [85]. They consider systems which are perturbations of Hamiltonian systems and which have a high degree of symmetry. By bifurcating limit cycles out of homoclinic and heteroclinic orbits, they construct systems with at least 11 limit cycles. It is well worth citing the examples as given by Tian:

$$\dot{x} = x(1 + 4x^2 - y^2) + \mu y(x^2 + \delta y^2 - \lambda)$$
$$\dot{y} = y(1 - x^2 - \tfrac{1}{2}y^2) + \mu y(x^2 + \delta y^2 - \lambda)\,,$$

where $\delta = 0.43$, and

$$\dot{x} = x(1 - y^2) + \mu x(x^2 - 3y^2 - \lambda)$$
$$\dot{y} = -x(1 - 2x^2) + \mu y(x^2 - 3y^2 - \lambda)\,.$$

In both systems, μ is small and λ is chosen appropriately.

Open questions. Further investigation of the properties of cubic systems in general is likely to be worthwhile. Initially, it might be a good idea to consider systems of the form (9), and discover whether there are counterparts of some of the known results for quadratic systems which were noted in Section 4. For example, is it true that there are no limit cycles if the origin is a fine focus of order 5 (in analogy with the theorem of Li Chenzhi [45])? One of my students is currently working on such questions, and, in particular, exploring the consequences of the existence of invariant curves of various kinds.

As far as small-amplitude limit cycles are concerned, a sensible strategy is to consider a variety of special cases of (10), in the expectation that systems will be discovered in which more than six limit cycles bifurcate from the origin. A related question is the co-existence of fine foci. Multiple bifurcation from infinity is likely to repay closer consideration (see [10] and [73]). In addition, a detailed study of bifurcation from separatrix cycles could well lead to interesting new developments and a better understanding of cubic systems in general - as well as being a fruitful source of examples.

6 LIENARD EQUATIONS

In this section we consider systems of the form

$$\dot{x} = y - F(x), \quad \dot{y} = -g(x)\,; \tag{12}$$

these are, of course, equivalent to second order equations of Liénard type:

$$\ddot{x} + f(x)\dot{x} + g(x) = 0\,,$$

where $f(x) = F'(x)$. In the context of this paper, we naturally suppose that F and g are polynomials. For the most part, we shall suppose that $g(x)$ represents a genuine restoring force, so that

$$g(x)\,\text{sign}\, x > 0 \qquad (x \neq 0)\,. \tag{13}$$

In particular, $x = y = 0$ is the only critical point. In many ways, little is lost even if it is supposed that $g(x) = x$.

There is an extensive literature on such systems, and a variety of criteria for the existence of limit cycles have been developed. These are usually proved in one of two ways: a bounded solution is shown to exist and the Poincaré-Bendixson theorem is used, or an 'a priori' bound for periodic solutions is derived and the methods of degree theory deployed.

The uniqueness of limit cycles has also attracted a great deal of interest, and though this has been less intensively studied than the question of existence, a variety of results have been proved which give conditions under which (12) has at most one limit cycle.

Uniqueness is often proved by showing that all limit cycles are stable (or all are unstable). For system (1) this is done by first obtaining information about the possible locations of limit cycles and then estimating

$$\Delta(L) = \oint_L \left[\frac{\partial P}{\partial x} + \frac{\partial Q}{\partial y} \right], \tag{14}$$

where L is a limit cycle. It is a fundamental result of Poincaré-Bendixson theory that L is stable if $\Delta(L) < 0$ and unstable if $\Delta(L) > 0$. An alternative way to establish uniqueness is to use the formulae given in [53] for the derivatives of the return map.

I shall make no attempt to summarise the literature on existence and uniqueness; such a survey would be an extensive project in its own right. Staude's survey article [83] is a very useful account up to about 1980, and Chapters 5 and 6 of Ye's book [89] are particularly helpful. However, there is one contribution to which I wish to refer, namely that of Zhang Zhifen [91], [92]. Her result was announced in 1958, but it is only recently that a proof has become available in English. She uses the technique described above to show that (12) has a unique limit cycle if

$$\int_0^x g(u)du \to \infty \quad \text{as} \quad |x| \to \infty,$$

$F(0) = 0$ and $F'(x)/g(x)$ is non-decreasing on $(-\infty, 0)$ and $(0, \infty)$. In fact, she is able to replace y in (12) by $\varphi(y)$, where $\varphi(y)$ is a suitable increasing function. Zhang's result has been generalised by Cherkas & Zhilevich [14]; details are given in Chapter 6 of [89]. Interestingly, these results have proved useful in the investigation of quadratic systems; such systems can sometimes be transformed into the form (12) (see [22]).

In contrast to the questions of existence and uniqueness, there are relatively few results on the possible number of limit cycles of Liénard systems. Theorems on the existence of a specified number of limit cycles have been given by, for example, Ding [26] and de Figueiredo [24], but the best known work is that of Lins, de Melo & Pugh [50]. They showed that, if $N = 2n+1$ or $2n+2$ and $m \leq n$, there are polynomials F of degree N such that the system

$$\dot{x} = y - F(x), \quad \dot{y} = -x \tag{15}$$

has exactly m limit cycles. They conjectured that

there can be no more than $k(F) = \left[\frac{1}{2}(\partial F - 1)\right]$ *limit cycles.* (16)

(Here $[\cdot]$ and ∂ denote 'integer part' and 'degree of' respectively.) This remains an open question. Lins, de Melo and Pugh deduced their theorem by using small parameter theory, having first established a number of results on the behaviour of trajectories in a neighbourhood of infinity.

A natural way to establish the existence of several limit cycles is to construct a sequence of closed curves across which the flow is alternately inwards and outwards. This is what has been done in [56], where classes of equations of the form (12) are described with at least $k(F)$ limit cycles. The conditions give lower bounds for the difference between the values taken by F at successive maxima (and minima) in terms of the distance between these turning points and the interlacing zeros of F. In the special case of F odd, the conditions require that the values of F at maxima form an increasing sequence and that the values at successive minima be decreasing.

Returning to the conjecture (16) the only cases which have been resolved to date are when $\partial F \leq 5$. That (15) has at most one limit cycle when F is cubic was in fact proved in [50] (the result is, of course, classical when F is an odd function). For $\partial F = 4$, Zeng [90] has shown that, under certain conditions on the critical values of F, (15) has at most one limit cycle. When F is quintic, it is known that (15) has at most two limit cycles provided that F is odd. This was proved by Rychkov [74], and his strategy is worth describing; it is a stability argument, using the formula (14). Suppose that $F'(0) > 0$, and that the zeros of F in $x > 0$ are α_1 and α_2; let $x = \beta_1 > 0$ be a maximum of F and $x = \beta_2 > \beta_1$ a minimum. It is first shown that at most one limit cycle can intersect Γ: $y = F(x)$ with $\alpha_1 < x < \beta_1$, and that any such limit cycle is unstable. It is then shown that if two limit cycles intersect Γ with $x > \beta_2$, the innermost is unstable and the other is stable. It follows that there can be no more than two such limit cycles, and that if there are two, no limit cycles can intersect Γ with $x < \beta_2$. The technical details are non-trivial, and a clear version has been prepared by James [40].

Though not directly relevant in the setting of this paper, it is nevertheless worth mentioning another paper of Zhang [93] in which she proves that the system

$$\dot{x} = y, \qquad \dot{y} = -x - \mu \sin y$$

has exactly n limit cycles in the strip $|y| \leq (n+1)\pi$. This is again proved by carefully considering the stability of limit cycles in appropriate regions of the phase plane.

It is thus clear that much remains to be done to resolve the full form of the conjecture of Lins, de Melo & Pugh. A restricted form of the question relates to the possible number of small-amplitude limit cycles of (12). In [9], Blows and myself used the algorithm described in Section 3 to prove that (15) has at most $k(F)$ small-amplitude limit cycles and that any number $m \leq k(F)$ can be attained, thus proving the conjecture in this special situation. We also showed that the same conclusion holds for (12) whenever g is an odd polynomial. When g is not odd, the number of small-amplitude limit cycles is still of interest, but we cannot expect the same upper bound. Recently, Stephen Lynch, one of my research students and I have solved the question in three particular cases [57]. Suppose that $g(x)$ has the sign of x for sufficiently small x, and let the maximum possible number of small-amplitude limit cycles be $\hat{H}$. Then

(i) if F is odd, $\hat{H} = k(F)$;
(ii) if $\partial F = 2$ and $\partial g = 2m$ or $2m+1$, then $\hat{H} = m$;
(iii) if F is even, $\partial F = 2n+2$, and $\partial g = 2m+2$ or $2m+3$, then $\hat{H} = \max(n,m)$.

The proofs of these results are essentially inductive in nature.

In addition to these particular classes of systems, the maximum number of small-amplitude limit cycles can be found directly if ∂F and ∂g are small by using the algorithm of Section 3 to compute focal values,and implementing it on a computer using symbolic manipulation techniques. Let $\hat{H}(m, n)$ be the value of $\hat{H}$ when $\partial F = m$ and $\partial g = n$. We have obtained the following values

n \ m	3	4	5	6	7
2	1	2	3	3	4
3	2	2	4		
4	3	4			
5	3				
6	4				

Open questions. The conjecture of Lins, de Melo and Pugh is obviously an unsolved question of considerable interest. All the available evidence suggests that it is true, but it is a genuinely difficult problem in general. However, it should be possible to resolve it in some particular cases by developing existing techniques: for example, (i) when $\partial F = 5$ and F is not necessarily odd, and (ii) when $\partial F = 6$. A somewhat easier problem is to extend the results of Lins, de Melo and Pugh (n=3), Zeng (n=4) and Rychkov (n=5) to cover the situation when $g(x)$ has the sign of x but $g(x) \not\equiv x$.

Progress should also be possible with regard to small-amplitude limit cycles. The inductive arguments of the kind used to prove the particular results already described become much more complicated in general. However, I anticipate that the work currently under way will lead to new results of interest. It is certainly possible to augment the table given above by computing more focal values. In general, the relationship between the results for small-amplitude limit cycles and the corresponding global results merits further exploration.

It would be interesting to understand more about the way in which the limit cycles of Liénard systems vary with a parameter, especially when there are several limit cycles for some values of the parameter and not for others. In Section 3, we mentioned the idea of tracking limit cycles, numerically if necessary. In the example (4), there are k_* and k^* such that there are two limit cycles if $0 < k < k_*$ or $k > k^*$, and none if $k_* < k < k^*$. Both k_* and k^* depend on μ, and it appears that the outer limit cycle expands monotonically as k increases. In general, a full investigation of the kinds of bifurcations which the limit cycles of Liénard systems can undergo is desirable.

7 SYSTEMS WITH HOMOGENEOUS NONLINEARITIES

In this, the final section, we consider systems of the form

$$\left.\begin{aligned} \dot{x} &= \lambda x + y + p_n(x, y) \\ \dot{y} &= -x + \lambda y + q_n(x, y)\,, \end{aligned}\right\} \qquad (17)$$

where p_n and q_n are homogeneous polynomials of degree n. In polar form, (17) is

$$\dot{r} = \lambda r + f(\theta) r^n\,, \quad \dot{\theta} = -1 + g(\theta) r^{n-1}$$

where f and g are homogeneous polynomials of degree $n+1$ in $\cos\theta$ and $\sin\theta$. Now let

$$\rho = r^{n-1}(1-r^{n-1}g(\theta))^{-1}; \tag{18}$$

a little calculation shows that ρ satisfies the first order non-autonomous equation

$$\frac{d\rho}{d\theta} = A(\theta)\rho^3 - B(\theta)\rho^2 - \lambda(n-1)\rho, \tag{19}$$

where

$$A(\theta) = -(n-1)g(\theta)\big(f(\theta) + \lambda g(\theta)\big)$$

and

$$B(\theta) = (n-1)f(\theta) + 2\lambda(n-1)g(\theta) - g'(\theta).$$

Thus $A(\theta)$ and $B(\theta)$ are homogeneous polynomials in $\cos\theta$ and $\sin\theta$ of degree $2(n+1)$ and $(n+1)$, respectively. The transformation (18) is defined in

$$D \equiv \{(r,\theta);\, r^{n-1}g(\theta) < 1\},$$

which is an open set containing the origin. Quadratic systems are obviously of the form (17), and in their case D extends at least as far as the outermost limit cycle surrounding the origin.

The limit cycles of (17) correspond to positive 2π-periodic solutions of (19). Considerable interest has been shown recently in systems of the form (17), and much of it has been stimulated by this relationship.

In the case $n = 2$, transformation (18) was introduced by Lins Neto [49]. The connection between (17) and (19) for $n > 2$ was explained in [54], where it was used to calculate the focal values for (17). The contribution of Cherkas [13] should also be noted.

In order to be able to keep track of the number of periodic solutions of almost any class of differential equations it is useful, if not essential, to work with the appropriate complexified form. This is because the number of zeros of a holomorphic function in a bounded region of the complex plane cannot be changed by small perturbations of the function. Since periodic solutions correspond to the zeros of the displacement function $q(c) = z(\omega; 0, c) - c$ (in the obvious notation), periodic solutions can be created or destroyed only at infinity. Thus we are led to consider equations of the form

$$\dot{z} = \alpha(t) z^3 + \beta(t) z^2 + \gamma(t) z, \tag{20}$$

where z is complex but t remains real, and the coefficients α, β, γ are real-valued functions. We specify $\omega \in \mathbb{R}$ and seek information about the number of solutions which satisfy the periodic boundary condition

$$z(0) = z(\omega); \tag{21}$$

we say that such solutions are 'periodic' whether or not the coefficients in (20) are themselves periodic. We are particularly interested, of course, in the case in which $\omega = 2\pi$ and α, β, γ have the forms arising in (19); in that case we say that (20) is of 'Hilbert type.'

Equation (20) is reminiscent of the equations

$$\dot{z} = p_0(t) z^N + p_1(t) z^{N-1} + \dots + p_N(t) \tag{22}$$

which were investigated in detail in [51] and [52] (also see Pliss [66]). In these papers the $p_i(t)$ were periodic functions, all of the same period, but there is no difficulty in adapting the techniques to apply to the case in which the p_k are not periodic and solutions satisfying (21) are sought. It was shown, in particular, that if $N = 3$, then there are precisely three periodic solutions (counting multiplicity) if p_0 is never zero, while there are at most three if p_0 does have zeros but does not change sign. There are also at most three periodic solutions when $\beta(t) \leq 0$ for all t (see [2]).

Returning to (20), suppose that α does not change sign. Since $z = 0$ is always a solution, there are at most two non-trivial periodic solutions. Complex periodic solutions occur in pairs: if $\varphi(t)$ is such a solution, then so is $\bar{\varphi}(t)$. If, moreover, (20) is of Hilbert type with n even, the real periodic solutions also occur in pairs: if $\psi(t)$ is a periodic solution, then so is $-\psi(t+\pi)$. Consequently, if the associated function α does not change sign, then (17) has at most one limit cycle in D when n is even and at most two when n is odd. In [55] it was further shown that there are no positive periodic solutions if n is even and $\lambda\alpha \leq 0$, while there can be no more than one if n is odd, $\alpha \geq 0$ and $\lambda > 0$.

Many of the recent results which exploit the connection between (17) and (19) depend on obtaining conditions under which α does not change sign. Systems of the form (17) in which the origin is the only critical point were considered in [55]. It was shown that α does not change sign if there are no critical points at infinity (in which case n is odd), or if there is only one pair of critical points at infinity (when n is even) and at these the function g changes sign. In [21], similar results were proved for quadratic systems and conditions given for the existence of a limit cycle. Both these contributions were stimulated by the papers of Chicone [15] and of Kotitschek & Narendra [43]. In [15], systems of the form (17) in which p_n and q_n have a common factor of degree $(n-1)$ were discussed, and [43] is concerned with quadratic systems in which p_2 and q_2 both split into linear factors. Both papers give conditions for the existence of a unique limit cycle. The methods used do not utilize the connection with (19); it must be said, however, that the approach via (19) is much simpler.

Quadratic systems have also been investigated by means of equation (19) by Coll, Gasull & Llibre [19]. They give a detailed discussion of systems with one finite critical point, and also of systems all of whose orbits are bounded and which have at most two finite critical points. They used the formulae given in [53] for the derivatives of the return map, and were able to show that limit cycles vary with a parameter in much the same manner as in semi-complete rotated vector fields [25]. The point of considering the derivatives of the return map h is that if $h'''(x) > 0$, say, for $x > 0$, then there are at most two limit cycles (as proved in [53]).

By using similar methods, Gasull, Llibre & Sotomayor [31,98] have obtained a lot of detailed information about systems of the form (17) in which p_n and q_n have a common factor of degree $(n-1)$. Recently, Carbonell & Llibre [11] have considered system (17) in the whole plane and not only inside the set D. They prove that any limit cycle surrounds the origin, and that there are at most two limit cycles when n is odd while there cannot be more than one when n is even - in other words, the results previously known to hold within D are in fact true without this restriction. Again, the formulae for the first three derivatives of the return map h play a major role.

All the contributions described above rely on finding conditions under which α in (19) does not change sign. In general, however, α does change sign, and consequently the results of [51] do not hold. Indeed, Lins Neto [49] has given examples which demonstrate that there is no upper bound for the number of periodic solutions of (20) - unless, that is, the coefficients are suitably restricted. The real question, therefore, is the maximum possible number of periodic solutions for various classes of coefficients. We discuss (20) in isolation from (17), but the connection with Hilbert's problem is self-evident.

In [1] we consider solutions of (20) which bifurcate out of the origin - a problem which directly parallels the investigation of small-amplitude limit cycles of (17). Following Lins Neto [49], we consider two types of coefficients: (i) polynomials in t, and (ii) polynomials in $\cos t$ and $\sin t$. The idea is to consider various classes $\mathfrak{C}$ of equations of the form (20), and for each to calculate the maximum possible multiplicity of the origin, which we denote by $\mu_{max}(\mathfrak{C})$.

The *multiplicity* of $z = 0$ as a solution of (20) is the multiplicity of $z = 0$ as a zero of the displacement function $q: c \mapsto z(\omega; 0, c) - c$, as usually defined in complex function theory. To compute the multiplicity (which we call μ), we write $z(t; 0, c) = \sum a_n(t) c^n$ and substitute directly into the equation. This gives a recursive set of linear differential equations for the $a_n(t)$. Let $\eta_i = a_i(\omega)$; then $\mu = k$ if $\eta_1 = 1$, $\eta_2 = ... = \eta_{k-1} = 0$ and $\eta_k \neq 0$. The η_i are the 'focal values' in this context. It is easily verified that if $\mu > 1$, then it may be supposed without loss of generality that $\lambda = 0$. The equations for the a_n are then

$$\dot{a}_n = \alpha \sum_{\substack{i,j,k \geq 1 \\ i+j+k=n}} a_i a_j a_k + \beta \sum_{\substack{i,j \geq 1 \\ i+j=n}} a_i a_j. \tag{23}$$

The expressions for the $a_i(t)$ are lengthy and their calculation tedious; the formulae for η_k with $k \leq 8$ are given in [1]. To be able to calculate the maximum multiplicity of the origin, criteria are required which ensure that the origin is a centre in the sense that all solutions in a neighbourhood are periodic. Several such conditions are presented in [1]; most of them are derived from the following: if there is a function σ such that $\sigma(0) = \sigma(\omega)$, and functions f and g such that

$$\alpha(t) = f(\sigma(t))\dot{\sigma}(t), \;\; \beta(t) = g(\sigma(t))\dot{\sigma}(t), \tag{24}$$

then $z = 0$ is a centre of

$$\dot{z} = \alpha(t)z^3 + \beta(t)z^2.$$

To illustrate the kind of results obtained in [1], we give three examples. (i) Let $\mathfrak{D}$ be the set of equations (20) in which α and β are both linear combinations of the same three powers of t; then $\mu_{max}(\mathfrak{D}) = 4$ whatever these powers are. (ii) Let $\mathfrak{C}_{p,q}$ be the class of equations in which α and β are polynomials in t of degree p and q, respectively; then $\mu_{max}(\mathfrak{C}_{3,2}) = 8$ and $\mu_{max}(\mathfrak{C}_{k,1}) = 3 + [\frac{1}{2}k]$ for $k < 5$. (iii) Let $\mathfrak{H}_k$ be the class of equations (20) of Hilbert type with α, β of degree $2k$, k respectively; then $\mu_{max}(\mathfrak{H}_k) = 5$ for $k=1, 2$.

The next task is obviously to investigate the recursive system of equations (23) in general. This leads to an algebraic structure involving three operations (addition, multiplication and composition) connected by the integration by parts formula. This work is continuing and is now at an interesting stage.

Except in the very special circumstances in which α does not change sign, the methods used in [51] to investigate (22) are not applicable to (20). There is one device, however, which enables a closer connection between (20) and (22) to be established. For sufficiently small ε, the equation

$$\dot{z} = \varepsilon z^4 + \alpha(t)z^3 + \beta(t)z^2 + \gamma(t)z \tag{25}$$

has at least as many periodic solutions as (20). After rescaling, (25) is of the form (22), and an upper bound for the number of periodic solutions of (22) with $n = 4$ is also an upper bound for the number of such solutions of (20).

The machinery of [51] is now available. The essential idea is that as a path is traversed in the space of coefficients, the number of periodic solutions can change only by encountering a solution which becomes unbounded both as t increases and as t decreases. Thus it can be proved that (25) has exactly four periodic solutions if $\beta \leq 0$, if α, β and γ are small enough, and in several other instances. Equation (25) differs from (20) in one significant respect - the origin is an isolated periodic solution.

In [2] we compute the maximum possible multiplicity of the origin for various classes of equation (25) - with $\varepsilon = 1$. The spirit of the work is similar to that of [1], though the calculations are somewhat more onerous. It is a definite advantage, however, that the origin cannot be a centre. We answer questions posed by Shahshahani in [75] and also by Lins Neto [49]. As an indication of the kind of results obtained, $\mu_{max} = 6$ for the class of equations in which α and β are homogeneous polynomials of degree 1 and 2.

Open questions. The number of periodic solutions of (20) which can bifurcate out of the origin continues to be investigated by computing further focal values and examining more classes of coefficients. Perhaps more interesting is the search for general conclusions by considering the recursive equations (23), an approach which has a distinctly combinatorial flavour. If the two operations 'multiply by α and integrate' and 'multiply by β and integrate' are denoted by ℓ and r, respectively, then each a_n can be expressed as a linear combination of words in ℓ and r, where the product operation is composition. The aim is to obtain bases in terms of which each word of a given length can be expressed.

Related to the investigation of particular classes of equation (20) is a problem which we note in [1]. It was shown that condition (24) is sufficient for the origin to be a centre; we conjecture that it is also a necessary condition, at least if appropriate restrictions are placed on α and β.

Several other questions about (20) also arise: for example, are there any non-trivial periodic solutions if the origin is of maximal multiplicity?

Equation (20) has not proved to be of as much value in the global study of (17) as might have been expected. A theory which parallels that expounded in [51] for (22) is required. A start has been made on this: what I have been able to prove so far is that there is an identifiable set B in (α, β) space, given by certain inequalities, such that along a path which avoids B, the number of periodic solutions does not change: recall that the number of periodic solutions can change only if they 'drift off' to infinity. Results such as that of Li [45] can be deduced.

A number of interesting issues concerning (25) come to mind. Let S be the set of pairs of periodic functions (α, β), and let A be the subset consisting of equations which have no singular periodic solutions (for definitions, see [51]). It was shown in [51] that equations in the same component of A have the same number of periodic solutions. For the class of equations in which α and β are of 'Hilbert type' of a fixed degree, the question of the maximum number of periodic solutions would be resolved if the number of components of the set A could be found. This was done in [52] for equations of Riccati type and requires conditions which exclude the possibility of singular periodic solutions. Such an outcome would, of course, have implications for the corresponding problem for (20) and consequently (17).

References

1. M.A.M. Alwash & N.G. Lloyd. Non-autonomous equations related to polynomial two-dimensional systems. *Proc. Roy. Soc. Edinburgh Sect. A*, **105** (1987), 129-152.

2. M.A.M. Alwash & N.G. Lloyd. Periodic solutions of a quartic non-autonomous equation. *Nonlinear Anal.*, to appear.

3. R. Bamón. A class of planar quadratic vector fields with a limit cycle surrounded by a saddle loop. *Proc. Amer. Math. Soc.* **88** (1983), 719-724.

4. R. Bamón. Quadratic vector fields in the plane have a finite number of limit cycles. *Inst. Hautes Etudes Sci. Publ. Math.* **64** (1987), 111-142.

5. N.N. Bautin. On the number of limit cycles which appear with the variation of coefficients from an equilibrium position of focus or centre type. Mat. Sb. **30** (1952), 181-196 (in Russian); *Amer. Math. Soc. Transl.* No.100 (1954).

6. T.R. Blows & N.G. Lloyd. A note on cubic systems in the plane. *Abstracts Amer. Math. Soc.*, January 1985.

7. T.R. Blows. Bifurcation of limit cycles from a degenerate focus. *Internal report*, Northern Arizona University, 1986.

8. T.R. Blows & N.G. Lloyd. The number of limit cycles of certain polynomial differential equations. *Proc. Roy. Soc. Edinburgh Sect. A* **98** (1984), 215-239.

9. T.R. Blows & N.G. Lloyd. The number of small-amplitude limit cycles of Liénard equations. *Math. Proc. Cambridge Philos. Soc.* **95** (1984), 359-366.

10. T.R. Blows, M.C. Kalenge & N.G. Lloyd. Limit cycles of some cubic systems. *Preprint*, University College of Wales, Aberystwyth.

11. M. Carbonell & J. Llibre. Limit cycles of a class of polynomial systems. *Preprint*, Universitat Autònoma de Barcelona, 1987.

12. Chen Lansun & Wang Mingshu. The relative position and number of limit cycles of the quadratic differential system. *Acta Math. Sinica* **22** (1979), 751-758.

13. L.A. Cherkas. Number of limit cycles of an autonomous second-order system. *Differencial'nye Uravneniya* **12** (1976), 666-668.

14. L.A. Cherkas & L.I. Zhilevich. Some tests for the absence or uniqueness of limit cycles. *Differencial'nye Uravneniya* **6** (1970), 1170-1178.

15. C. Chicone. Limit cycles of a class of polynomial vector fields in the plane. *J. Differential Equations* **63** (1986), 68-87.

16. C. Chicone & D.S. Shafer. Separatrix and limit cycles of quadratic systems and Dulac's theorem. *Trans. Amer. Math. Soc.* **278** (1983), 585-612.

17. C. Chicone & Tian Jinghuang. On general properties of quadratic systems. *Amer. Math. Monthly* **89** (1982), 167-179.

18. C. Chicone & J. Sotomayor. On a class of complete polynomial vector fields in the plane. *J. Differential Equations* **61** (1986), 398-418.

19. B. Coll, A. Gasull & J. Llibre. Some theorems on the existence, uniqueness and non-existence of limit cycles for quadratic systems. *J. Differential Equations* **67** (1987), 372-399.

20. W.A. Coppel. A survey of quadratic systems. *J. Differential Equations* **2** (1966), 293-304.

21. W.A. Coppel. A simple class of quadratic systems. *J. Differential Equations* **64** (1986), 275-282.

22. W.A. Coppel. The limit cycle configurations of quadratic systems. *Proceedings of the Ninth Conference on Ordinary and Partial Differential Equations, University of Dundee, 1986*, (Longman), to appear.

23. J. Cronin. The point at infinity and periodic solutions. *J. Differential Equations* **1** (1965), 156-170.

24. Rui J.P. de Figueiredo. On the existence of N periodic solutions of Liénard's equation. *Nonlinear Anal.* **7** (1983), 483-499.

25. R.J. Dickson & L.M. Perko. Bounded quadratic systems in the plane. *J. Differential Equations* **7** (1970), 251-273.

26. Ding Sunhong. Theorem of existence of n limit cycles for Liénard's equation. *Sci. Sinica Ser. A* **26** (1983), 449-459.

27. H. Dulac. Sur les cycles limites. *Bull. Soc. Math. France* **51** (1923), 45-188.

28. J.-P. Françoise. Cycles limites, étude locale. *Report,* I.H.E.S., 1983.

29. J.-P. Françoise & C. Pugh. Keeping track of limit cycles. *J. Differential Equations* **65** (1986), 139-157.

30. A. Gasull, Sheng Liren & J. Llibre. Chordal quadratic systems. *Rocky Mountain J. Mathematics*, **16** (1986) 751-782.

31. A. Gasull, J. Llibre & J. Sotomayor. Limit cycles of vector fields of the form X(v) = Av+f(v)Bv. *J. Differential Equations* **67** (1987), 90-110.

32. F. Göbber & K.-D. Willamowski. Liapunov approach to multiple Hopf bifurcation. *J. Math. Anal. Appl.* **71** (1979),333-350.

33. E.A. Gonzales Velasco. Generic properties of a polynomial vector field at infinity. *Trans. Amer. Math. Soc.* **143** (1969) 201-222.

34. D. Hilbert. Mathematical problems. *Bull. Amer. Math. Soc.* **8** (1902), 437-479.

35. C.A. Holmes. Some quadratic systems with a separatrix cycle surrounding a limit cycle. *J. London Math. Soc.*, to appear.

36. Yu. S. Il'yashenko. The appearance of limit cycles under a perturbation of the equation $dw/dz = R_z/R_w$, where R(z, w) is a polynomial. *Math. Sb. (N.S.)* **78** (120) (1969), 360-373.

37. Yu. S. Il'yashenko. An example of equations $dw/dz = P_n(z,w)/Q_n(z,w)$ having a countable number of limit cycles and arbitrarily high Petrowskii-Landis genus, *Mat. Sb. (N.S.)* **80** (122) (1969), 388-404.

38. Yu. S. Il'yashenko. Limit cycles of polynomial vector fields with nondegenerate singular points on the real plane. *Funkcional. Anal. i Prilozen* **18** (1984), 32-42.

39. Yu. S. Il'yashenko. Dulac's memoir "On limit cycles" and related problems of the local theory of differential equations. *Uspekhi Mat. Nauk* **40:6** (1985), 41-78; *Russian Math. Surveys* **40:6** (1985), 1-49.

40. E.M. James. Liénard equations with at most two limit cycles. *Internal report,* University College of Wales, Aberystwyth, 1987.

41. J.P. Jouanolou. *Equations de Pfaff algebriques.* Lecture Notes in Mathematics Vol.708, Springer Verlag, Berlin/New York, 1979.

42. M.C. Kalenge. On some polynomial systems in the plane. *PhD. thesis*, University College of Wales, Aberystwyth, 1986.

43. D.E. Koditschek & K.S. Narendra. Limit cycles of planar quadratic differential equations. *J. Differential Equations* **54** (1984), 181-195.

44. Li Chengzhi. Two problems of planar quadratic systems. *Sci. Sinica Ser. A* **26** (1983), 471-481.

45. Li Chengzhi. Non-existence of limit cycle around a weak focus of order three for any quadratic system. *Chinese Ann. Math.* **7B** (1986), 174-190.

46. Li Jibin & Huang Q-M. Bifurcations of limit cycles forming compound eyes in the cubic system (Hilbert number $H_3 \geq 11$). *J. Yunnan University* **1** (1985), 7-16.

47. Li Jibin & Li Chunfu. Global bifurcations of planar disturbed Hamiltonian systems and distributions of limit cycles of cubic systems. *Acta. Math. Sinica* **28** (1985), 509-521.

48. Li Jibin, Tian Jinghuang & Xu S-L. A survey of cubic systems. *Sichuan Shiyuan Xuebao* (1983), 32-48.

49. A. Lins Neto. On the number of solutions of the equations $dx/dt = \sum_{j=0}^{n} a_j(t)t^j$, $0 \leq t \leq 1$, for which $x(0) = x(1)$. *Invent. Math.* **59** (1980), 67-76.

50. A. Lins, W. de Melo & C.C. Pugh. On Liénard's equation. *Geometry and Topology (Rio de Janeiro, 1976)* Lecture Notes in Mathematics, No.597 (Springer-Verlag, 1977), 335-357.

51. N.G. Lloyd. The number of periodic solutions of the equation $\dot{z} = z^N + p_1(t) z^{N-1} + \ldots + p_N(t)$. *Proc. London Math. Soc.* (3) **27** (1973), 667-700.

52. N.G. Lloyd. On a class of differential equations of Riccati type. *J. London Math. Soc.* (2) **10** (1975), 1-10.

53. N.G. Lloyd. A note on the number of limit cycles in certain two-dimensional systems. *J. London Math. Soc.* (2) **20** (1979), 277-286.

54. N.G. Lloyd. Small amplitude limit cycles of polynomial differential equations. *Ordinary Differential Equations and Operators*, eds. W.N. Everitt & R.T. Lewis. Lecture Notes in Mathematics, No.1032 Springer-Verlag, (1982), 346-357.

55. N.G. Lloyd. Limit cycles of certain polynomial systems, *Nonlinear Functional Analysis and its Applications*, ed. S.P. Singh, NATO ASI Series C, Vol. 173 (1986), 317-326.

56. N.G. Lloyd. Liénard systems with several limit cycles. *Math. Proc. Camb. Phil. Soc.*, to appear.

57. N.G. Lloyd & S. Lynch. Small-amplitude limit cycles of certain Liénard systems. *Preprint*, University College of Wales, Aberystwyth (1987).

58. F.W. Long & I. Danicic. Algebraic manipulation of polynomials in several indeterminates. *Proceedings of the Conference on Applications of Algol 68* (University of East Anglia: March 1976), 112-115.

59. V.V. Nemytskii & V.V. Stepanov. *Qualitative Theory of Differential Equations* (Princeton University Press, 1960).

60. Tyre A. Newton. Two dimensional homogeneous quadratic differential systems. *SIAM Rev.* **20** (1978), 120-138.

61. L.M. Perko. Centre-foci for analytic systems. *Preprint*, Northern Arizona University, 1986.

62. L.M. Perko. On the accumulation of limit cycles. *Proc. Amer. Math. Soc.*, **99** (1987), 515-526.

63. I.G. Petrovskii & E.M. Landis. On the number of limit cycles of the equation $dy/dx = P(x,y)/Q(x,y)$ where P and Q are polynomials of the second degree. *Mat. Sb. N. S.* **37** (79) (1955), 209-250; *Amer. Math. Soc. Transl.* (2) **16** (1958), 177-221.

64. I.G. Petrovskii & E.M. Landis. On the number of limit cycles of the equation $dy/dx = P(x,y)/=Q(x,y)$, where P and Q are polynomials. *Mat. Sb. N. S.* 43 (85) (1957), 149-168; *Amer. Math. Soc. Transl.* (2) **14** (1960), 181-200.

65. I.G. Petrovskii & E.M. Landis. Corrections to the articles "On the number of limit cycles of the equation $dy/dx = P(x,y)/Q(x,y)$ where P and Q are polynomials of the second degree" and "On the number of limit cycles of the equation $dy/dx = P(x,y)/Q(x,y)$, where P and Q are polynomials". *Mat. Sb. N.S.* **48** (90) (1959), 263-255.

66. V.A. Pliss. *Non-local problems of the theory of oscillations.* (Academic Press, New York, 1966.)

67. Qin Yuanxun, Shi Songling & Cai Suilin. On limit cycles of planar quadratic systems. *Sci Sinica Ser. A* **25** (1982), 41-50.

68. Qin Yuanxun. On surfaces defined by ordinary differential equations - a new approach to Hilbert's 16th problem. *J. Northwest Univ.* **1** (1984), 1-15.

69. R.H. Rand & W.L. Keith. Determinacy of degenerate equilibria with linear part $x'=y$, $y'=0$ using MACSYMA. *Appl. Math. Comput.* **21** (1987), 1-19.

70. J.W. Reyn. Phase portraits of a quadratic system of differential equations occurring frequently in applications. *Nieuw. Arch. Wisk.*, to appear.

71. C. Rousseau. Example of a quadratic system with 2-cycles appearing in a homoclinic loop bifurcation. *J. Differential Equations*, **66** (1987), 140-150.

72. C. Rousseau. Bifurcation methods in quadratic systems. *Preprint*, University of Montreal, 1986.

73. C. Rousseau, Bifurcation of limit cycles at infinity in polynomial vector fields. *Preprint*, University of Montreal, 1986.

74. G.S. Rychkov. The maximum number of limit cycles of the system $\dot{y} = x$, $\dot{x} = y - \sum_{i=0,1,2} a_i x^{2i+1}$ is two. *Differencial'nye Uravneniya*, **11** (1973), 380-391.

75. S. Shahshahani. Periodic solutions of polynomial first order differential equations. *Nonlinear Anal.* **5** (1981), 157-165.

76. Shi Songling. Example of five limit cycles for cubic systems. *Acta Math. Sinica* **18** (1975), 300-304.

77. Shi Songling. A concrete example of the existence of four limit cycles for plane quadratic systems. *Sci. Sinica Ser. A* **23** (1980), 153-158.

78. K.S. Sibirskii. The number of limit cycles in the neighbourhood of a singular point. *Differencial'nye Uravneniya* **1** (1965), 53-66.

79. K.S. Sibirskii & V.A. Lunkevich. On the conditions for a centre. *Differencial'nye Uravneniya* **1** (1965), 176-181.

80. K.S. Sibirskii. On dynamical systems which are close to being Hamiltonian. *Differencial'nye Uravneniya* **3** (1967), 2177-2178.

81. B.D. Sleeman. The number of limit cycles of polynomial autonomous systems in the plane and Hilbert's 16th problem. *Report* AA/862, University of Dundee, 1986.

82. J. Sotomayor & R. Paterlini. Quadratic vector fields with finitely many periodic orbits. *International Symposium on Dynamical Systems.* I.M.P.A., Rio de Janeiro,(1983).

83. U. Staude. Uniqueness of Periodic Solutions of the Liénard Equation. *Recent Advances in Differential Equations.* (Academic Press,1981).

84. F. Takens. Singularities of vector fields. *Inst. Hautes Etudes Sci. Publ. Math.* **43** (1974), 47-100.

85. Tian Jinghuang. A survey of Hilbert's problem 16. *Research report* 19, Institute of Mathematical Sciences, Acadmia Sinica, Chengdu, 1986.

86. Wang Mingshu & Luo Dingjun. Global bifurcation of some cubic planar systems. *Nonlinear Anal.* **8** (1984), 711-722.

87. Ye Yanqian. Some problems in the qualitative theory of ordinary differential equations. *J. Differential Equations* **46** (1982), 153-164.

88. Ye Yanqian. On the impossibility of (2,2) distribution of limit cycles of any real quadratic system. *J. Nanjing University* (1985), 161-182.

89. Ye Yanqian & others. *Theory of limit cycles.* Trans. Math. Monographs, 66 (Amer. Math. Soc., 1986).

90. Zeng Xianwu. Remarks on the uniqueness of limit cycles. *Kexue Tongbao* **28** (1983), 452-455.

91. Zhang Zhifen. On the uniqueness of limit cycles of certain equations of nonlinear oscillations. *Dokl. Akad. Nauk. SSSR* **119** (1958), 659-662.

92. Zhang Zhifen. Proof of the uniqueness theorem of limit cycles of generalised Liénard equations. *Preprint*, Instituto Matematica 'U. Dini', Florence, 1985.

93. Zhang Zhifen. Theorem of existence of exactly n limit cycles in $|\dot{x}| \leq (n+1)R$ for the differential equation $\ddot{x} + \mu \sin \dot{x} + x = 0$. *Sci. Sinica* **23** (1980), 1502-1510.

94. W. A. Coppel. Some quadratic systems with at most one limit cycle. *Research report* No. 14 (1987), Mathematical Sciences Research Centre, The Australian National University, Canberra.

95. B. Drachman, S. van Gils & Zhang Zhiffen, Abelian integrals for quadratic vector fields. *Preprint*, Michigan State University, 1987.

96. Jean Ecalle, Jean Martinet, Robert Moussu & Jean-Pierre Ramis. Non-accumulation des cycles-limites (I). *C. R. Acad. Sci. Paris Sér. I Math.* **304** (1987), 375-377.

97. Jean Ecalle, Jean Martinet, Robert Moussu & Jean-Pierre Ramis. Non-accumulation des cycles-limites (II). *C. R. Acad. Sci. Paris Sér. I Math.* **304** (1987), 431-434.

98. A. Gasull, J. Llibre & J. Sotomayor. Further considerations on the number of limit cycles of vector fields of the form X(t)=Av+f(v)Bv. *J. Differential Equations* **68** (1987), 36-40.

99. John Guckenheimer, Richard Rand & Dana Schlomiuk. Degenerate homoclinic cycles in perturbations of quadratic Hamiltonian systems. *Preprint*, Cornell University, 1987.

100. R. Moussu. Développement asymptotique de l'application retour d'un polycycle. *Preprint*, University of Dijon, 1987.

101. J.W. Reyn & P. de Jager. Phase portraits for quadratic systems with higher order singularity I. Third and fourth order points with two zero eigenvalues. *Report* 86-52, Delft university of Technology, 1986.

102. R. Roussarie. On the number of limit cycles which appear by perturbation of a separatrix loop of planar vector fields. *Bol. Soc. Brasil. Mat.*, **17** (1986), 67-101.

BIFURCATIONS WITH SYMMETRY

Ian Stewart

Mathematics Institute
University of Warwick
Coventry CV4 7AL, U.K.

0. INTRODUCTION

Symmetries abound in nature, in technology, and - especially - in the simplified mathematical models that we study so assiduously. Symmetries complicate things and simplify them. They complicate them by introducing exceptional types of behaviour, increasing the number of variables involved, and making vanish things that usually do not vanish. They simplify them by introducing exceptional types of behaviour, increasing the number of variables involved, and making vanish things that usually do not vanish. They violate all the hypotheses of our favourite theorems, yet lead to natural generalizations of those theorems. It is now standard to study the 'generic' behaviour of dynamical systems. Symmetry is not generic. The answer is to work within the world of symmetric systems and to examine a suitably restricted idea of genericity.

The pioneering work of Sattinger [1979, 1983], Vanderbauwhede [1982] and others opened up the possibility of a systematic theory, and during the past decade understanding of the bifurcation of dynamical systems with symmetry has developed into a recognizable subject with its own distinctive identity: Equivariant Bifurcation Theory. It is not just a tactical development: it embodies a general strategy for tackling the bifurcations of symmetric nonlinear systems. The technical machinery is extensive - Lie theory, representation theory, invariant theory, dynamical systems, and topology, for example - and the literature has grown to the point where the details can obscure the broader principles of the subject.

Symmetries are often exploited without being made explicit. For example, symmetry often forces multiple eigenvalues; but in any computation of those eigenvalues their multiplicity will emerge in due course. 'Accidental' simplifications in calculations can be exploited without tracing their origins. However, to formulate a general context and understand the ways in which symmetry influences behaviour, it is best to render the symmetry explicit. This introduces the language of group representation theory - a powerful tool, but not one that is universally familiar. Fortunately its most pertinent results can easily be summarized.

The main message is that the range of 'typical' behaviour that can occur in a symmetric system is model-independent; that is, it is largely conditioned by the symmetry and not by the precise details of the system. The behaviour that actually occurs from within this range is much more sensitive to the details of the model. The suggested strategy is to deal with the problem in these two steps, and this order. This lets us disentangle the model-independent phenomena, due purely to symmetry, from those that depend on the precise system concerned.

This paper is based on the material in Golubitsky and Schaeffer [1984] and Golubitsky, Stewart, and Schaeffer [1987], and these sources should be consulted for proofs and further details. They also include a much more detailed analysis of the stability of bifurcating branches of solutions to equations with symmetry. This is a highly important topic, but in the interests of brevity it will not be discussed in detail in this paper. This survey is not exhaustive and many important contributions are not mentioned. See also Golubitsky [1983] and Stewart [1981, 1982, 1983a,b,c].

1. SYSTEMS WITH SYMMETRY

For simplicity, our main object of study will be a system of ordinary differential equations (ODEs)

$$\dot{x} + f(x,\lambda) = 0 \tag{1}$$

where $x \in \mathbb{R}^n$, $\lambda \in \mathbb{R}$ is a bifurcation parameter, and $f:\mathbb{R}^n\times\mathbb{R} \to \mathbb{R}^n$ is a smooth (C^∞) mapping defined near $(0,0)$. We emphasize that in general f is nonlinear, and that we study only local behaviour near the origin. We also assume that there is a trivial solution, so that $f(0,\lambda) \equiv 0$. The methods apply more generally, for example to certain partial differential equations, but it helps to focus on the main ideas if we work in this restricted setting. We concentrate on classical 'prechaotic' phenomena such as Hopf bifurcation. A theory of symmetric chaos is one long-term aim, but the more classical groundwork needs to be put in first.

Let Γ be a Lie group of linear transformations of $\mathbb{R}^n$. We say that Γ is a group of symmetries of (1) if

$$f(\gamma x,\lambda) = \gamma f(x,\lambda) \tag{2}$$

for all $\gamma \in \Gamma$. Another term for this is that f is Γ-equivariant. For technical reasons we assume that Γ is compact, although some of the more elementary results hold more generally and it would be interesting to know more about the non-compact case.

To set the scene we first describe, very broadly, what happens when symmetries are absent from (1). Generically there are two types of bifurcation:

(a) Steady-state bifurcation, when an eigenvalue of $(df)_{0,\lambda}$ passes through 0 (without loss of generality at $\lambda = 0$).

(b) Hopf bifurcation, when a pair of conjugate complex eigenvalues of $(df)_{0,\lambda}$ crosses the imaginary axis with nonzero speed at $\pm\omega i$, $\omega \neq 0$.

Generically these eigenvalues are simple. In case (a), subject to suitable nondegeneracy conditions, there will be a branch of steady-state solutions bifurcating from the origin. In case (b) a branch of periodic solutions, of period near $2\pi/\omega$, will bifurcate.

When the initial solution is not an isolated point but lives in a group orbit of solutions, other phenomena can occur. For example, a single real eigenvalue passing through zero can create a periodic solution from a circle of equilibria. We will not discuss this point further, but it illustrates the need for caution.

The above are the generic cases, likely to happen in a single 'typical' ODE. If there are parameters in the equation, then at certain parameter values a degeneracy may occur. The number of parameters needed for a given degeneracy to occur 'typically' is called its codimension; and the behaviour near such a parameter value can be found by 'unfolding' the degeneracy via a universal family of perturbations. This technique is valuable because it can detect quasiglobal behaviour by a local analysis: for a survey see Guckenheimer [1984].

We distinguish two principal types of degeneracy.

(a) Linear degeneracy. The Jacobian $(df)_{0,0}$ has a multiple eigenvalue (0 or $\pm\omega i$) or several distinct such eigenvalues.

(b) Nonlinear degeneracy. Degeneracies in higher order terms of f lead to 'atypical' branching behaviour. (For example consider a Hopf-type bifurcation where the eigenvalue meets the imaginary axis with zero speed.) The eigenfunctions associated with the linearization of (1) at (0,0) are often called modes; and in case (a) we speak of new types of solution being created by mode interaction. There are three basic types of mode interaction: steady-state/steady state, Hopf/steady-state, and Hopf/Hopf.

This classification is not exhaustive, especially as regards more complicated dynamic behaviour. A case in point is the Takens-Bogdanov bifurcation, where

$$(df)_{0,0} = \begin{pmatrix} 0 & 1 \\ 0 & 0 \end{pmatrix},$$

see Guckenheimer[1984], Guckenheimer and Holmes [1983]. Homoclinic orbits can arise near such a system. The study of dynamic bifurcations of this type is new even in the asymmetric case; but a start on the symmetric case has been made by Dangelmayr and Knobloch [1986a,b].

Now we turn to symmetric systems. The main technical feature is that multiple eigenvalues arise generically. The correct analogue of a simple eigenvalue is one whose eigenspace is Γ-irreducible (see below for clarification and a caveat). It follows that steady-state, Hopf, and mode interaction behaviour must be indexed by the irreducible representations of Γ, leading to a great diversity of behaviour.

2. EXAMPLES

The ideas, methods, and phenomena that occur are most easily grasped by studying simple examples. The role of the examples is to motivate and illustrate general principles which can then be applied to more complicated situations. Here we discuss two archetypes for static and Hopf bifurcation: deformation of an elastic cube, and oscillations of a circular hosepipe.

(a) Deformation of an elastic cube

Consider an incompressible elastic cube, subjected to a uniform tension λ perpendicular to each face. As λ increases the undeformed cubic shape loses stability, and we ask what new equilibrium shapes can arise. For pedagogical purposes we consider only deformations for which the cube becomes a rectangular parallelepiped, of sides v_1, v_2, v_3 where $v_1v_2v_3 = 1$ by incompressibility. Any reasonable model of this system will be unchanged under permutations of these lengths, so the group of symmetries is $\mathbf{S}_3$, the symmetric group of degree 3, consisting of all permutations on three symbols. By applying such permutations we may assume that $v_1 \leq v_2 \leq v_3$, leading us to distinguish the four cases illustrated in Figure 1:

(1) $v_1 = v_2 = v_3$ [cube] (2) $v_1 = v_2 < v_3$ [rod]

(3) $v_1 < v_2 = v_3$ [slab] (4) $v_1 < v_2 < v_3$ [box].

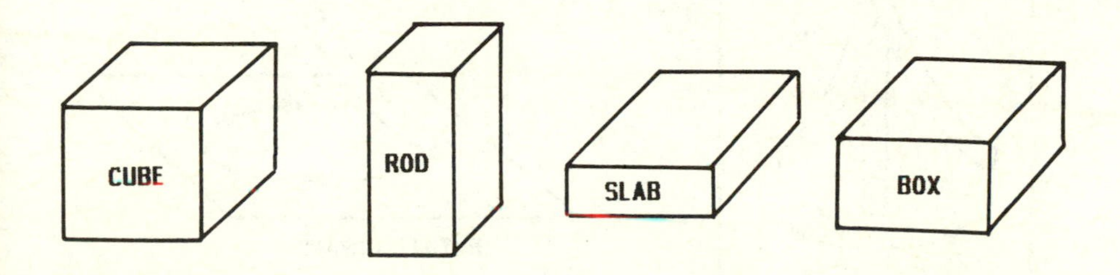

Figure 1. Some possible equilibria of a deformed elastic cube.

These types of possible equilibrium configuration are distinguished by their symmetries. The original undeformed cube has the full $\mathbf{S}_3$ symmetry. The slab and rod solutions have symmetry $\mathbf{S}_2$, permuting only the two equal sides. The unequal-sided box has only the trivial permutational symmetry $\mathbb{1}$. (By focussing on the permutational symmetry we are ignoring the various reflectional and rotational symmetries of these shapes.)

A naive view would be that the box (4) is the generic possibility: in the space of possible (v_1,v_2,v_3), almost all points satisfy (4). This approach, however, applies genericity arguments in the wrong place. In fact - as we show later - the bifurcating equilibria are the rod and plate solutions with $\mathbf{S}_2$ symmetry. This is an example of the general phenomenon of spontaneous symmetry-breaking. When a system with symmetry group Γ bifurcates, the bifurcating solutions have a symmetry group Δ which is a subgroup of Γ - and usually a 'large' one.

(b) The hosepipe

Consider a circular hosepipe hanging vertically, with water running through it at a speed λ. If λ is small, the hosepipe will stay in equilibrium in the vertical position; but as λ increases, a Hopf-type bifurcation occurs and the pipe begins to oscillate. Two types of primary oscillation can be observed: see e.g. Bajaj and Sethna [1984]. The first is a standing wave, where the pipe remains in a vertical plane and swings rather like a pendulum. The other is a rotating wave, where the free end describes circles in a horizontal plane. See Figure 2.

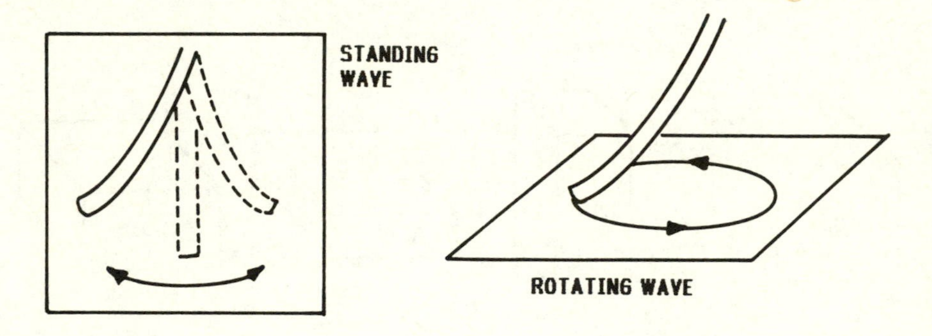

Figure 2. Oscillations of a circular hosepipe.

Again these solutions are distinguished by their symmetries. The original system has symmetry group $\mathbf{O}(2)$, the group of rotations and reflections of a circle. If we consider only these spatial symmetries then the standing wave has symmetry $\mathbf{Z}_2$ (reflection in its plane) and the rotating wave has symmetry $\mathbb{1}$. But this is not the best way to think about the symmetries of periodic states. For example, it fails to pick up the 'rotating wave' structure. The temporal phase-shift symmetries must also be introduced. These form a circle group $\mathbf{S}^1 \equiv \mathbb{R}/T\mathbf{Z}$ where T is the period (which we usually normalize to 2π). The mixed spatio-temporal symmetries are elements of $\mathbf{O}(2)\times\mathbf{S}^1$. The rotating wave then has symmetries of the form $(\theta,-\theta) \in \mathbf{O}(2)\times\mathbf{S}^1$. That is, a phase shift of θ is equivalent to a spatial rotation of the solution by θ. (The standing wave also has a further mixed spatio-temporal symmetry (π,π).)

3. A Package Tour of Representation Country

We now introduce some mathematical precision. A group Γ _acts_ on a (real) vector space V if there is a map

$$\Gamma \times V \to V$$
$$(\gamma, v) \mapsto \gamma.v$$

which is linear in v and satisfies

$$\gamma.(\delta.v) = (\gamma\delta).v \quad (\forall \gamma, \delta \in \Gamma)$$
$$1.v = v$$

where 1 is the identity element of Γ. We call V a Γ-_space_. A subspace W $\subset$ V is Γ-_invariant_ if $\gamma.w \in W$ for all $\gamma \in \Gamma$, $w \in W$. If V has no Γ-invariant subspaces other than 0, V then it is Γ-_irreducible_.

For each $\gamma \in \Gamma$ the map

$$\rho_\gamma : V \to V$$
$$\rho_\gamma(v) = \gamma.v$$

is an invertible linear mapping. Let **GL**(V) be the group of invertible linear transformations of V. The map

$$\rho : \Gamma \to \mathbf{GL}(V)$$
$$\rho(\gamma) = \rho_\gamma$$

is a _representation_ of Γ on V. Actions and representations are two technically different ways to describe the identical idea and we will tend to use the terms interchangeably.

The main result on representations of compact Lie groups is:

Theorem 3.1 (_Complete reducibility_) If Γ is a compact Lie group acting on V, then V decomposes as

$$V = V_1 \oplus \ldots \oplus V_r$$

where each V_j is Γ-irreducible.

A Γ-map between Γ-spaces V and W is a linear map $\theta: V \to W$ such that

$$\theta(\gamma.v) = \gamma.(\theta(v)) \quad (\forall \gamma \in \Gamma, v \in V).$$

Define $Hom_\Gamma(V,W)$ to be the space of all such Γ-maps. This is the space of linear maps $V \to W$ that commute with Γ. Say that V and W are Γ-isomorphic if there exists an invertible Γ-map $V \to W$, and write $V \cong W$.

There are very few Γ-maps between irreducible Γ-spaces:

Theorem 3.2 (Schur's Lemma) Let V, W be Γ-irreducible. Then

(a) If $V \ncong W$ then $Hom_\Gamma(V,W) = 0$.

(b) If $V \cong W$ then $Hom_\Gamma(V,W) \cong \mathbb{R}, \mathbb{C}$, or $\mathbb{H}$.

Here $\mathbb{R}$ is the reals, $\mathbb{C}$ the complex numbers, and $\mathbb{H}$ the quaternions. Their dimensions over $\mathbb{R}$ are 1, 2, and 4. The distinction between these three types of irreducible representation is fundamental to real representation theory.

Examples

1. $\Gamma = \mathbf{SO}(2)$, the group of orientation-preserving symmetries of a circle. This acts naturally on $\mathbb{R}^2 \equiv \mathbb{C}$ by

$$\theta.z = e^{i\theta}z \quad (\theta \in [0,2\pi)).$$

The action is irreducible. $Hom_{\mathbf{SO}(2)}(\mathbb{C},\mathbb{C})$ is isomorphic to $\mathbb{C}$ since any $w \in \mathbb{C}$ commutes with the $\mathbf{SO}(2)$-action:

$$\theta.(wz) = e^{i\theta}wz = we^{i\theta}z = w(\theta.z).$$

2. $\Gamma = \mathbf{O}(2)$, generated by $\mathbf{SO}(2)$ together with the 'flip' κ, acting by complex conjugation $\kappa.z = \bar{z}$. Again $\mathbb{C}$ is an irreducible $\mathbf{O}(2)$-space, but now $Hom_{\mathbf{O}(2)}(\mathbb{C},\mathbb{C}) \cong \mathbb{R}$. For w to commute with κ we need $\kappa.(wz) = w(\kappa.z)$, but

$$\kappa.(wz) = \overline{wz} = \overline{w}\overline{z},$$
$$w(\kappa.z) = w\overline{z}.$$

Thus $w = \overline{w}$ so $w \in \mathbb{R}$.

3. Γ = $\mathbf{SU}(2)$, the unit quaternions $\{a+bi+cj+dk : a^2+b^2+c^2+d^2 = 1\}$ acts on $\mathbb{H}$ by left multiplication $q.h = qh$ $(q \in \mathbf{SU}(2), h \in \mathbb{H})$. The space $\mathbb{H}$ is irreducible. $\mathrm{Hom}_{\mathbf{SU}(2)}(\mathbb{H},\mathbb{H}) \cong \mathbb{H}$ since multiplication on the right by any $h' \in \mathbb{H}$ commutes with $\mathbf{SU}(2)$:

$$(q.h)h' = (qh)h' = q(hh') = q.(hh').$$

4. $\Gamma = \mathbf{D}_n$, the dihedral group of order $2n$, is the subgroup of $\mathbf{O}(2)$ generated by $\zeta = 2\pi/n$ and κ. For $n \geq 3$ it acts irreducibly on $\mathbb{C}$ with the natural action

$$\zeta.z = e^{2\pi i/n}z \qquad \kappa.z = \overline{z}.$$

$\mathrm{Hom}_{\mathbf{D}_n}(\mathbb{C},\mathbb{C}) \cong \mathbb{R}$.

5. $\Gamma = \mathbf{Z}_n$, the cyclic group of order n, is the subgroup of $\mathbf{SO}(2)$ generated by ζ. For $n \geq 3$ it acts irreducibly on $\mathbb{C}$ with the natural action, and $\mathrm{Hom}_{\mathbf{Z}_n}(\mathbb{C},\mathbb{C}) \cong \mathbb{C}$.

4. Eigenstructure

If $L:V \to V$ is a linear map and $\lambda \in \mathbb{C}$, define the (real) eigenspace E_λ and generalized eigenspace G_λ to be

$$\begin{aligned}
E_\lambda &= \{x \in V: (L-\lambda I)x = 0\} && (\lambda \in \mathbb{R})\\
& \{x \in V: (L-\lambda I)(L-\overline{\lambda}I)x = 0\} && (\lambda \notin \mathbb{R})\\
G_\lambda &= \{x \in V: (L-\lambda I)^n x = 0\} && (\lambda \in \mathbb{R})\\
& \{x \in V: (L-\lambda I)^n(L-\overline{\lambda}I)^n x = 0\} && (\lambda \notin \mathbb{R})
\end{aligned}$$

where $n = \dim V$. Clearly $V = \bigoplus_\lambda G_\lambda$.

Proposition 4.1 If Γ acts on V and L commutes with Γ then its eigenspaces E_λ and generalized eigenspaces G_λ are Γ-invariant.

Proof Let $p(L)$ be any polynomial in L. We claim that $\ker p(L)$ is Γ-invariant. Let $x \in \ker p(L)$ so that $p(L)x = 0$. If $\gamma \in \Gamma$ then $p(L)\gamma.x = \gamma.p(L)x = 0$. So $\gamma.x \in p(L)$.

Irreducible Γ-spaces usually have dimension ≥ 1. Thus Proposition 4.1 explains why multiple eigenvalues tend to occur in symmetric systems.

There is a 'dual' result. If V is a Γ-space and X is an irreducible Γ-space we define the isotypic component of V of type X to be

$$W_X(V) = \sum\{W : W \subset V \text{ is } \Gamma\text{-invariant}, W \cong X\}.$$

This is unique, contains every Γ-invariant subspace of V that is Γ-isomorphic to X, and can be written as $W_X(V) = X_1 \oplus \ldots \oplus X_k$ where each $X_j \cong X$. We have $V = \bigoplus_X W_X(V)$ as X runs through all possible (isomorphism types of) irreducible Γ-spaces.

Proposition 4.2 If L commutes with Γ, then L leaves each $W_X(V)$ invariant.

Remark This means that L possesses a block matrix structure

$$\begin{matrix} W_{X_1}(V) \\ \cdot \\ \cdot \\ \cdot \\ W_{X_p}(V) \end{matrix} \quad \begin{pmatrix} L_1 & & & 0 \\ & \cdot & & \\ & & \cdot & \\ 0 & & & \cdot \\ & & & L_p \end{pmatrix}$$

and this fact can often simplify the analysis of L.

5. FIXED-POINT YOGA

As the examples in §2 illustrate, one of the most striking phenomena in the bifurcation of symmetric systems is spontaneous symmetry-breaking. Formulating this idea in a group-theoretic manner leads to two important concepts: the isotropy subgroup of a solution, and the corresponding fixed-point subspace.

Suppose that Γ acts on V. If $x \in V$ then the orbit of x under Γ is

$$\Gamma x = \{\gamma x \mid \gamma \in \Gamma\}.$$

If $f:V \to V$ is Γ-equivariant, the the zero-set of f is a union of orbits. For if $f(x) = 0$ then $f(\gamma x) = \gamma f(x) = 0$. In schematic bifurcation diagrams, and in discussing particular bifurcation problems, it is convenient to consider solutions in the same Γ-orbit as being the same.

The isotropy subgroup of $x \in V$ is

$$\Sigma_x = \{\gamma \in \Gamma \mid \gamma x = x\}.$$

Isotropy subgroups of points on the same orbit are conjugate. Indeed $\Sigma_{\gamma x} = \gamma \Sigma_x \gamma^{-1}$. We therefore tend not to distinguish between isotropy subgroups and their conjugates.

If $H \subset \Gamma$ is any subgroup, we define the fixed-point subspace

$$\mathrm{Fix}(H) = \{x \in V \mid \gamma x = x \text{ for all } \gamma \in H\}.$$

We have $\mathrm{Fix}(\gamma H \gamma^{-1}) = \gamma \mathrm{Fix}(H)$. The importance of fixed-point subspaces is:

Proposition 5.1 If f is Γ-equivariant and $H \subset \Gamma$ then f leaves Fix(H) invariant.

Proof Let $\gamma \in H$, $x \in \mathrm{Fix}(H)$. Then $\gamma f(x) = f(\gamma x) = f(x)$ so $f(x) \in \mathrm{Fix}(H)$.

Despite its trivial proof, the above fact is very useful. For suppose we are seeking a branch of solutions to a Γ-equivariant bifurcation problem $f(x,\lambda) = 0$, breaking symmetry to H. Then $x \in \mathrm{Fix}(H)$, and it suffices to solve $f|\mathrm{Fix}(H) = 0$. Since dim Fix(H) is generally smaller than dim V, this is a simpler problem.

The isotropy lattice of Γ acting on V is the set of conjugacy classes of isotropy subgroups, with partial ordering induced by inclusion. Strictly speaking it is not a lattice but a finite partially ordered set. Successive bifurcations tend to break symmetry further and further down this lattice, and it plays an important role in organizing calculations.

6. THE EQUIVARIANT BRANCHING LEMMA

In the case where dim Fix(H) is as small as possible, we have the following theorem of Vanderbauwhede [1980] and Cicogna [1981]:

Theorem 6.1 (Equivariant Branching Lemma). Let $f(x,\lambda) = 0$ be a Γ-equivariant bifurcation problem where $\mathrm{Fix}(\Gamma) = 0$. Let Σ be an isotropy subgroup such that $\dim \mathrm{Fix}(\Sigma) = 1$. If $(df_\lambda)_{0,0}(v_0) \neq 0$, $0 \neq v_0 \in \mathrm{Fix}(\Sigma)$, then near $(0,0)$ there exists a branch of solutions $(tv_0, \lambda(t))$ to $f(x,\lambda) = 0$.

If $\dim \mathrm{Fix}(\Sigma) = 1$ then it is easy to see that Σ is a maximal isotropy subgroup, that is, one not properly contained in an isotropy subgroup $\neq \Gamma$. However, there exist maximal isotropy subgroups Δ with $\dim \mathrm{Fix}(\Delta)$ arbitrarily large, see Ihrig and Golubitsky [1984]. The important question of which isotropy subgroups generically lead to branches is open, although there are numerous special results. Here we note only one simple proposition:

Proposition 6.2 For Γ-equivariant bifurcation problems g, generically the 0-eigenspace E of $(dg)_{0,0}$ is Γ-irreducible, and $\mathrm{Hom}_\Gamma(E,E) \cong \mathbb{R}$.

Remark By 'generically' we mean that the situation cannot be destroyed by a small perturbation of the vector field.

Example

Consider the deformation of an elastic cube as in §2a. Here the group $\mathbf{S}_3$ acts on the space of all (v_1,v_2,v_3) such that $v_1v_2v_3 = 1$. Letting $w_i = \log v_i$ we may work on the plane $w_1+w_2+w_3 = 0$. The subgroup $\mathbf{S}_2$ that interchanges w_1 and w_2 but fixes w_3, corresponding to rods and slabs, has the fixed-point subspace $\{(w, w, -2w)\}$ which is 1-dimensional. Hence by the Equivariant Branching Lemma rod and slab solutions exist.

Note that other solutions in the same $\mathbf{S}_3$-orbit correspond to the fixed-point subspaces $\{(w, -2w, w)\}$ and $\{(-2w, w, w)\}$ of conjugates of $\mathbf{S}_2$ in $\mathbf{S}_3$. Physically these are rods (or slabs) in different orientations.

Theorem 6.1, while apparently rather restrictive, is widely applicable. For example, many of the existence results of Busse [1975] on the spherical Bénard problem can be recovered from it. Melbourne [1987a] has applied it to bifurcations with the symmetry group of the cube acting on $\mathbb{R}^3$.

7. HOPF BIFURCATION

The situation for Hopf bifurcation is closely analogous to that for static bifurcation, but with the symmetry group Γ replaced by $\Gamma\times\mathbf{S}^1$. Consider a system of ODEs

$$\dot{v} + f(v,\lambda) = 0 \tag{3}$$

where $v \in \mathbb{R}^n$, λ is a bifurcation parameter, $f:\mathbb{R}^n\times\mathbb{R} \to \mathbb{R}^n$ is Γ-equivariant, and $f(0,\lambda) \equiv 0$. The main hypothesis of the classical Hopf bifurcation theorem is that $(df)_{0,0}$ should have a pair of purely imaginary eigenvalues. In the equivariant case this already places restrictions on the action of Γ: the corresponding eigenspace of $(df)_{0,0}$ must contain a Γ-invariant subspace that is Γ-simple, i.e. Γ-isomorphic to $V\oplus V$, where V is irreducible and $\mathrm{Hom}_\Gamma(V,V) \cong \mathbb{R}$, or to W where W is irreducible and $\mathrm{Hom}_\Gamma(W,W) \cong \mathbb{C}$ or $\mathbb{H}$. Generically we may assume that the entire imaginary eigenspace is Γ-simple.

This is the caveat promised in §1: generically the imaginary eigenspaces is 'as irreducible as possible subject to being an imaginary eigenspace'.The essential reason for this is that $(df)_{0,0}$ commutes with Γ. If X is a Γ-irreducible subspace such that $\mathrm{Hom}_\Gamma(X,X) \cong \mathbb{R}$, and X is invariant under $(df)_{0,0}$, then the eigenvalues of $(df)_{0,0}$ are *real* on X. There are two ways out: either $\mathrm{Hom}_\Gamma(X,X)$ must be $\cong \mathbb{C}$ or $\mathbb{H}$, or X must occur with multiplicity at least 2. These turn out to be the generic cases. By Centre Manifold or Liapunov-Schmidt reduction methods, we can assume without loss of generality that $\mathbb{R}^n$ itself is Γ-simple. From now on we work in this context. We also scale time so that the imaginary eigenvalues are $\pm i$, and the period of oscillations is therefore near 2π.

As observed in §2, temporal symmetries are important in Hopf bifurcation. Let $\mathbf{S}^1 = \mathbb{R}/2\pi\mathbf{Z}$, and make $\Gamma\times\mathbf{S}^1$ act on 2π-periodic functions $u(t)$ by $(\gamma,\theta)u(t) = \gamma u(t+\theta)$. We can seek periodic solutions to (3) of period near 2π as follows. We rescale time by $s = (1+\tau)t$ for a period-scaling parameter τ near 0. This yields the system

$$(1+\tau)du/ds + f(u,\lambda) = 0 \tag{4}$$

where $u(s) = v((1+\tau)t)$. Then 2π-periodic solutions to (4) correspond to $2\pi/(1+\tau)$-periodic solutions to (3). Define the operator

$$\Phi:\mathfrak{C}^1{}_{2\pi}\times\mathbb{R}\times\mathbb{R}\to\mathfrak{C}^0{}_{2\pi}$$
$$\Phi(u,\lambda,\tau) = (1+\tau)du/ds + f(u,\lambda),$$

where $\mathfrak{C}^k{}_{2\pi}$ is the Banach space of 2π-periodic C^k maps $\mathbb{R}\to\mathbb{R}^n$. Let $\mathcal{L} = (d\Phi)_{0,0,0} = d/ds + L$ where $L = (df)_{0,0}$. Then Liapunov-Schmidt reduction lets us find solutions to $\Phi = 0$ by solving a reduced set of equations

$$\varphi:\ker\mathcal{L}\times\mathbb{R}\times\mathbb{R}\to\operatorname{coker}\mathcal{L}. \tag{5}$$

In fact $\mathcal{L}$ is almost self-adjoint: coker $\mathcal{L}$ = ker $\mathcal{L}$. It can be shown that φ commutes with an induced action of $\Gamma\times\mathbf{S}^1$, in which

$$\theta x = e^{-L\theta}x \qquad (\theta\in\mathbf{S}^1,\ x\in\mathbb{R}^n). \tag{6}$$

Also, τ may be eliminated from the reduced equation. In other words, Hopf bifurcation with symmetry group Γ can be reduced to static bifurcation with symmetry group $\Gamma\times\mathbf{S}^1$. In a sense the $\mathbf{S}^1$-action is a static substitute for the periodic part of the dynamic.

We then have the following analogue of the Equivariant Branching Lemma, due to Golubitsky and Stewart [1985]:

Theorem 7.1 (Equivariant Hopf Theorem) With the above notation, suppose that the eigenvalues of df cross the imaginary axis with nonzero speed as λ passes through 0. Let Σ be an isotropy subgroup of $\Gamma\times\mathbf{S}^1$ acting on $\mathbb{R}^n$, such that dim Fix $(\Sigma) = 2$. Then there exists a unique branch of small-amplitude periodic solutions to (3) with period near 2π, having Σ as their group of symmetries.

Note that when interpreting the symmetries Σ for a given solution, we think of the $\mathbf{S}^1$-action as phase shift; but when calculating the dimension of the fixed-point subspace Fix(Σ) we think of it as the action by $e^{-L\theta}$ in (6). In this context the condition dim Fix$(\Sigma) = 2$ implies that Σ is a maximal isotropy subgroup. Recently Fiedler [1986] has proved a generalization of Theorem 7.1 in which Σ can be any maximal isotropy subgroup.

Example

The hosepipe. Under suitable conditions on the model we can reduce to the simplest case, where $\mathbf{O}(2)\times\mathbf{S}^1$ acts on $\mathbb{R}^4 \equiv \mathbb{C}^2$ as follows:

$$\theta(z_1, z_2) = (e^{i\theta}z_1, e^{i\theta}z_2) \qquad (\theta \in \mathbf{S}^1)$$
$$\varphi(z_1, z_2) = (e^{-i\varphi}z_1, e^{i\varphi}z_2) \qquad (\varphi \in \mathbf{SO}(2))$$
$$\kappa(z_1, z_2) = (z_2, z_1).$$

The orbit data are:

ORBIT REPRESENTATIVE	ISOTROPY SUBGROUP Σ	Fix(Σ)	dim Fix(Σ)
(0,0)	$\mathbf{O}(2)\times\mathbf{S}^1$	$\{0\}$	0
(a,0) a>0	$\widetilde{\mathbf{SO}}(2) = \{(\theta,\theta)\}$	$\{(z,0)\}$	2
(a,a) a>0	$\mathbf{Z}_2\oplus\mathbf{Z}_2^c = \{(0,0), \kappa, (\pi,\pi), (\kappa\pi,\pi)\}$	$\{(z,z)\}$	2
(a,b) a>b>0	$\mathbf{Z}_2^c = \{(0,0),(\pi,\pi)\}$	$\mathbb{C}^2$	4

Therefore Theorem 7.1 yields two branches of solutions, with isotropy subgroups $\widetilde{\mathbf{SO}}(2)$ and $\mathbf{Z}_2\oplus\mathbf{Z}_2^c$. These correspond precisely to the rotating waves and standing waves, respectively.

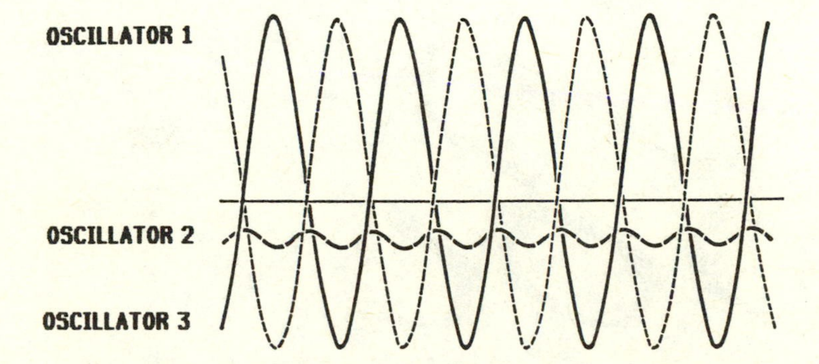

Figure 3. One type of generic symmetry-breaking in a system of three coupled identical nonlinear oscillators. From Golubitsky and Stewart [1986a].

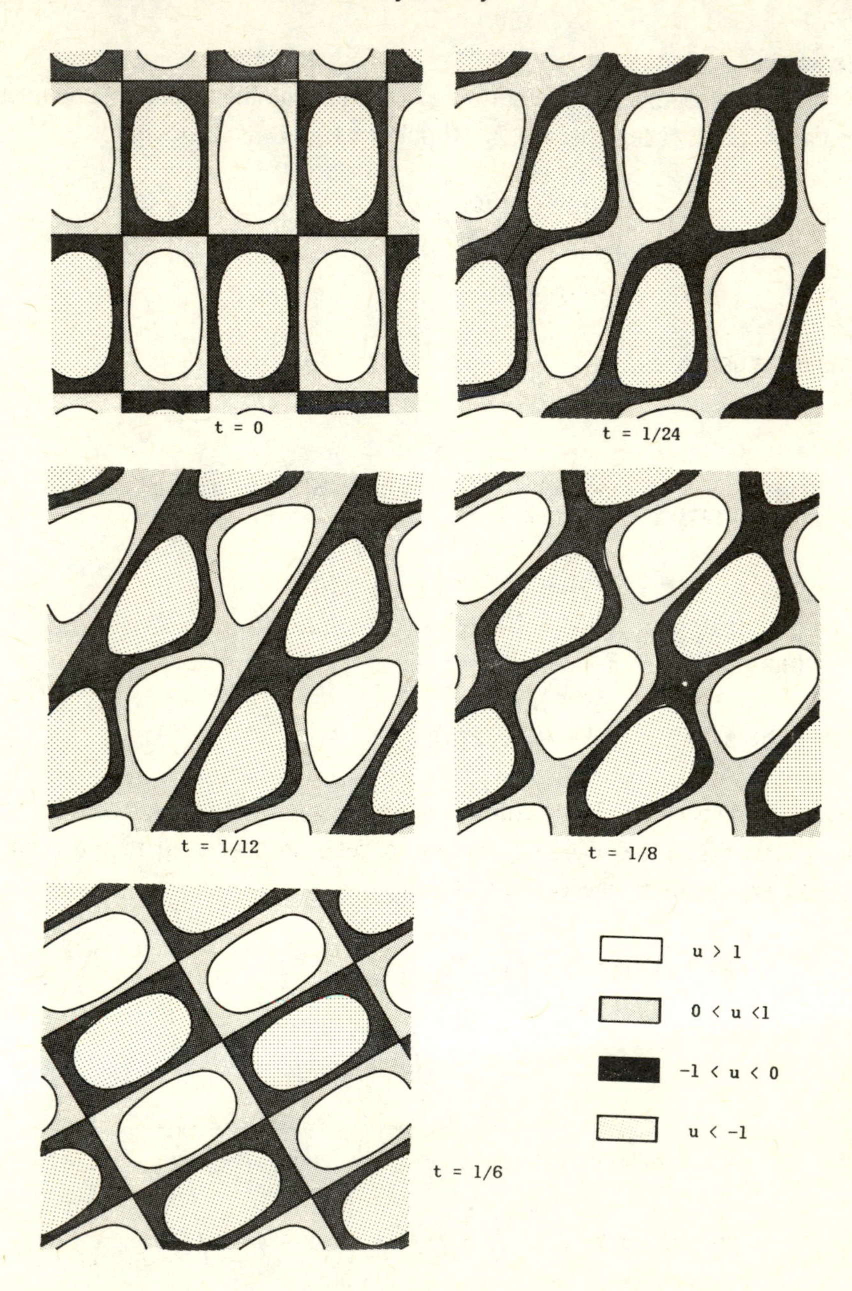

Figure 4: Wavy rolls in doubly diffusive flow on a hexagonal lattice.

Other cases have been analysed. The dihedral groups $\mathbf{D}_n$ are studied in Golubitsky and Stewart [1986b] who find three primary branches: one discrete rotating wave and two standing waves. They also apply the results to rings of identical coupled oscillators. Perhaps the most interesting oscillation is shown in Figure 3: two oscillators are 180° out of phase, while the third has half the period.

Exceptional features found by them in the case n = 4 have been the subject of further study by Swift [1986]. Roberts, Swift, and Wagner [1986] study Hopf bifurcation on the hexagonal lattice, which is relevant to certain doubly diffusive systems. Figure 4 shows one of the solutions that they find. Golubitsky and Stewart [1985] classify isotropy subgroups Σ of $\mathbf{O}(3)\times\mathbf{S}^1$ such that dim Fix(Σ) = 2, for all irreducible representations of $\mathbf{O}(3)$. These representations are on the space of spherical harmonics of degree ℓ. Predicted types of periodic solution, for ℓ = 0,1,2, are illustrated in Figures 5 and 6, which are taken from Montaldi, Roberts, and Stewart [1986]. Nagata [1986] has applied $\mathbf{O}(2)$- and $\mathbf{SO}(2)$-symmetric Hopf bifurcation to magnetoconvection. Cowan [1982] suggests a connection with visual hallucinations.

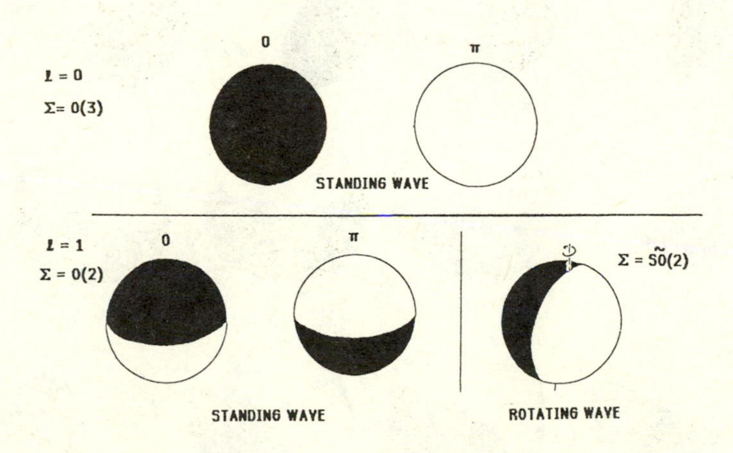

Figure 5. Hopf bifurcation in $\mathbf{O}(3)$-symmetric systems on spherical harmonics V_ℓ, for ℓ = 0,1. Black regions are below the surface of the sphere, white regions above.

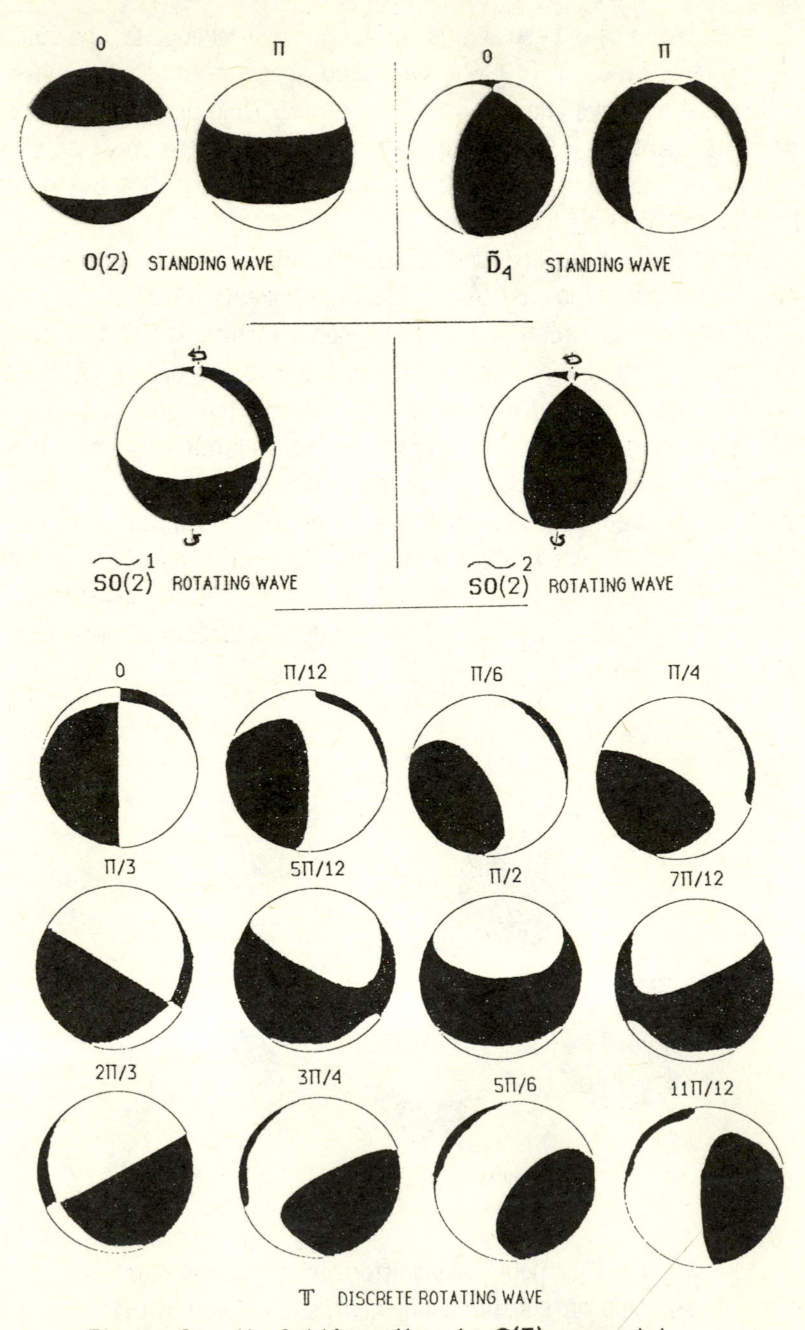

Figure 6. Hopf bifurcation in **O**(3)-symmetric systems on spherical harmonics V_ℓ, for $\ell = 2$. Black regions are below the surface of the sphere, white regions above.

8. BIRKHOFF NORMAL FORM

It is easier to study the stability of the solutions obtained by Hopf bifurcation if the system of ODEs is put into a more convenient form, known as <u>(Poincaré-) Birkhoff normal form</u>. We discuss this method first. The idea is to simplify the form of a system of ODEs

$$\dot{x} + f(x) = 0, \quad f(0) = 0, \quad x \in \mathbb{R}^n \tag{7}$$

by using successive polynomial changes of coordinates to set to zero many terms in the Taylor expansion of f at degree k. More precisely, let

$$f(x) = \tilde{f}_k(x) + h_k(x) + \ldots \tag{8}$$

where $\tilde{f}_k$ is a polynomial mapping of degree < k, h_k is a homogeneous polynomial mapping of degree k, and ... indicates terms of degree k+1 or higher. The main question is: how can we simplify h_k by changing coordinates in (7) while leaving $\tilde{f}_k$ unchanged? Since the terms of degree > k impose no constraint, the process of putting a system of ODEs into Birkhoff normal form is a recursive one.

We motivate the theorem by performing a calculation that also proves it. Consider coordinate changes of the form

$$x = y + P_k(y) \tag{9}$$

where P_k is a homogeneous polynomial of degree k. Now $\dot{x} = \dot{y} + (dP_k)_y \dot{y}$, so in the new coordinates (7) takes the form

$$\dot{y} = (I + (dP_k)_y)^{-1} f(y + P_k(y)). \tag{10}$$

Modulo higher order terms (10) is

$$\dot{y} = (I - (dP_k)_y) f(y + P_k(y)) + \ldots . \tag{11}$$

Let

$$\tilde{f}_k(x) = Lx + f_k(x) \tag{12}$$

where L is the linear part of f, and f_k consists of terms of degree between 2 and $k-1$. Since P_k contains only terms of degree k,

$$\text{(a)}\quad f_k(y+P_k(y)) = f_k(y) + \ldots \qquad (13)$$
$$\text{(b)}\quad h_k(y+P_k(y)) = h_k(y) + \ldots .$$

Substitute (12,13) into (7) to obtain

$$\begin{aligned}\dot{y} &= (I-(dP_k)_y)[L(y+P_k(y))+f_k(y)+h_k(y)] + \ldots \qquad (14)\\ &= Ly + f_k(y) + h_k(y) + LP_k(y) - (dP_k)_yLy + \ldots .\end{aligned}$$

Thus any homogeneous term of degree k in f of the form

$$LP_k(y) - (dP_k)_yLy \qquad (15)$$

may be eliminated from (7) by a change of coordinates of the type (9). Moreover, this change of coordinates does not disturb terms of degree $< k$.

More abstractly, let $\mathcal{P}_k$ be the space of homogeneous polynomial mappings of degree k on $\mathbb{R}^m$. Then ad_L, defined by

$$ad_L(P_k)(y) = LP_k(y) - (dP_k)_yLy$$

as in (15), is a linear map $\mathcal{P}_k \to \mathcal{P}_k$. Thus the terms in the Taylor expansion of f that can be eliminated by this process are precisely those in the subspace ad $L(\mathcal{P}_k) \subset \mathcal{P}_k$. For each k we choose a complementary subspace $\mathcal{G}_k \subset \mathcal{P}_k$, so that

$$\mathcal{P}_k = \mathcal{G}_k \oplus \operatorname{Im} ad_L. \qquad (16)$$

Then we have proved the following:

Theorem 8.1 Poincaré-Birkhoff Normal Form Theorem) Let $L = (df)_0$ and choose a value of k. Then there exists a polynomial change of coordinates of degree k such that in the new coordinates the system (7) has the form

$$\dot{y} = Ly + g_2(y) + \ldots + g_k(y) + \ldots$$

where $g_j \in \mathfrak{G}_j$, and ... indicates terms of degree at least k+1.

We call the system

$$\dot{x} = Lx + g_2(x) + \ldots + g_k(x) \tag{17}$$

the (k^{th} order) <u>truncated Birkhoff normal form</u> of (7).

The dynamics of the truncated Birkhoff normal form are related to, but not identical with, the local dynamics of the system (7) around the equilibrium point $x = 0$. The question of determining precisely which qualitative features of the dynamics are preserved in the kth order truncated Birkhoff normal form for some k, is still open. The technical problem is essentially that the process of putting a vector field f into Birkhoff normal form is a formal one. Suppose that f is C^∞ or analytic. Although there exists a change of coordinates that puts f in normal form for each k, it is not possible to find a single change of coordinates that puts f into normal form to all orders. The problem is one of 'small divisors'. The power series defined by successive coordinate changes may have zero radius of convergence, see for example Siegel and Moser [1971]. As we show below, despite this truncation problem, the Birkhoff normal form is an important tool for discussing Hopf bifurcation.

There are many possible choices for the complement $\mathfrak{G}_k$ in (16). Elphick <u>et al</u>. [1986] have shown that there is a canonical choice, in which the elements of $\mathfrak{G}_k$ commute with a 1-parameter group **S** of mappings defined in terms of the linear part L of f. The group **S** effectively introduces extra symmetry into the mathematical analysis, with important consequences.

The linear part $L = (df)_0$ acts on $\mathbb{R}^n$ and we have a 1-parameter group of transformations $\mathbf{R} = \{\exp(sL^t) : s \in \mathbb{R}\}$. The group **R** is not necessarily a Lie subgroup since it may not be closed. Let **S** be the closure of this group in **GL**(n),

$$\mathbf{S} = \bar{\mathbf{R}} = \overline{\{\exp(sL^t)\}}.$$

Then **S**, being closed, is a Lie subgroup of **GL**(n).

Remarks 8.2

(a) **R** is an abelian group since the exponentials commute. Hence its closure **S** is also an abelian group.

(b) An example where **R** is not closed is where, say,

$$L = \begin{pmatrix} 0 & -1 & 0 & 0 \\ 1 & 0 & 0 & 0 \\ 0 & 0 & 0 & -\sqrt{2} \\ 0 & 0 & \sqrt{2} & 0 \end{pmatrix}.$$

Then

$$\exp(sL^t) = \begin{pmatrix} R_s & 0 \\ 0 & R_{\sqrt{2}s} \end{pmatrix}$$

which is dense on the torus $\mathbf{T}^2 = \mathbf{S}^1 \times \mathbf{S}^1 \subset \mathbb{R}^2 \times \mathbb{R}^2 = \mathbb{R}^4$, where $\mathbf{S}^1$ is the unit circle in $\mathbb{R}^2$. Thus **R** is not closed. In this case we have $\mathbf{S} = \mathbf{T}^2$.

(c) **S** need not in general be compact. For example, suppose that

$$L = \begin{pmatrix} 0 & 1 \\ 0 & 0 \end{pmatrix}.$$

Then

$$\exp(sL^t) = \begin{pmatrix} 1 & 0 \\ s & 1 \end{pmatrix}$$

which is unbounded for $s \in \mathbb{R}$.

We can now state the main result of Elphick <u>et al</u>. [1986]:

Theorem 8.3 Let **S** be as above. Define

$$\mathcal{G}_k(S) = \{p \in \mathcal{P}_k : p \text{ commutes with } \mathbf{S}\}.$$

Then for $k \geq 2$,

$$\mathcal{P}_k = \mathcal{G}_k(S) \oplus \operatorname{Im} \operatorname{ad}_L .$$

If L and L^t commute, then the full Birkhoff normal form, including the linear term Lx, commutes with **S**. For example if (the important case for Hopf bifurcation)

$$L = \begin{pmatrix} 0 & -I \\ I & 0 \end{pmatrix}$$

then the Birkhoff normal form commutes with **S**, which here is the circle group $\mathbf{S}^1$ of matrices

$$\begin{pmatrix} \cos\theta\, I & -\sin\theta\, I \\ \sin\theta\, I & \cos\theta\, I \end{pmatrix}.$$

There is an equivariant generalization of Theorem 8.3:

Theorem 8.4 If $k \geq 2$ then $\mathcal{P}_k(\Gamma) = \mathcal{G}_k(\Gamma\times\mathbf{S}) \oplus \operatorname{ad}_L(\mathcal{P}_k(\Gamma))$.

In other words, we can make the Birkhoff normal form (up to given order and perhaps omitting the linear part) commute not just with Γ, but with $\Gamma\times\mathbf{S}$. In equivariant Hopf bifurcation, we can make the truncated Birkhoff normal form, to any order, commute with $\Gamma\times\mathbf{S}^1$.

9. FLOQUET THEORY AND ASYMPTOTIC STABILITY

We now apply this idea to compute the stability of the solutions found in Theorem 7.1. To study the asymptotic stability of a periodic solution u(s) to (4) we employ Floquet theory. The <u>Floquet equation</u> for u(s) is the linearisation

$$\frac{dz}{ds} + \frac{1}{1+\tau}(df)_{u(s),\lambda} z = 0 \qquad (18)$$

The <u>Floquet operator</u> $M_u : V\oplus V \to V\oplus V$ (often called the <u>monodromy operator</u>) is defined as follows. Let $z_0 \in V\oplus V$ and let z(s) be a solution to (18) such that $z(0) = z_0$. Define

$$M_u z_0 = z(2\pi). \tag{19}$$

Since (18) is linear, M_u is linear. The eigenvalues of M_u are the Floquet multipliers of u(s).

Because the periodic solution u(s) is 'neutrally stable' to phase shift, at least one Floquet multiplier is always forced to be 1. The standard Floquet Theorem states that if the remaining eigenvalues of M_u lie strictly inside the unit circle, then u(s) is an asymptotically stable periodic solution to (4). This is intuitively clear since the eigenvalues measure the amount of contraction or expansion in the flow of (18) near the periodic orbit u(s).

Symmetry may force many eigenvalues of M_u to 1. More precisely, let $\Sigma \subset \Gamma\times\mathbf{S}^1$ be the isotropy subgroup of u(s). Define $d_\Sigma = \dim \Gamma + 1 - \dim \Sigma$. Then M_u has at least d_Σ eigenvalues equal to 1, corresponding to neutral stability within the group orbit of the solution u(s). We say that u is orbitally asymptotically stable if the remaining eigenvalues of M_u lie strictly inside the unit circle.

It is in general not easy to compute eigenvalues of M_u analytically. However, there is an explicit formula if f if in Birkhoff normal form. Write (4) as

$$du/ds + \tfrac{1}{1+\tau} f(u,\lambda) = 0. \tag{20}$$

The $\mathbf{S}^1$-action in Birkhoff normal form is by e^{-Lt}. But this identifies with the phase-shift action on $\mathcal{C}^0{}_{2\pi}$. Hence when f is in Birkhoff normal form the 2π-periodic solutions of (20) are of the form

$$u(s) = e^{-Ls}u(0). \tag{21}$$

We seek conditions on u(0) so that (21) is a solution, and we find that u(0) satisfies the 'steady-state' equation

$$-Lu(0) + \tfrac{1}{1+\tau} f(u(0),\lambda) = 0. \tag{22}$$

One way to express this is that by changing to a rotating coordinate frame, we convert (20) to a steady-state bifurcation: see Renardy [1982]. We can now solve the Floquet equation (19) explicitly, obtaining

$$z(s) = \exp((\tfrac{1}{1+\tau}(df)_{u(0)} - L)s)w(0).$$

Since L has purely imaginary eigenvalues we have:

Theorem 9.1 (<u>Principle of Equivariant Reduced Stability</u>) If f is in Birkhoff normal form, then u(s) is orbitally asymptotically stable if all eigenvalues of $(df)_{u(0)}$ have positive real part, with exactly d_Σ of them being at 0.

In practice this theorem leads to stability criteria involving the Taylor coefficients of f when in Birkhoff normal form. There are similar results with Birkhoff normal form replaced by the Liapunov-Schmidt reduced function (which has almost the same coefficients).

Example

Consider **O**(2) Hopf bifurcation in the coordinates of §7. The general Birkhoff normal form of f is

$$f = (p+iq)\begin{pmatrix} z_1 \\ z_2 \end{pmatrix} + (r+is)\delta \begin{pmatrix} z_1 \\ -z_2 \end{pmatrix}$$

where p, q, r, s are functions of $N = |z_1|^2+|z_2|^2$, $\Delta = \delta^2$, and $\delta = |z_2|^2-|z_1|^2$. Further $p(0) = 0 = q(0)$, $p_\lambda(0) \neq 0$. Applying Theorem 9.1 we find the eigenvalue signs

$r(0), -p_N(0)+r(0)$ for the rotating wave,
$-r(0), p_N(0)$ for the standing wave.

We deduce that <u>in order for either solution to be stable, both must be supercritical; and then exactly one of them is stable</u>. Versions of this result have been obtained independently by numerous authors over the last fifteen years, in a variety of contexts.

Taliaferro [1986] has shown that the stability criteria found in this way remain valid even when f is not in Birkhoff normal form, <u>provided that</u>

it is transformed into Birkhoff normal form up to an order higher than that of any of the coefficients involved. That is, the truncation involved in the Birkhoff normal form does not alter the stability (near the bifurcation point) provided it is at high order.

10. NONLINEAR COMMUTING MAPPINGS

We now begin to look at nonlinear degeneracies, that is, the effect of higher order terms. Let us return to the example of the elastic cube. The action of $\mathbf{S}_3$ on the space $\mathbb{R}^2 = \{(w_1,w_2,w_3) \mid w_1+w_2+w_3 = 0\}$ is isomorphic to one that is more convenient for our purposes, namely the standard action of the dihedral group $\mathbf{D}_3$ on $\mathbb{C} \equiv \mathbb{R}^2$. This action is generated by the transformations

$$z \to \bar{z}$$
$$z \to \omega z \quad (\omega = e^{2\pi i/3}).$$

A $\mathbf{D}_3$-equivariant mapping $g:\mathbb{R}^2 \to \mathbb{R}^2$ has the general form

$$g(z) = p(u,v)X(z) + q(u,v)Y(z)$$

where

$$u = z\bar{z} \qquad v = \mathrm{Re}(z^3)$$
$$X(z) = z \qquad Y(z) = \bar{z}^2.$$

Hence the bifurcation equation for the cube under traction reduces to

$$0 = g(z,\lambda) = p(u,v,\lambda)X(z) + q(u,v,\lambda)Y(z),$$

the trivial cubic shape corresponding to $z = 0$. If $\lambda = 0$ is a bifurcation point then $p(0,0,0) = 0$. Generically we have $q(0,0,0) \neq 0$. If z and $\bar{z}^2$ are linearly independent (as vectors in $\mathbb{R}^2$) then $g = 0$ has a solution only when $p = q = 0$, contrary to this genericity assumption. Hence all bifurcating solutions have z a real multiple of $\bar{z}^2$, that is, $\mathrm{Im}(z^3) = 0$. Such points are precisely those that have $\mathbf{Z}_2$ (or a conjugate) as isotropy subgroup, corresponding to the isotropy subgroup $\mathbf{S}_2 \subset \mathbf{S}_3$. These are the rod and slab solutions, and we have explained why generically the symmetry breaks not to the trivial group $\mathbb{1}$, but to Z_2.

However, a separate analysis shows that all such branches are unstable. To find stable solutions it is therefore necessary to consider also the degenerate case $q(0,0,0) = 0$. But now it is much harder to solve $g = 0$. It is here that the power of singularity theory can be exploited: it reduces this problem to the solution of a polynomial equation, as we shall see below.

11. SINGULARITY THEORY

Singularity Theory arose from the work of Thom, Mather, Malgrange, Arnold, and others in the 1960s. In the present context it may be described as the local theory of nonlinear degeneracies of mappings $\mathbb{R}^n \to \mathbb{R}^p$. It was adapted to bifurcation problems with a distinguished parameter λ by Golubitsky and Schaeffer [1979a] and extended to the equivariant case in Golubitsky and Shaeffer [1979b].

The singularity-theoretic approach to static bifurcation problems $g(x,\lambda) = 0$, or to problems that can be reduced to this form, is to introduce an equivalence relation preserving the main qualitative features of the bifurcation diagram. Associated with each equivalence class is a normal form: a bifurcation problem $\hat{g}$ in that class having an especially simple expression. The main objectives are:

(a) The Recognition Problem. For a given normal form $\hat{g}$, find necessary and sufficient conditions on the Taylor coefficients of g for g to be equivalent to $\hat{g}$.

(b) The Unfolding Problem. For each normal form $\hat{g}$, what is the structure of all small perturbations $\hat{g}+\varepsilon p$?

(c) The Classification Problem. Up to a given level of complexity, list all possible equivalence classes of bifurcation problems. 'Complexity' here is usually measured by an invariant known as the codimension, see below.

All three problems can be solved, at least in principle, by algorithmic methods. For example Keyfitz [1986] has classified bifurcation problems without symmetry in one state variable up to codimension 7. In practice the computations can become very difficult, especially in the equivariant case; nonetheless singularity theory provides rigorous proofs of many results that have not, as yet, been obtained by other means.

The rigorous setting is as follows. Suppose that a compact Lie group Γ acts linearly on $V = \mathbb{R}^n$. Let $\vec{\mathcal{E}}(\Gamma)$ be the set of (germs at 0 of) equivariant mappings $V \to V$. Let $\mathcal{E}(\Gamma)$ be the set of Γ-invariant (germs of) functions $V \to \mathbb{R}$, that is, functions f such that $f(\gamma x) = f(x)$ for all $x \in V$. The set $\mathcal{E}(\Gamma)$ is closed under sums and products, that is, it forms a ring. The set $\vec{\mathcal{E}}(\Gamma)$ is closed under sums, and under multiplication by elements of $\mathcal{E}(\Gamma)$, hence it is a module over $\mathcal{E}(\Gamma)$. The classical Theorem of Hilbert and Weyl states that the polynomial functions in $\mathcal{E}(\Gamma)$ are generated by a finite subset $\rho_1,\dots,\rho_r$. Schwarz [1975] proved that every element of $\mathcal{E}(\Gamma)$ is of the form $h(\rho_1,\dots,\rho_r)$ for smooth h. Again, classically it is known that all polynomial mappings in $\vec{\mathcal{E}}(\Gamma)$ are generated over polynomial functions in $\mathcal{E}(\Gamma)$ by some finite subset $X_1,\dots,X_s$. Poénaru [1976] showed that the same X_j generate $\vec{\mathcal{E}}(\Gamma)$ over $\mathcal{E}(\Gamma)$.

For example when $\Gamma = \mathbf{D}_3$ acting on $\mathbb{R}^2$ as in §8, we have $\rho_1 = u$, $\rho_2 = v$, $X_1 = X$, $X_2 = Y$. The important point is that for all compact Γ these sets of generators are always finite.

Introduce a new parameter λ and let $\mathcal{E}_{x,\lambda}(\Gamma)$ be the set of invariant functions on $\mathbb{R}^n \times \mathbb{R}$, $\vec{\mathcal{E}}_{x,\lambda}(\Gamma)$ the set of equivariant mappings on $\mathbb{R}^n \times \mathbb{R}$. The above results generalize to the parametrized setting.

Next we define equivalence of bifurcation problems. Let $\mathcal{L}_\Gamma(V)^o$ be the connected component of $\mathrm{Hom}_\Gamma(V,V) \cap \mathbf{GL}(V)$ that contains the identity. For example if Γ acts irreducibly on V and $\mathrm{Hom}_\Gamma(V,V) \cong \mathbb{R}$ then $\mathcal{L}_\Gamma(V)^o = \{cI \mid c > 0\}$. We say that $g,h \in \vec{\mathcal{E}}_{x,\lambda}(\Gamma)$ are Γ-equivalent if there exists a change of coordinates $(x,\lambda) \mapsto (X(x,\lambda),\Lambda(\lambda))$ and a matrix-valued map $S(x,\lambda)$ such that

$$g(x,\lambda) = S(x,\lambda)h(X(x,\lambda),\Lambda(\lambda))$$

where

$$X(0,0) = 0,\ \Lambda(0) = 0,$$

and for all $\gamma \in \Gamma$

$$X(\gamma x,\lambda) = \gamma X(x,\lambda),\quad S(\gamma x,\lambda) = \gamma S(x,\lambda),$$
$$S(0,0) \text{ and } (dX)_{0,0} \in \mathcal{L}_\Gamma(V)^o.$$

The equivalence is strong if $\Lambda(\lambda) \equiv \lambda$.

The motivation for this definition is that we want the bifurcation diagrams $g = 0$ and $h = 0$ to have the same qualitative structure, including its symmetry aspects.

Conditions for the equivalence of two bifurcation problems can be given in terms of the tangent space - obtained by considering 'infinitesimal' equivalences. First we define the Γ-equivariant restricted tangent space of g to be

$$RT(g,\Gamma) = \{S.g+(dg)X\}$$

where S and X are equivariant and $X(0,0) = 0$. The full tangent space is more complicated. First let

$$\vec{\mathcal{M}}_{x,\lambda}(\Gamma) = \{g \in \vec{\mathcal{E}}_{x,\lambda}(\Gamma) : g(0,0) = 0\}$$

and let $Y_1,\ldots,Y_m$ be a basis for $\vec{\mathcal{E}}_{x,\lambda}(\Gamma)$ modulo $\vec{\mathcal{M}}_{x,\lambda}(\Gamma)$, so that

$$\vec{\mathcal{E}}_{x,\lambda}(\Gamma) = \vec{\mathcal{M}}_{x,\lambda}(\Gamma) + \mathbb{R}\{Y_1,\ldots,Y_m\}.$$

Then we define

$$T(g,\Gamma) = RT(g,\Gamma) + \mathbb{R}\{(dg)_{x,\lambda}Y_1,\ldots,(dg)_{x,\lambda}Y_m\} + \mathcal{E}(\lambda)g_\lambda.$$

The codimension of g (in the sense of singularity theory) is defined to be

$$\mathrm{cod}_\Gamma(g) = \dim_{\mathbb{R}} \vec{\mathcal{E}}_{x,\lambda}(\Gamma)/T(g,\Gamma).$$

In fact the term 'codimension' has the same intuitive intepretation as in §1, by Theorem 12.1 below.

The space $RT(g,\Gamma)$ is an $\mathcal{E}_{x,\lambda}(\Gamma)$-module, giving it more pleasant algebraic properties than $T(g,\Gamma)$. For this reason, much of the theory concentrates on $RT(g,\Gamma)$. A basic result is a condition for strong equivalence:

Proposition 11.1 Suppose that $h, p \in \vec{\mathcal{E}}_{x,\lambda}(\Gamma)$ and $RT(h+tp,\Gamma) = RT(h,\Gamma)$ for all $t \in [0,1]$. Then h+tp is strongly equivalent to h for all $t \in [0,1]$. In particular h+p is equivalent to h.

Next we define

$$\overleftrightarrow{\mathcal{E}}_{x,\lambda}(\Gamma) = \{n{\times}n \text{ matrices } S(x,\lambda) \mid S(\gamma x,\lambda) = \gamma S(x,\lambda)\}.$$

Each of $\vec{\mathcal{M}}_{x,\lambda}(\Gamma)$ and $\overleftrightarrow{\mathcal{E}}_{x,\lambda}(\Gamma)$ is a finitely generated module over $\mathcal{E}_{x,\lambda}(\Gamma)$, and we denote sets of generators respectively by $X_1,\dots,X_s$ and $S_1,\dots,S_t$. The next result lets us compute $RT(h,\Gamma)$.

Proposition 11.2 $RT(h,\Gamma)$ is generated by $S_1h,\dots,S_th$; $(dh)X_1,\dots,(dh)X_s$.

Γ	V	ACTION	GENERATORS FOR $\mathcal{E}(\Gamma)$	$\vec{\mathcal{E}}(\Gamma)$	$\overleftrightarrow{\mathcal{E}}(\Gamma)$	$RT(h)$
$\mathbf{Z}_2$ $\begin{pmatrix}1&0\\0&-1\end{pmatrix}$	$\mathbb{R}^2$ $x=(y,z)$	$\begin{pmatrix}1&0\\0&-1\end{pmatrix}\begin{pmatrix}y\\z\end{pmatrix}$	$u=y$ $v=z^2$	$g_1=(1,0)$ $g_2=(0,z)$ $h=(p,qz)$ $\equiv[p,q]$	$S_1=\begin{pmatrix}1&0\\0&0\end{pmatrix}$ $S_2=\begin{pmatrix}0&z\\0&0\end{pmatrix}$ $S_3=\begin{pmatrix}0&0\\z&0\end{pmatrix}$ $S_4=\begin{pmatrix}0&0\\0&1\end{pmatrix}$	$[p,0],[vq,0]$ $[0,p],[0,q]$ $[up_u,uq_u],[vp_u,vq_u]$ $[\lambda p_u,\lambda q_u],[vp_v,vq_v]$
$\mathbf{O}(2)$ $0\le\theta\le 2\pi$ κ	$\mathbb{C}$	$\theta.z=e^{i\theta}z$ $\kappa.z=\bar{z}$	$u=z\bar{z}$	$g_1=z$ $h=pz$ $\equiv[p]$	$w\in\mathbb{C}$ $S_1(z)w=w$ $S_2(z)w=z^2\bar{w}$	$[p],[up_u]$
$\mathbf{D}_n$ $(n\ge 3)$ $\theta=2k\pi/n$ κ	$\mathbb{C}$ $x=z$	$\theta.z=e^{i\theta}z$ $\kappa.z=\bar{z}$	$u=z\bar{z}$ $v=z^n+\bar{z}^n$	$g_1=z$ $g_2=\bar{z}^{n-1}$ $h=pz+q\bar{z}^{n-1}$ $\equiv[p,q]$	$w\in\mathbb{C}$ $S_1(z)w=w$ $S_2(z)w=z^2\bar{w}$ $S_3(z)w=\bar{z}^{n-2}\bar{w}$ $S_4(z)w=z^n w$	$[p,q],[2up+vq,0]$ $[u^{n-2}q,p]$, $[vp+2u^{n-1}q,0]$, $[2up_u+nvp_v,$ $nq+(n-1)uq_u+nvq_v]$ $[vp_u+2nu^{n-1}p_v$ $+(n-2)u^{n-1}q$ $+(n-1)u^{n-1}q_u,$ $vq_u+2nu^{n-1}q_v]$

Table 11.1. Data for Γ-equivariant restricted tangent spaces, $\Gamma = \mathbf{Z}_2, \mathbf{O}(2), \mathbf{D}_n$

Table 11.1 shows the relevant algebraic data for the standard actions of $\mathbf{Z}_2$, $\mathbf{O}(2)$, and $\mathbf{D}_n$, using the notation $[p_1,...,p_s] = p_1X_1 + ... + p_sX_s$ for equivariants, the p_j being functions of the invariant generators $\rho_1,...,\rho_r$ listed in the column headed $\mathcal{E}(\Gamma)$.

For example in the case $\Gamma = \mathbf{D}_3$, which is relevant to the elastic cube, we can solve the recognition problem for the two simplest cases:

Proposition 11.3 Let g be a $\mathbf{D}_3$-equivariant bifurcation problem. Assume

$$\begin{aligned} &(a) \quad p(0,0,0) = 0 \\ &(b) \quad p_\lambda(0,0,0) \neq 0 \\ &(c) \quad q(0,0,0) \neq 0. \end{aligned} \tag{23}$$

Then g is $\mathbf{D}_3$-equivalent to

$$h(z,\lambda) = \varepsilon\lambda z + \delta\bar{z}^2 \tag{24}$$

where $\varepsilon = \text{sgn}\, p_\lambda(0,0,0)$ and $\delta = \text{sgn}\, q(0,0,0)$.

For the second example we begin by specifying the lower order terms in p and q as follows:

$$\begin{aligned} p(u,v,\lambda) &= Au + Bv + \alpha\lambda + ... \\ q(u,v,\lambda) &= Cu + Dv + \beta\lambda + ... \end{aligned} \tag{25}$$

We call any $\mathbf{D}_3$-equivariant bifurcation problem g satisfying $p(0,0,0) = 0 = q(0,0,0)$ <u>nondegenerate</u> if

$$\alpha \neq 0,\ A \neq 0,\ \alpha C-\beta A \neq 0,\ AD-BC \neq 0. \tag{26}$$

This terminology is slightly confusing, since we have already described the bifurcation problem as 'degenerate'. In this context, 'nondegenerate' means 'as little degenerate as possible'.

Theorem 11.4 Let g be a $\mathbf{D}_3$-equivariant bifurcation problem. Assume that $p(0,0,0) = 0 = q(0,0,0)$ and that g is nondegenerate. Then g is $\mathbf{D}_3$-equivalent to the normal form

$$\text{(a)} \qquad N(z,\lambda) = (\varepsilon u + \delta\lambda)z + (\sigma u + mv)\bar{z}^2 \qquad (27)$$

where $\varepsilon = \text{sgn } A$, $\delta = \text{sgn } \alpha$, $\sigma = \text{sgn}(\alpha C - \beta A).\text{sgn } \alpha$, and

$$\text{(b)} \qquad m = \text{sgn}(A).(AD-BC)\alpha^2/(\alpha C-\beta A)^2.$$

The parameter m in (b) above is said to be modal. This means that it affects the differentiable, but not the topological, structure of the bifurcation diagram. (Strictly speaking, even the topological structure may change, but only at certain special values of m, finite in number.)

Before exhibiting the bifurcation diagrams in these cases we develop the theory of equivariant unfoldings. We remark that singularities of $\mathbf{D}_n$-equivariant bifurcation problems have been studied in Buzano, Geymonat, and Poston [1985] in connection with buckling rods of regular polygonal cross-section.

12. UNFOLDING THEORY

Let $g(x,\lambda) \in \vec{\mathcal{E}}_{x,\lambda}(\Gamma)$ be a Γ-equivariant bifurcation problem. A k-parameter unfolding of g is a Γ-equivariant map germ $G(x,\lambda,\alpha) \in \vec{\mathcal{E}}_{x,\lambda,\alpha}(\Gamma)$ where $\alpha \in \mathbb{R}^k$ and $G(x,\lambda,0) = g(x,\lambda)$. More precisely, $G \in \vec{\mathcal{E}}_{x,\lambda,\alpha}(\Gamma)$ if it is the germ of a mapping $G: V\times\mathbb{R}\times\mathbb{R}^k \to V$ satisfying the equivariance condition

$$G(\gamma x,\lambda,\alpha) = \gamma G(x,\lambda,\alpha) \qquad (28)$$

for all $\gamma \in \Gamma$. That is, we assume that Γ acts trivially on all parameters λ,α.

Of course, we use Γ-unfoldings to represent families of perturbations of g that preserve the symmetries Γ. In the qualitative theory defined by Γ-equivalences we need to know when the perturbations in a Γ-unfolding H are contained in another unfolding G. We codify this idea using the notion of 'factoring'. Let $H(x,\lambda,\beta)$ be an λ-parameter Γ-unfolding of $g(x,\lambda)$, and let $G(x,\lambda,\alpha)$ be a k-parameter Γ-unfolding of $g(x,\lambda)$. We say that H factors through G if

$$H(x,\lambda,\beta) = S(x,\lambda,\beta)G(X(x,\lambda,\beta),\Lambda(\lambda,\beta),A(\beta)) \tag{29}$$

where $S(x,\lambda,0) = I$, $X(x,\lambda,0) = x$, $\Lambda(\lambda,0) = \lambda$ and $A(0) = 0$. Equation (29) states that for each β, $H(\cdot,\cdot,\beta)$ is equivalent to some member $G(\cdot,\cdot,A(\beta))$ of the family G. We also assume that the equivalence of g with g when $\beta = 0$ is the identity. Moreover, we want these equivalences to be Γ-equivalences. We ensure this by demanding that

$$\begin{aligned} &\text{(a)} \quad S \in \overleftrightarrow{\mathcal{E}}_{x,\lambda,\alpha}(\Gamma), \text{ i.e. } S(\gamma x,\lambda,\alpha)\gamma = \gamma S(x,\lambda,\alpha), \\ &\text{(b)} \quad X \in \vec{\mathcal{E}}_{x,\lambda,\alpha}(\Gamma), \text{ i.e. } X(\gamma x,\lambda,\alpha) = \gamma X(x,\lambda,\alpha). \end{aligned} \tag{30}$$

As expected we define a Γ-unfolding G of g to be <u>versal</u> if every Γ-unfolding H of g factors through G. A versal Γ-unfolding G of g is <u>universal</u> if G depends on the minimum number of parameters needed for a versal unfolding. This minimum number is equal to the Γ-codimension of g.

The main theoretical result is:

Theorem 12.1 (Γ-<u>equivariant Universal Unfolding Theorem</u>). Let Γ be a compact Lie group acting on V. Let $g \in \vec{\mathcal{E}}_{x,\lambda}(\Gamma)$ be a Γ-equivariant bifurcation problem and let $G \in \vec{\mathcal{E}}_{x,\lambda,\alpha}(\Gamma)$ be a k-parameter unfolding of g. Then G is versal if and only if

$$\vec{\mathcal{E}}_{x,\lambda}(\Gamma) = T(g,\Gamma) + \mathbb{R}\{\partial G/\partial\alpha_1(x,\lambda,0),\ldots,\partial G/\partial\alpha_k(x,\lambda,0)\}.$$

For example, the normal form (24) has codimension 0 and hence is its own universal unfolding, whereas (27) has codimension 2 and universal unfolding $H(z,\lambda,\mu,\alpha) = (\varepsilon u+\delta\lambda)z+(\sigma u+\mu v+\alpha)\bar{z}^2$ for α near 0 and μ near m.

The bifurcation diagrams for these two normal forms, together with the associated stabilities, are shown in Figures 7 and 8. They arise also in the spherical Bénard problem, see Busse [1975], Chossat [1979], and Golubitsky and Schaeffer [1982]. Other applications of equivariant singularity theory to static bifurcation theory include:

(a) The planar Bénard problem: Buzano and Golubitsky [1983], Golubitsky, Swift, and Knobloch [1984].

(b) Mode-jumping in the buckling of a rectangular plate, Schaeffer and Golubitsky [1979].

(c) Deformation of an elastic cube, Ball and Schaeffer [1982].

(d) Reaction-diffusion equations, Schaeffer and Golubitsky [1981], Keyfitz and Kuiper [1984]. See Fiedler [1986a] for an application of global Hopf bifurcation.

(d) Deformation of crystals under heat, Melbourne [1987b,c].

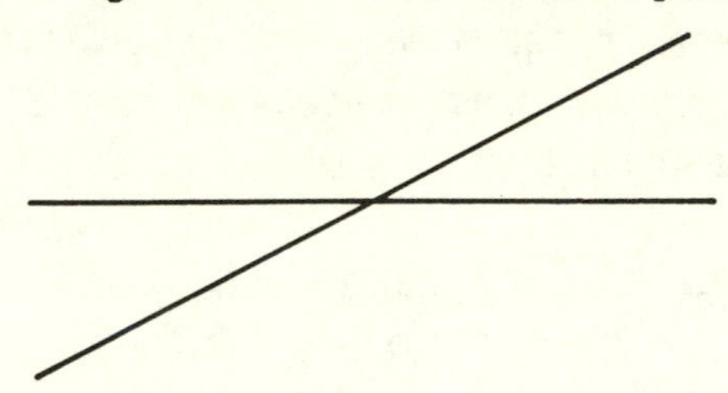

Figure 7: Bifurcation diagram for normal form (24).

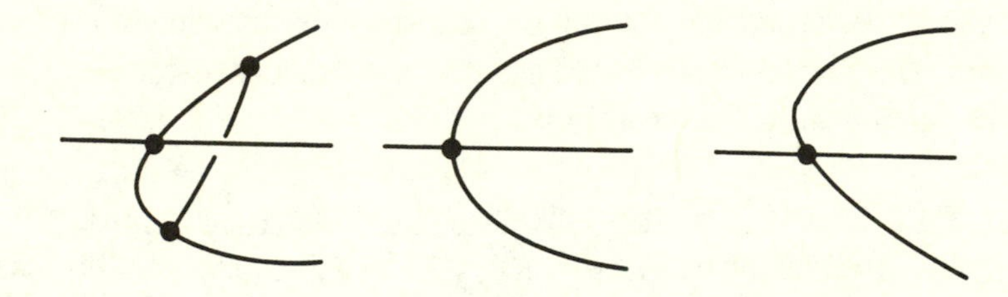

Figure 8: Bifurcation diagrams for normal form (27).

13. DEGENERATE HOPF BIFURCATIONS WITH SYMMETRY

Degenerate Hopf bifurcation, which occurs when some hypotheses of the standard Hopf theorem are false, has been studied by Golubitsky and Langford [1981] using singularity-theoretic methods. Liapunov-Schmidt reduction converts the problem to one in $\mathbf{S}^1$-equivariant singularity theory, much as in §7. There is a further reduction to $\mathbf{Z}_2$-symmetric singularities, and a relatively complete theory exists.

One might imagine that degenerate Γ-equivariant Hopf bifurcation ought, by the same token, to reduce to $\Gamma\times\mathbf{S}^1$-equivariant singularity theory, but this is not straightforward. In particular the usual equivalence relation appears unsuitable. At present it remains an open problem, how best to proceed. However, in one important case, a great deal can be achieved by special methods. This is $\mathbf{O}(2)$-symmetric Hopf bifurcation. The analysis, due to Golubitsky and Roberts [1986], proceeds in three stages. First the system is put into Birkhoff normal form to some high order k and truncated at that order. Then it is separated into phase and amplitude equations.

Finally it is observed that the amplitude equations have $\mathbf{D}_4$ symmetry and can be classified by equivariant singularity theory, and the results are interpreted back in the original system. Because of the reliance on Birkhoff normal form, the results are only formal; but additional analysis suggests that most of the phenomena found persist in the original system.

As we saw in §9, in Birkhoff normal form a general $\mathbf{O}(2)$-equivariant Hopf bifurcation takes the form of an $\mathbf{O}(2)\times\mathbf{S}^1$-equivariant vector field

$$X = (p+iq)\begin{pmatrix} z_1 \\ z_2 \end{pmatrix} + (r+is)\begin{pmatrix} z_1 \\ -z_2 \end{pmatrix} \tag{31}$$

in coordinates $(z_1,z_2) \in \mathbb{C}^2 \equiv \mathbb{R}^4$. Here p, q, r, s are functions of $N = |z_1|^2+|z_2|^2$, $\Delta = (|z_1|^2-|z_2|^2)^2$, and λ, and $p(0) = 0$, $q(0) = 1$.

The Birkhoff normal form (31) lets us separate the 4×4 system of ODEs into amplitude and phase equations, which in turn permit a simple analysis of the stability of the rotating and standing waves. Write $z_1 = xe^{i\psi_1}$, $z_2 = ye^{i\psi_2}$. Then (31) implies that the system takes the form

$$\begin{aligned} \dot{z}_1 &= (p+iq+(r+is)\delta)z_1 \\ \dot{z}_2 &= (p+iq-(r+is)\delta)z_2. \end{aligned} \tag{32}$$

In amplitude/phase variables these become

$$\begin{aligned} &\text{(a)} \qquad \dot{x} + (p+r\delta)x = 0 \qquad \dot{y} + (p-r\delta)y = 0 \\ &\text{(b)} \qquad \dot{\psi}_1 + (q+s\delta) = 0 \qquad \dot{\psi}_2 + (q-s\delta) = 0 \end{aligned} \tag{33}$$

where now p, q, r, s are functions of $N = x^2+y^2$, $\Delta = (y^2-x^2)^2$, and λ; and $\delta = y^2-x^2$. Nontrivial zeros of the amplitude equations correspond to <u>invariant circles</u> and <u>invariant tori</u>. In particular when $y = 0$ (the rotating waves branch) zeros correspond to circles; and when $x = y$ (the standing waves branch) zeros correspond to invariant 2-tori in the original 4-dimensional system. On such a torus we see from (33b) that $\dot{\psi}_1 = \dot{\psi}_2$ since $\delta = 0$. It follows that trajectories on this 2-torus are all circles. This qualitative feature of the flow persists even when the vector field is not in Birkhoff normal form, see Chossat [1986]. Finally, a zero $x > y > 0$ of the amplitude equations corresponds to an invariant 2-torus with linear flow. This linear flow is, in general, quasiperiodic. Generically $r(0) \neq 0$, so there are (locally) no zeros of the amplitude equations with $x > y > 0$. Therefore such tori do not occur in $\mathbf{O}(2)$-symmetric nondegenerate Hopf bifurcation. It is not surprising, however, that when considering degenerate cases such as $r(0) = 0$, such tori do occur. This observation was made by Erneux and Matkowsky [1984].

14. $\mathbf{D}_4$-SYMMETRY

The amplitude equations inherit some of the $\mathbf{O}(2)\times\mathbf{S}^1$ symmetry. Let the dihedral group $\mathbf{D}_4$ act on $\mathbb{C}$ as symmetries of the square. That is, the action is generated by

$$z \mapsto \bar{z} \qquad \text{and} \qquad z \mapsto iz. \tag{34}$$

The general $\mathbf{D}_4$-equivariant mapping has the form $pz+q\bar{z}^3$ where p and q are functions of $u = z\bar{z}$ and $v = \mathrm{Re}(z^4)$. If we let $z = x+iy$, then we have $u = x^2+y^2 = N$ and $v = x^4-6x^2y^2+y^4 = -(x^2+y^2)^2+2(y^2-x^2)^2 = 2\Delta-N^2$. Moreover, as mappings $\mathbb{R}^2 \to \mathbb{R}^2$, the mappings $z \mapsto z$ and $z \mapsto \bar{z}^3$ correspond to

$$\begin{pmatrix} x \\ y \end{pmatrix} \mapsto \begin{pmatrix} x \\ y \end{pmatrix} \quad \text{and} \quad \begin{pmatrix} x \\ y \end{pmatrix} \mapsto \begin{pmatrix} x^3-3xy^2 \\ y^3-3x^2y \end{pmatrix} = -N\begin{pmatrix} x \\ y \end{pmatrix} -2\delta\begin{pmatrix} x \\ -y \end{pmatrix}$$

Therefore <u>the general form of the amplitude equations is exactly the same as the general form of the $\mathbf{D}_4$-equivariant mappings</u> $\mathbb{R}^2 \to \mathbb{R}^2$. The $\mathbf{D}_4$ symmetry may be thought of as what remains of the original $\mathbf{O}(2)\times\mathbf{S}^1$ symmetry after reduction to the amplitude equations.

Suppose we let

$$h(x,y,\lambda) = p(N,\Delta,\lambda)\begin{pmatrix} x \\ y \end{pmatrix} + r(N,\Delta,\lambda)\delta\begin{pmatrix} x \\ -y \end{pmatrix}. \tag{35}$$

Then the amplitude equations (33) have the form

$$\begin{pmatrix} \dot{x} \\ \dot{y} \end{pmatrix} + h(x,y,\lambda) = 0. \tag{36}$$

The $\mathbf{D}_4$ symmetries restrict the form of dh at solutions, and it is possible to classify the <u>degenerate</u> $\mathbf{O}(2)$-equivariant Hopf bifurcations, by classifying the $\mathbf{D}_4$-equivariant germs h up to $\mathbf{D}_4$-equivalence. This is done by Golubitsky and Roberts [1986]. The solutions so obtained contain not only the periodic solutions of rotating and standing waves, but also the invariant 2-tori with linear flow.

This kind of reduction (Γ-equivariant Hopf bifurcation $\to$ $\Gamma\times\mathbf{S}^1$ Birkhoff normal form $\to$ Σ-equivariant amplitude equations) seldom happens in such a nice way, but when it does, we have a method for studying the degenerate Γ-equivariant Hopf bifurcations. It can be placed in a more general setting by considering the way group orbits of solutions evolve in time; or equivalently how the invariant generators evolve.

15. MODE INTERACTIONS

As remarked in the introduction, there are three distinct types of mode interaction: steady-state/steady-state, Hopf/steady-state, and Hopf/Hopf. Each has its own distinctive features, and so far only the $\mathbf{O}(2)$ case has been treated systematically. Even there, the Hopf/Hopf case has not been fully analysed. For the current state of knowledge see Dangelmayr and Armbruster [1983, 1986], Armbruster, Dangelmayr, and Güttinger [1985], Dangelmayr [1987], Langford [1986], and Chossat, Golubitsky, and Keyfitz [1987]. For treatments using different methods, see Langford [1979], Shearer [1981], and Guckenheimer and Holmes [1983].

In the Hopf/steady-state case, the problem reduces, via amplitude equations, to the classification of bifurcation problems in two state variables with $\mathbf{Z}_2$ symmetry $(x,y) \mapsto (x,-y)$. This is all that remains, after reduction, of the $\mathbf{S}^1$ symmetry of Hopf bifurcation. The singularity theory mechanics for this classification were set up in Golubitsky and Schaeffer [1979]; the actual classification - up to topological codimension 5 - was completed by Dangelmayr and Armbruster [1983].

16. APPLICATIONS OF MODE INTERACTIONS WITH SYMMETRY

We briefly describe two applications: one to combustion and one to fluid dynamics.

Keyfitz, Golubitsky, Gorman and Chossat [1985] use equivariant bifurcation theory to explore the behaviour of mesh-stabilized flames on a porous plug burner with circular $\mathbf{O}(2)$ symmetry. Here the steady state planar flame can become unstable to various modes. One is a pulsating flame with $\mathbf{O}(2)$ symmetry, another is a standing wave, and a third is a spiral wave. Assuming that the former may be modelled by an $\mathbf{O}(2)$ Hopf bifurcation with trivial $\mathbf{O}(2)$-action and the latter two may by an $\mathbf{O}(2)$ Hopf bifurcation with nontrivial $\mathbf{O}(2)$-action, they propose an $\mathbf{O}(2)$ Hopf/Hopf mode interaction as a model. After reduction to amplitude equations they obtain a $\mathbf{Z}_2\times\mathbf{D}_4$-equivariant singularity, selected from the classification due to Chossat, Golubitsky and Keyfitz [1987]. The isotropy lattice is

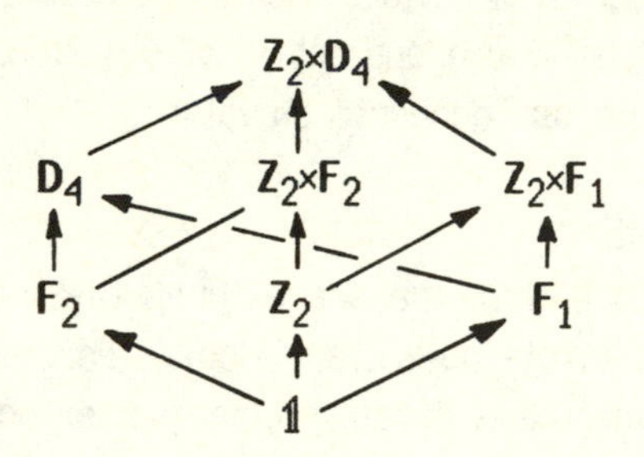

where $\mathbf{F}_1$ and $\mathbf{F}_2$ are two particular subgroups of $\mathbf{D}_4$ of order 2. A sample bifurcation diagram, occurring for particular ranges of coefficients, is shown in Figure 9. Here the primary branch is a Hopf bifurcation that is $\mathbf{O}(2)$-invariant (a circular pulsating flame). There is a secondary bifurcation to a doubly periodic mixed-mode solution with symmetry group $\mathbf{F}_2$, and a tertiary torus bifurcation from the mixed mode. This torus branch terminates in a homoclinic orbit.

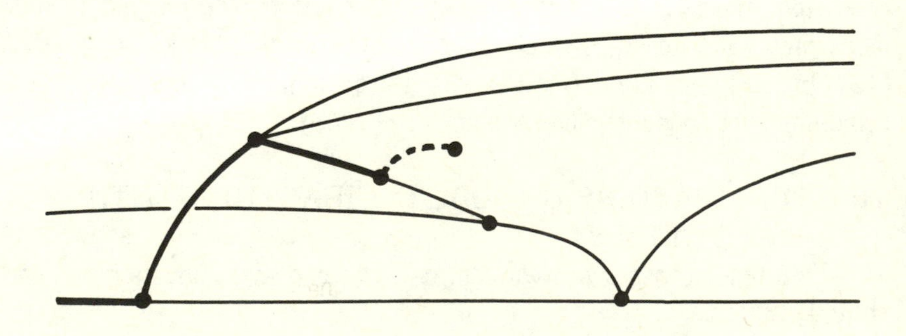

Figure 9. Sample bifurcation diagram for mode interaction in a porous plug burner with $\mathbf{O}(2)$ symmetry.

In Taylor-Couette flow a fluid is placed between two coaxial rotating cylinders. Depending on the speeds of the two cylinders, numerous flow-patterns are observed. This problem has been investigated from the viewpoint of equivariant bifurcation theory by a number of authors, including Chossat and Iooss [1985], Iooss [1985], Golubitsky and Stewart [1986a], Chossat, Demay and Iooss [1986], Iooss, Coullet and Demay [1986], Iooss [1986]. In most cases the approach is to analyse the interaction of a small number of modes (usually two although some progress is being made on three). The symmetry group for the interaction of Taylor vortex flow and helices is $\mathbf{O}(2)\times\mathbf{SO}(2)$, analysed by Golubitsky and Stewart [1986a]. Other groups, such as $\mathbf{O}(2)\times\mathbf{SO}(2)\times\mathbf{T}^2$, occur in interactions of periodic modes. The results obtained classify a number of prechaotic flows and qualitative agreement with experiments is quite good. Figure 10 shows a sample bifurcation diagram from Golubitsky and Stewart [1986a], corresponding to the 'main sequence' of bifurcations from Couette flow to Taylor vortices to wavy vortices to modulated wavy vortices. (A computational error, pointed out by Troger, occurs in Golubitsky and Stewart [1986a] on pp. 268, 270, yielding incorrect eigenvalues on the wavy vortices and twisted vortices branches. The two eigenvalues of multiplicity 1 should be replaced by the trace and determinant of a suitable 2×2 matrix. This opens up new possibilities for torus bifurcations and implies that two different kinds of modulation may occur. See Golubitsky, Stewart and Schaeffer [1987].)

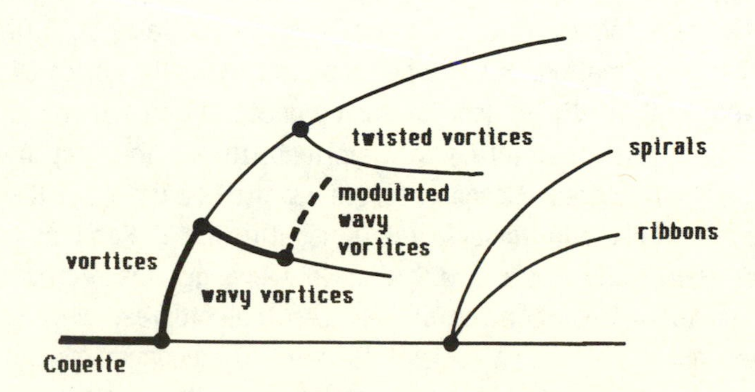

Figure 10: The 'main sequence' in Taylor-Couette flow, seen as a mode interaction with symmetry.

17. HAMILTONIAN SYSTEMS

Very recently it has become clear that many of the results obtained for bifurcation of symmetric dissipative systems have analogues in conservative and Hamiltonian systems. In the absence of symmetry, generic bifurcation of Hamiltonian systems was studied using singularity-theoretic methods by Meyer [1975]. Golubitsky and Stewart [1986] obtain similar results for the bifurcation of equilibria of symmetric Hamiltonian systems.

Montaldi, Roberts, and Stewart [1987] have obtained equivariant analogues of the Liapunov Centre Theorem and the Weinstein-Moser Theorem about the existence of periodic solutions near equilibria of Hamiltonian systems. We sketch their results. First suppose that no symmetries are present. Let $\mathcal{H}$ be a Hamiltonian function on a 2n-dimensional symplectic manifold (phase space) P, and let p be a nondegenerate minimum of $\mathcal{H}$, that is, $D\mathcal{H}_p = 0$ and $D^2\mathcal{H}_p$ is positive definite. The eigenvalues of the linearisation L of the vector field at p are purely imaginary pairs $\{\pm\lambda_1,\ldots,\pm\lambda_n\}$. Liapunov [1907] proved the Centre Theorem: if some λ_i is <u>nonresonant</u> then there exists a smooth 2-dimensional submanifold of P, passing through p, and intersecting each energy level near p in a periodic trajectory with period near $2\pi/|\lambda_i|$, see Abraham and Marsden [1983]. By 'nonresonant' we mean that λ_j is not an integer multiple of λ_i for any $j \neq i$. In a celebrated generalisation, Weinstein [1973] and Moser [1976] proved that even when there is resonance, there still exist n periodic trajectories on each energy level near p, each having period near $2\pi/|\lambda_i|$ for some i.

Let Γ be a compact Lie group acting symplectically on P, that is, by 'canonical transformations'. Let $\mathcal{H}$ be invariant under the action of Γ and let p be a fixed point for Γ. In general such a group action will force some of the λ_i to be equal, restricting the applicability of the Liapunov Centre Theorem. The Weinstein-Moser Theorem can still be used, but the estimate of the number of periodic trajectories is usually weak. For instance, in the system of Hénon and Heiles [1964] at low energy, the Weinstein-Moser Theorem predicts two families of periodic trajectories, whereas in fact there are eight. The reason for this is that the standard Weinstein-Moser Theorem does not (directly) give information about the symmetry groups of periodic trajectories.

For simplicity suppose that $\lambda_1 = \ldots = \lambda_n = \lambda$. (The actual results are more general.) Then the trajectories of the linearised vector field at p, defined on the tangent space T_pP, are all periodic with period $2\pi/|\lambda|$. This flow therefore defines an $\mathbf{S}^1$-action which commutes with the linearisation of the Γ-action, the two together yielding an action of $\Gamma\times\mathbf{S}^1$ on T_pP. Let Σ be <u>any</u> subgroup of $\Gamma\times\mathbf{S}^1$. The main existence theorem is the equivariant Weinstein-Moser Theorem: there exist at least $\frac{1}{2}$ dim Fix(Σ) periodic trajectories of the nonlinear system, with periods near $2\pi/|\lambda|$ and symmetry groups containing Σ, on each energy level sufficiently close to p. There is also a version of the Liapunov Centre Theorem: if dim Fix (Σ) = 2 then there exists a family of periodic trajectories, with symmetry groups equal to Σ, forming a smooth 2-dimensional submanifold through p. Examples suggest that in 'generic' systems with symmetry, the result gives sharp estimates for the number of periodic trajectories with periods near those of the linearised system.

The stability of these periodic trajectories is also strongly constrained by symmetry - much more so than in the dissipative case. First recall some definitions. Let $\sigma:P\times\mathbb{R}\to P$ denote the flow generated by $\mathcal{H}$ and for fixed $t\in\mathbb{R}$ write $\sigma_t = \sigma(\cdot,t)$. If u(t) is a T-periodic trajectory then its Floquet operator is

$$(D\sigma_T)_{u(0)}:T_{u(0)}P \to T_{u(0)}P.$$

We need the following definitions:

(a) A T-periodic trajectory u(t) is <u>Liapunov stable</u> if for all $\varepsilon > 0$ there exists $\delta > 0$ such that if v(t) is a trajectory satisfying $\|u(0)-v(0)\| < \delta$, then $\inf_s \|u(s)-v(t)\| < \varepsilon$ for all $t > 0$ and some metric $\|\cdot\|$ on P.

(b) A T-periodic trajectory u(t) is <u>linearly stable</u> if it is Liapunov stable as a trajectory of the linearisation of the vector field about u(t).

(c) A T-periodic trajectory u(t) is <u>spectrally stable</u> if the eigenvalues of $(D\sigma_T)_{u(0)}$ all lie on the unit circle in $\mathbb{C}$.

Linear stability is neither necessary nor sufficient for Liapunov stability, but spectral stability is a consequence of either.

Montaldi, Roberts and Stewart [1987] give a set of sufficient representation-theoretic conditions for a periodic trajectory near an equilibrium point to be spectrally stable. Perhaps the most surprising result is that for some isotropy subgroups Σ these conditions are satisfied <u>automatically</u>, independent of the coefficients in the Hamiltonian. Examples show that these conditions are satisfied sufficiently often for the theorem to be useful. Of perhaps wider significance is that in proving this theorem they develop machinery that describes precisely the constraints put on Floquet operators by group actions.

Let p be an equilibrium point and let u(t) be a periodic trajectory near p with symmetry group Σ and period T near $2\pi/|\lambda|$, where $\lambda = \lambda_1$, say. In The Floquet operator $M_u:T_{u(0)}P \to T_{u(0)}P$ is close to the operator $e^{-2\pi L/|\lambda|}T_pP \to T_pP$, which has eigenvalues $e^{\pm 2\pi\lambda_i/|\lambda|}$, on the unit circle. By using 'Krein theory' it can be shown that if neither λ_i nor $\lambda_i-\lambda_j$ $(j\neq i)$ is an integer multiple of λ (so that in particular $e^{\pm 2\pi\lambda_i/|\lambda|} \neq 1$) then the eigenvalues of M_u near $e^{\pm 2\pi\lambda_i/|\lambda|}$ remain on the unit circle for u near p. The eigenvalues of M_u near 1 fall into two classes: those that remain at 1 because of the group action, and the rest. Those forced to equal 1 can be computed from the the momentum mapping, or 'constants of motion', associated to the Γ-action. Dividing out those parts of $T_{u(0)}P$ corresponding to eigenvalues of M_u forced to 1, we obtain a residual Floquet operator N_u whose eigenvalues are the remaining eigenvalues of M_u.

To describe the approach to the eigenvalues of N_u again suppose, for simplicity only, that $\lambda_1= \dots =\lambda_n=\lambda$. Then N_u is a linear symplectic map which commutes with an action of Σ derived from the restriction of the $\Gamma\times\mathbf{S}^1$-action on T_pP. (This is not obvious.) It is then possible to apply general results about equivariant symplectic maps. These include a representation-theoretic characterisation of cyclospectral representations: representations of Σ for which every Σ-equivariant symplectic linear map has all its eigenvalues on the unit circle .

The results can be applied to a number of examples, including equilibria fixed by actions of $\mathbf{O}(3)$ (such as occur in a vibrating liquid drop). Near these, under suitable conditions, there exists a variety of periodic trajectories with prescribed symmetries. They can be approximated by certain linear combinations of rotating spherical harmonics. Figures 4 and 5 above illustrate the form of these solutions in low-dimensional cases: they have exactly the same symmetries that are predicted for $\mathbf{O}(3)$ Hopf bifurcation. For low order spherical harmonics many of these trajectories are forced by the group action to be spectrally stable. As the order increases this is no longer true, although there are always two types of rotating wave that are spectrally stable. However, even here the group action limits the possible forms that the Floquet operator can take.

18. CONCLUSIONS AND SUGGESTIONS FOR FURTHER RESEARCH

Symmetry imposes strong constraints on the bifurcations of vector fields, which usually take the form of degeneracies. If the symmetry is not exploited these degeneracies make analysis extremely difficult and the results obtained are necessarily weak. However, by exploiting the symmetries it is often possible to prove much stronger theorems, taking the form of equivariant analogues of the standard results. In other words, when operating within the world of symmetric systems, all the machinery should - as far as possible - be made symmetric too. Most of the classical prechaotic phenomena can be so generalized.

The influence of symmetries there is so great that it would be surprising if symmetry did not carry implications for chaotic systems. However, little work has been done in this area. In part this is because certain key results, that are trivial in the non-symmetric case, do not yet have good equivariant analogues. A case in point is the existence of a Poincaré section for a periodic orbit. Is there a good existence theorem for an equivariant Poincaré section? Until such a theorem exists, the usual approach to chaos by way of mappings is unable to get off the ground in the equivariant case.

There are still a great many open problems, and the diversity of known results in dynamical systems theory opens up broad vistas for equivariant generalization. A sample of specific questions follows. It could easily have been made longer.

1. Develop a good theory of degenerate equivariant Hopf bifurcation.

2. Characterize those isotropy subgroups that occur generically in spontaneous symmetry-breaking. Examples of Chossat [1983], Lauterbach [1986], and Swift [1986] show that non-maximal isotropy subgroups can occur.

3. Develop efficient machinery to solve the recognition problem for equivariant bifurcation problems. Some progress has been made by Gaffney [1986] and Melbourne [1987b] building on results of Bruce, Du Plessis, and Wall [1985].

4. Classify equivariant bifurcation problems for crystallographic groups, and apply the results to the Landau theory of phase transitions in crystals.

5. Complete the analysis of **O**(2) Hopf/Hopf mode interactions.

6. Study specific groups, such as **O**(3) or **O**(4).

7. Study the simplest mode interactions for **O**(3). Apply the results to spherical Bénard convection.

8. Use bifurcation theory with torus group symmetry to analyse instabilities in plasmas.

9. Prove or disprove the existence of an equivariant Poincaré section near a periodic orbit of a dynamical system.

10. Develop a theory of mode interactions in the Hamiltonian case.

REFERENCES

R.Abraham and J.E.Marsden [1978]. *Foundations of Mechanics* (2nd. ed.), Benjamin/Cummings, Reading, Mass.

D.Armbruster, G.Dangelmayr, and W.Güttinger [1985]. Interacting Hopf and steady-state bifurcations, Preprint, Univ. of Tübingen.

A.K.Bajaj and P.R.Sethna [1984]. Flow induced bifurcations to three-dimensional oscillatory motions in continuous tubes, *SIAM J. Appl. Math.* **44** 270-286.

J.Ball and D.Schaeffer [1982]. Bifurcation and stability of homogeneous equilibrium configurations of an elastic body under dead-load traction, *Math. Proc. Camb. Phil. Soc.* **94** 315-339.

J.W.Bruce, A.A. Du Plessis, and C.T.C.Wall [1985]. Determinacy and unipotency, preprint, Univ. of Liverpool.

F.Busse [1975]. Patterns of convection in spherical shells, *J. Fluid Mech.* **72** 65-88.

E.Buzano, G.Geymonat, and T.Poston [1985]. Post-buckling behavior of a nonlinearly hyperelastic thin rod with cross-section invariant under the dihedral group $\mathbf{D}_n$, *Arch. Rational Mech. Anal.* **89** 307-388.

P.Chossat [1979]. Bifurcation and stability of convective flows in a rotating or not rotating shell, *SIAM J. Appl. Math.* **37** 624-647.

P.Chossat [1983]. Solutions avec symétrie diédrale dans les problèmes de bifurcation invariantes par symétrie spherique, *C. R. Acad. Sci. Paris* **297** 639-642.

P.Chossat [1986]. Bifurcation secondaire de solutions quasi périodiques dans un problème be bifurcation de Hopf invariant par symétrie **O**(2), *C.R. Acad. Sci. Paris* **302** 539-541.

P.Chossat, M.Golubitsky, B.L.Keyfitz [1987]. Hopf-Hopf mode interactions with **O**(2) symmetry, *Dynamics and Stability of Systems,* to appear.

P.Chossat and G.Iooss [1985]. Primary and secondary bifurcations in the Couette-Taylor problem, *Japan. J. Appl. Math.* **2** 37-68.

P.Chossat, Y.Demay, and G.Iooss [1986]. Interactions de modes azimutaux dans le problème de Couette-Taylor, preprint, Univ. of Nice.

G.Cicogna [1981]. Symmetry breakdown from bifurcation, *Lett. Nuovo Cimento* **31** 600-602.

J.D.Cowan [1982]. Spontaneous symmetry breaking in large scale nervous activity, *Int. J. Quantum Chem.* **22** 1059-1082.

G.Dangelmayr [1987]. Steady-state mode interactions in the presence of **O**(2) symmetry, *Dynamics and Stability of Systems* (to appear).

G.Dangelmayr and D.Armbruster [1983]. Classification of Z(2)-equivariant imperfect bifurcations with corank 2, *Proc. London Math. Soc.* **46** 517-546.

G.Dangelmayr and D.Armbruster [1986]. Steady-state mode interactions in the presence of **O**(2) symmetry and in non-flux boundary value problems, in *Multiparameter bifurcation theory* (ed. M.Golubitsky and J.Guckenheimer), *Contemporary Math.* **56**, Amer. Math. Soc., Providence RI. 53-68.

G.Dangelmayr and E.Knobloch [1986a]. The Takens-Bogdanov bifurcation with **O**(2) symmetry, *Proc. Roy. Soc. Lond. A* (to appear).

G.Dangelmayr and E.Knobloch [1986b]. Interaction between standing and travelling waves and steady states in magnetoconvection, *Phys. Lett. A* **117** 394-398.

C.Elphick, E.Tirapegui, M.E.Brachet, P.Coullet, and G.Iooss [1986]. A simple global characterization for normal forms of singular vector fields, preprint, Nice.

T.Erneux and B.J.Matkowsky [1984]. Quasi-periodic waves along a pulsating propagating front in a reaction-diffusion system, *SIAM J. Appl. Math.* **44** 536-544.

B.Fiedler [1986a]. Global Hopf bifurcation in reaction diffusion systems with symmetry, in *Dynamics of Infinite Dimensional Systems* (ed. J.K.Hale et al.), Lisbon, to appear.

B.Fiedler [1986]. Habilitationsschrift, Heidelberg.

T.Gaffney [1986]. Some new results in the classification theory of bifurcation problems, in *Multiparameter bifurcation theory* (ed. M.Golubitsky and J.Guckenheimer), *Contemporary Math.* **56**, Amer. Math. Soc., Providence RI. 97-118.

M.Golubitsky [1983]. The Bénard problem, symmetry, and the lattice of isotropy subgroups, in *Bifurcation Theory, Mechanics, and Physics* (ed. C.P.Bruter et al.), Reidel, Dordrecht.

M.Golubitsky and R.M.Roberts [1986]. A classification of degenerate Hopf bifurcations with **O**(2) symmetry, preprint, Univ. of Warwick.

M.Golubitsky and W.F.Langford [1981]. Classification and unfoldings of degenerate Hopf bifurcations, *J. Diff. Eqn.* **41** 375-415.

M.Golubitsky and D.G.Schaeffer [1979a]. A theory for imperfect bifurcation via singularity theory, *Commun. Pure Appl. Math.* **32** 21-98.

M.Golubitsky and D.G.Schaeffer [1979b]. Imperfect bifurcation in the presence of symmetry, *Commun. Math. Phys.* **67** 205-232.

M.Golubitsky and D.G.Schaeffer [1984]. *Singularities and Groups in Bifurcation Theory* vol. I, Springer, Berlin, Heidelberg, New York.

M.Golubitsky and D.G.Schaeffer [1982]. Bifurcations with **O**(3) symmetry including applications to the Bénard problem, *Commun. Pure Appl. Math.* **35** 81-111.

M.Golubitsky, I.N.Stewart, and D.G.Schaeffer [1987]. *Singularities and Groups in Bifurcation Theory* vol. II, Springer, Berlin, Heidelberg, New York, to appear.

M.Golubitsky and I.N.Stewart [1985a]. Hopf bifurcation in the presence of symmetry, *Arch. Rational Mech. Anal.* **87** 107-165.

M.Golubitsky and I.N.Stewart [1986a]. Symmetry and Stability in Taylor-Couette flow, *SIAM J. Math. Anal.* **17** 249-288.

M.Golubitsky and I.N.Stewart [1986b]. Hopf bifurcation with dihedral group symmetry: coupled nonlinear oscillators, in *Multiparameter bifurcation theory* (ed. M.Golubitsky and J.Guckenheimer), *Contemporary Math.* **56** 131-173, Amer. Math. Soc., Providence RI.

M.Golubitsky and I.N.Stewart [1986c]. Generic bifurcation of Hamiltonian systems with symmetry, *Physica D: Nonlinear Systems,* in press.

J.Guckenheimer [1984]. Multiple bifurcation of codimension two, *SIAM J. Math. Anal.* **15** 1-49.

J.Guckenheimer and P.Holmes [1983]. *Nonlinear Oscillations, Dynamical Systems, and Bifurcation of Vector Fields,* Springer, Berlin, Heidelberg, New York.

M.Hénon and C.Heiles [1964]. The applicability of the third integral of motion; some numerical experiments, *Astronom. J.* **69** 73-79.

E.Ihrig and M.Golubitsky [1984]. Pattern selection with **O**(3) symmetry, *Physica D: Nonlinear Phenomena* **13** 1-33.

G.Iooss [1985]. Recent progresses in the Couette-Taylor problem, preprint, Univ. of Nice.

G.Iooss [1986]. Secondary bifurcations of Taylor vortices into wavy inflow or outflow boundaries, *J. Fluid Mech.* **173** 273-288.

G.Iooss, P.Coullet, Y.Demay [1986]. Large scale modulations in the Taylor-Couette problem with counterrotating cylinders, preprint, Univ. of Nice.

B.L.Keyfitz, M.Golubitsky, M.Gorman and P.Chossat, [1985]. The use of symmetry and bifurcation techniques in studying flame stability, preprint, Univ. of Houston.

B.L.Keyfitz and H.J.Kuiper [1984]. Bifurcation resulting from changes in domain in a reaction diffusion equation, Tech. Report **60**, Arizona State Univ.

B.L.Keyfitz [1986]. Classification of one-state-variable bifurcation problems up to codimension seven, *Dynamics and Stability of Systems* **1** 1-42.

W.F.Langford [1979]. Periodic and steady-state mode interactions lead to tori, *SIAM J. Appl. Math.* **37** 22-48.

W.F.Langford [1986]. A review of interactions of Hopf and steady-state bifurcations, in *Nonlinear Dynamics and Turbulence,* (eds. G.I.Barenblatt, G.Iooss, D.D.Joseph), Pitman, London.

R.Lauterbach [1986]. An example of symmetry-breaking with submaximal isotropy subgroup, in *Multiparameter bifurcation theory* (ed. M.Golubitsky and J.Guckenheimer), *Contemporary Math.* **56**, Amer. Math. Soc., Providence RI. 217-222.

A.M.Liapunov [1907]. Problème générale de la Stabilité du Mouvement, *Ann. Fac. Sci. Toulouse* **9**, reprinted by Princeton Univ. Press 1947. (Russian original 1895.)

I.Melbourne [1987a]. A singularity theory analysis of bifurcation problems with octahedral symmetry, *Dynamics and Stability of Systems*, to appear.

I.Melbourne [1987b]. New methods in the recognition problem for equivariant singularities, to appear.

I.Melbourne [1987c]. The classification up to low codimension of bifurcation problems with octahedral symmetry, to appear.

K.Meyer [1975]. Generic bifurcations in Hamiltonian systems, in *Dynamical Systems - Warwick 1974* (ed. A.K.Manning), Lecture notes in Math. **468**, Springer, Berlin, Heidelberg, New York 62-70.

J.Montaldi, R.M.Roberts, and I.N.Stewart [1986]. Periodic solutions near equilibria of symmetric Hamiltonian systems, preprint, Univ. of Warwick.

J.Moser [1976]. Periodic orbits near equilibrium and a theorem by Alan Weinstein, *Commun. Pure Appl. Math.* **29** 727-747.

W.Nagata [1986]. Symmetric Hopf bifurcations and magnetoconvection, in *Multiparameter bifurcation theory* (ed. M.Golubitsky and J.Guckenheimer), *Contemporary Math.* **56**, Amer. Math. Soc., Providence RI. 237-265.

V.Poénaru [1976]. *Singularités C^∞ en présence de symétrie*, Lecture Notes in Math. **510**, Springer, Berlin, Heidelberg, New York.

M.Renardy [1982]. Bifurcation from rotating waves, *Arch. Rational Mech. Anal.* **75** 49-84.

M.Roberts, J.W.Swift, and D.Wagner [1986]. The Hopf bifurcation on a hexagonal lattice, in *Multiparameter bifurcation theory* (ed. M.Golubitsky and J.Guckenheimer), *Contemporary Math.* **56**, Amer. Math. Soc., Providence RI.

D.H.Sattinger [1979]. *Group Theoretic Methods in Bifurcation Theory*, Lecture Notes in Math. **762**, Springer, Berlin, Heidelberg, New York.

D.H.Sattinger [1983]. *Branching in the presence of symmetry*, CBMS-NSF Regional Conf. Series **40** SIAM, Philadephia.

D.G.Schaeffer and M.Golubitsky [1979]. Boundary conditions and mode jumping in the buckling of a rectangular plate, *Commun. Math. Phys.* **69** 209-236.

D.G.Schaeffer and M.Golubitsky [1981]. Bifurcation analysis near a double eigenvalue of a model chemical reaction, *Arch. Rational Mech. Anal.* **75** 315-347.

G.Schwarz [1975]. Smooth functions invariant under the action of a compact Lie group, *Topology* **14** 63-68.

M.Shearer [1979]. Coincident bifurcations of equilibrium and periodic solutions of evolution equations, MRC. Tech. Report.

C.L.Siegel and J.Moser [1971]. *Lectures on Celestial Mechanics,* Springer, Berlin, Heidelberg, New York.

I.N.Stewart [1981]. Applications of catastrophe theory to the physical sciences, *Physica D: Nonlinear Phenomena* **2** 245-305.

I.N.Stewart [1982]. Catastrophe theory in physics, *Rep. Prog. Phys.* **45** 185-221.

I.N.Stewart [1983a]. Elementary catastrophe theory, *IEEE Trans. Circuits and Systems* **30** 578-586.

I.N.Stewart [1983b]. Nonelementary catastrophe theory, *IEEE Trans. Circuits and Systems* **30** 663-670.

I.N.Stewart [1984]. Applications of nonelementary catastrophe theory, *IEEE Trans. Circuits and Systems* **31** 165-174.

J.W.Swift [1984]. *Bifurcation and symmetry in convection,* Ph.D. Thesis, U.C.Berkeley.

J.W.Swift [1986]. Four coupled oscillators: Hopf bifurcation with the symmetry of a square, preprint, King's College Research Centre, Cambridge.

S.D.Taliaferro [1986]. Stability of bifurcating solutions in the presence of symmetry, preprint, Texas A&M Univ., College Station.

A.L.Vanderbauwhede [1980]. *Local Bifurcation and Symmetry,* Habilitation Thesis, Rijksuniversiteit Gent.

A.L.Vanderbauwhede [1982]. *Local Bifurcation and Symmetry,* Research Notes in Math. **75**, Pitman, London, Boston.

A.L.Vanderbauwhede [1986]. Center Manifolds, Normal Forms, and Elementary Bifurcations, preprint, Gent.

A.Weinstein [1973]. Normal modes for nonlinear Hamiltonian systems, *Invent. Math.* **20** 47-57.